Advance praise for Digital Fandom 2.0

"Paul Booth's original *Digital Fandom* bridges between fandom studies, internet studies, and game studies, to explore the ways that digital media has impacted the ways contemporary audiences engage with popular texts. With this new edition, Booth revisits and revises his earlier work, focusing attention onto newer forms of digital expression—from Kickstarter to Tumblr. Booth's characteristic concepts—such as the digi-gratis economy and Hybrid fandom—stress the complex ways fans have become entangled in, sometimes empowered by, sometimes exploited by, new media industry practices, even as he holds onto the messiness and playfulness of these new relationships."

—Henry Jenkins, Author of *Textual Poachers: Television Fans and Participatory Culture*

"The second edition of Paul Booth's *Digital Fandom* [*Digital Fandom 2.0*] offers a perceptive reconsideration of fandom past and present, and of the evolving field of fan studies: its stakes, limits, and potential. Booth observes that contemporary digital fandom is multivocal, and, by engaging with a wide range of perspectives on fandom, he succeeds in making *Digital Fandom 2.0* dynamically multivocal as well. *Digital Fandom 2.0* is a thought-provoking work for fans, fan scholars, and everyone in between."

—Louisa Stein, Middlebury College

"Rather than attempting to definitively map the moving target of digital fan culture, Paul Booth embraces its complexities to pose important questions about the future(s) of fan studies. Booth's ongoing applications of game studies to theorize the 'playful' nature of fan culture, and his nuanced consideration of hybrid fan practices, make the second edition of *Digital Fandom* [*Digital Fandom 2.0*] required reading."

—Suzanne Scott, the University of Texas at Austin

"Fan scholars will squee for Paul Booth's *Digital Fandom 2.0*! *Digital Fandom* (2010) is one of the most important books in fan studies, and version 2.0 will be equally influential to discussions of fan cultures. By significantly reworking old chapters and adding new ones, 2.0 is not just an update of the original, it's an entirely new contribution to the field."

—Casey McCormick, Co-founder, *Fan and Audience Studies*, Society for Media and Cinema Studies

"*Digital Fandom 2.0* resourcefully positions fandom as a means for interrogating notions of hybridity and participatory culture. Situated within an important discussion of digital-era media studies, Paul Booth illustrates that fan studies is not merely a specific area of scholarship but a worthwhile frame for a broad examination of popular culture."

—Candice Roberts, St. John's University

Digital Fandom 2.0

Digital Formations

Steve Jones
General Editor

Vol. 114

The Digital Formations series is part of the Peter Lang Media and Communication list.
Every volume is peer reviewed and meets
the highest quality standards for content and production.

PETER LANG
New York • Bern • Frankfurt • Berlin
Brussels • Vienna • Oxford • Warsaw

Digital Fandom 2.0

Steve Jones
General Editor

Vol. 114

The Digital Formations series is part of the Peter Lang Media and Communication list.
Every volume is peer reviewed and meets
the highest quality standards for content and production.

PETER LANG
New York • Bern • Frankfurt • Berlin
Brussels • Vienna • Oxford • Warsaw

Paul Booth

Digital Fandom 2.0

New Media Studies

SECOND EDITION

PETER LANG
New York • Bern • Frankfurt • Berlin
Brussels • Vienna • Oxford • Warsaw

Library of Congress Cataloging-in-Publication Data

Names: Booth, Paul, author.
Title: Digital fandom 2.0: new media studies / Paul Booth.
Other titles: Digital fandom.
Description: Second edition. | New York: Peter Lang, 2016.
Earlier edition published as: Digital fandom: new media studies.
Series: Digital formations, vol. 114 | ISSN 1526-3169
Includes bibliographical references and index.
Identifiers: LCCN 2016033979 | ISBN 978-1-4331-3150-9 (paperback: alk. paper)
ISBN 978-1-4539-1654-4 (ebook pdf) | ISBN 978-1-4331-3707-5 (epub)
ISBN 978-1-4331-3708-2 (mobi)
Subjects: LCSH: Digital media—Social aspects. | Fans (Persons)—Psychology.
Online social networks—Social aspects. | Technological innovations—Social aspects.
Classification: LCC HM851 .B675 2017 | DDC 303.48/33—dc23
DOI: 10.3726/978-1-4539-1654-4
LC record available at https://lccn.loc.gov/2016033979

Bibliographic information published by **Die Deutsche Nationalbibliothek.**
Die Deutsche Nationalbibliothek lists this publication in the "Deutsche Nationalbibliografie"; detailed bibliographic data are available on the Internet at http://dnb.d-nb.de/.

29 Broadway, 18th floor, New York, NY 10006
www.peterlang.com

I dedicate this book to my past, present, and future students
in my fandom classes.
I also dedicate this book to Kate, again.

TABLE OF CONTENTS

FOREWORD

The "Imaginary Opponents" of Digital Fandom (and Fan Studies)

Matt Hills

> "It was you who filled in the blanks on your own. … The real culprit is your excessive intertextual reading and linking of literary references. … And that's where the danger lies: an excess of references may have caused you to create the wrong opponent, or an imaginary opponent."
>
> —Pérez-Reverte, *The Club Dumas*, p. 335

In this second edition of *Digital Fandom* (just as in its predecessor) Paul Booth uses *The Club Dumas* as a fertile source of chapter epigraphs. When I was very kindly invited to write this short piece, I rapidly decided that I'd better follow suit—not so much to blend in, but more to see what other, remaining Pérez-Reverte quotes I could mine for Dumasian significance. And nearing the end of this twisting, turning literary-fan-thriller, our hero of sorts, Corso, confronts the idea that he may have been too (fannishly) keen to spot patterns and too enthusiastic to trace intertextual webs of meaning. The result, perhaps, is that he has created his own opponents, and narrated his own story. And herein lies one danger of penning a foreword to a rich and multi-layered academic book like this one: I might spot the wrong patterns, make the wrong links, and thus imagine an idiosyncratic set of interlocutors or opponents to the vital arguments contained within. But I'll have to take that chance. So, here goes …

Forewords are a curious thing; although they supposedly precede and contextualize a book, they can only properly be written after reading it. A sense of following afterwards thus awkwardly interacts with their presumed precedence: Forewords are really afterwords stuck in the wrong place. And given this twisted temporality, can we count on readers looking at a foreword before they dive into a favored chapter, or begin a book's own Introduction? In short, these brief pieces of writing act as what Jacques Derrida (1987) has termed the "parergon"; they work to frame a book, yet it is unclear whether such a framing leaves them inside the text looking out, or paratextually outside its borders and looking in.

In terms of their uneasy spatiotemporalities, forewords are, surprisingly, not unlike digital fandom itself. Representing one person's response to a text might color, guide, or agitate the responses of others; Forewords are not so very different to forum postings, comments threads, Tumblr notes, or livetweets. They may lack the visual recontextualizations of a shared GIF or a meme, but they remain part of a work in progress, mediating between the book's author and its readers, interpreting in ways that later readers might share, reject, or carefully mull over and negotiate.

The (para)textual blurring of the foreword also mimics, however fleetingly, a sense of digital fandom's productivity, proliferation, and invention. For the amazing array of texts, images, and alternatives created by and within digital fandom means that we cannot quite be sure where media texts begin and end anymore. Should analysis of a media franchise stop at the "primary" texts of a film sequence or a TV series? Or should it take in all the trailers, and the teasers, where these are typically leaked or posted online? And if we move our analytical borders outwards, then aren't fans' trailer mash-ups, hoaxes, and remixes (along with speculations, rumors, and spoilers) also vital parts of the build-up to an anticipated media text, or part of the textual experience itself? Although "official" parameters can be reestablished around the latest *Game of Thrones* or *Doctor Who* episode, much of the brand value and cultural meaning surrounding these (and other) shows surely emanates from all the digital fan activity clustering around them. They have so many, many forewords and afterwords, all teeming with ideas. Where analogue media studies could rather more feasibly isolate out its objects of study—what William Merrin (2014) has dubbed "media studies 1.0"—new media studies, or 2.0, confronts a mediatized world where texts have exploded into paratextual shards and gems. Such texts become media(tized) and digitized events, acting as anticipated, valued moments in a timeline of media industry, fan, and even "fan industry" practices. Treating media texts as *spaces of meaning* and consumption has been supplanted by an alternative way of doing things where media texts act, instead, within *timeframes of prosumption*.

Digital Fandom 2.0 is carefully attentive to such developments in media culture and scholarship. By updating the previous (2010) version, Paul Booth has also focused on a few key concepts that are further unfolded and developed across three new chapters (6, 7, and 8): the role of "demediation" in relation to digital fan practices, the "Digi-gratis economy" characteristic of digital fandom, and the "play" of fandom that has so much to offer fan studies' scholarship. As Booth has chosen to affirmationally and transformationally re-broach these concepts through new case studies, and in new contexts, I will endeavor to say a little about each here.

Demediation, as a term, demonstrates digital fandom's deconstruction of hypermediacy and immediacy, as fans shape media about media, and yet also desire to step outside realms of mediation and into domains of "immediate" experience. But these desired immediacies in fact presume and give rise to the creativity and activity of additional (social) media. Demediation is central to understanding contemporary fan practices. Digital fandom's hypermediacy now increasingly means that fans can be *fans of being fans*, with fannish memes and GIFs taking center-stage alongside all manner of multifannish engagements. But the coexistence of valorized immediacy suggests that sections of digital fandom simultaneously wish to break out of incessant mediation and into the co-presence of "being there" at a convention, a premiere, or a signing, despite such co-presence itself then being experienced and disseminated via shared photos, updates, and tweets. And in the communal domains of gaming, "being there" can be entirely inseparable from mediation, with demediation emerging through the collapse between hypermediation and immediacy.

In the (2010) first edition, Booth took the Alternate Reality Game or ARG as a metaphorical model for thinking about contemporary fandom. He returns to this in chapter 6, albeit focusing much more centrally on what he now calls "hybrid fandom," i.e., that which crosses affirmational and transformational positions, as well as liminally bridging offline and online experiences. In a sense, then, this latest upgrade of *Digital Fandom* actually has a secret, hidden subtitle that only emerges when you highlight the relevant text: *Digital Fandom … is Hybrid Fandom … is Material Fandom*. All too often, online/Web 2.0/social media-enabled fandoms have been studied in isolation from offline fan experiences and identities, treated as separate objects of study, partly because they are often easier to study online as naturally occurring data, rather than observing or interviewing fans in specific sites, spaces, and cultural geographies.

Writing in *The Official World*, Mark Seltzer (2016) suggests that rather than analyzing different cultural spaces as utopian or heterotopic—that is, as idealized zones of community or as alternative, other spaces resisting cultural norms or powers—we might now identify "isotopias." An isotopia is "a self-conditioned reenactment space, among a proliferation of synonymous but formally demarcated spaces" (17), and is thus a bounded, contained way of (re-)performing cultural identity. Booth's "hybrid fandom," we might say, traverses and occupies a number of "isotopias"; it can find a re-enactment, performance space in the "hotel-world" of conventions (Seltzer 2016, 18–19) just as much as in the Twitter-verse or via Tumblr's affordances.

By focusing on demediation and hybrid fandom, *Digital Fandom 2.0* engages with the cultural normalization of our digital lives, and with the mainstreaming of fandom. Digital and material aspects of fan practice are typically strongly interwoven. For instance, Abigail De Kosnik et al. have scraped data from major archives of fan fiction such as Archive of Our Own (AO3) and FanFiction.net, demonstrating pronounced spikes in fanfic uploading activity shortly after the premieres of major blockbusters such as *The Avengers* or the films in Christopher Nolan's *Batman* trilogy. However, De Kosnik et al.'s work also shows that in relation to a stand-alone film such as *Inception*, the productivity of fans can seemingly engender further digital engagement and production, rather than industrial promotion and marketing having this impact. De Kosnik et al. (2015) view digital fan practices as one indicator of the "mindshare" that fans give to media properties and franchises, i.e., "how long a certain text lives in the imaginations of viewers. How long does it hold on to the audiences' thoughts and emotions? How much mindshare do people grant to different media products that they consume?" (147). This mindshare is a finite cognitive and affective resource, but there is no reason to assume that it is restricted to a gift economy only, or to digital fandom in isolation. Rather, the very notion of "mindshare," I would say, resonates powerfully with Booth's emphasis here on hybrid fandom, since it implies an underlying engagement, preoccupation or attachment that can then be articulated across digital/material fandom, as well as across offline and online interactions. Fan studies may need to become more vigorously hybrid itself, not only moving across the identities of fan and scholar, but refusing to split mediatized fandom into categories such as affirmational/transformational or online/offline. Approached as multi-sited and as emerging out of multiple practices (even if fans have their niche specialisms such as fanfic creation or fanart), hybrid fandom suggests that another opponent can be discerned in relation to better understanding today's digital fandom, even if it is potentially my own "imagined opponent." This would be scholarship seeking to recuperate or re-install problematic binaries of fandom (or binaries where one term is insistently valorized over its marginalized other).

Where *Digital Fandom 2.0*'s brand new chapter 6 illuminates developments in demediation, chapter 7 re-emphasizes Booth's concept of the "Digi-gratis economy." The Digi-gratis economy offers a way of thinking about how fandom and industry intersect in ever more complex ways. It has become a tradition of fan studies to posit fandom as part of a "gift economy," strongly contrasting this to the capitalist economy of big media. Booth retains this "gifting" stance, whilst noting how digital culture plays with the situation. If fans upload fanfic or digital fanart, for instance, then unlike hard copy fanzines or printed artworks, these can be digitally "given" as gifts to a potentially infinite number of fans, in principle, without the quality of the digital artefact ever being degraded, and without the gift-giver sacrificing any further resources. The gifts of digital fandom keep on giving, then, but they also alter norms of reciprocation. Material gifts tend to evoke an obligation: One should offer a comparable gift in return at a suitable later date. But in the "Digi-gratis economy" one's counter-gift can instead be presented in the form of online attention, "likes," or posted comments and praise.

Booth discusses how this re-versioning of fandom's gift economy coexists with a capitalist economy; fans' textual productivity can be appropriated by social media platforms, and used to generate corporate revenue. The result is an unpredictable terrain where interference patterns can emerge between very different, co-present, and yet non-hybridized economic modes. Ramon Lobato and Julian Thomas (2015), in their book *The Informal Media Economy*, likewise note how formal and informal economies, corporate and grey markets, can interact in a wide variety of ways. The "Digi-gratis," that is to say, may not always be wholly "gratis"; fans can self-commodify, or aim to create fanworks within entrepreneurial models, as well as using their fan knowledge to produce commodified goods for a range of websites from Red Bubble to Etsy to Teefury and beyond. And, of course, fans can set up their own Kickstarters or assorted crowdfunders, hoping to evade copyright lawsuits, or to edge around copyrighted materials. Digital fandom offers a way of retooling the "gift economy," to be sure, but in some quarters it can also generate and sustain increasingly neoliberal or individualized versions of entrepreneurial fandom, where fan passion is integrated with a career path and even a will-to-profit. Karen Hellekson, who has been a vocal champion of fandom being viewed as a "gift economy," has latterly observed that

> some fannish groups have managed to construct a system that permits payment [for fan creativity – MH], but such attempts must be initiated by and embedded within the fandom in question. Such profiting must be a form of making use of; it must grow out of the fandom's community and its relationships with the powers that be, the copyright holders. It cannot be unilaterally imposed. (Hellekson 2015, 126)

The "Digi-gratis economy" thus includes expectations not only of "free content" and non-commercial reciprocity, but also of digital labor supposedly seamlessly uniting fan and business interests (and as I write this, Legion M has emerged as a putatively "fan-run entertainment company"). If there is an opponent here—and I am not convinced that, as posited in *The Club Dumas*, it is an "imaginary opponent"—then it may be the impact of individualizing "passion-work" or neoliberal structures and strictures on fandom. Rhiannon Bury (2016) has suggested that, at least for some fans familiar with older modes of digital fan community linked to Usenet or Live Journal, more recent platforms such as Facebook, Twitter, and Tumblr do not offer a strong sense of community: "online community will not disappear, but it may well become *residual*. … Further empirical investigation will be needed before we know if a new iteration of participatory fandom, one that is not organized around community, … is on the horizon" (14). It may be no accident that as fan community seems to be potentially eroding, at least in certain quarters, a "new economy of fandom" (Galuszka 2015) based on the co-creation of brand value by fans and rights-owners is simultaneously emerging.

Lastly, in his new closing chapter, Booth returns to the playfulness of fandom, pondering whether fan studies would in fact be best served by becoming a fully-fledged academic discipline, being disciplined and formalized in a process that is no doubt already underway in the academy. His response is to argue passionately for the value of a more playful and more affectively engaged fan studies—closer to lived experiences of hybrid fandom—which could reach outside academia. Such a version of fan studies would simultaneously avoid foreclosing the kinds of fan activity that might deserve study. Such openness and playfulness—whether related to Tumblr or not, given chapter 8's framing metaphor—can surely only serve the analysis of digital fandom, and hybrid fandom, well.

But if I were to sound one note of caution, it would be to wonder whether the formalization of an academic discipline and its legitimation are one and the same thing. Television (to think for a moment about a leading cultural form and technology of the pre-Tumblr era) existed as an economically and culturally potent force in varied societies for many years before discourses of legitimation (often modeled on other more "high cultural" media such as film) began to circulate around its dramas, even then relating only to some kinds of TV output and not others (Newman and Levine 2012). To compare fan studies to old-school television as well as Tumblr, I wonder whether fan studies' rise in terms of journals, networks, symposia—and a Special Interest Group within the Society of Cinema and Media Studies—really means that the area has been univocally or monolithically "legitimated." Speaking from my own experiences, and from those of many other fan studies' scholars that I have heard about personally, professionally, and anecdotally, fan studies remains delegitimated—or very uneasily legitimated—in some areas of fandom, as well as in long-established traditions of disciplinary practice within academia, whether these are potentially departments of English Literature, Sociology, Psychology, or Religious Studies. And in a neoliberal higher education context of shrinking research funding for increasing numbers of applicants, I also suspect that for certain funding bodies or disciplinary committees, fan studies may still seem like an "improper" arena. Perhaps this is a generational concern, but academic cultures can be tenacious in terms of their cultural reproduction, and so I doubt that we can simply place our hope in the passing of time, assuming that a "media generation" (Barker et al. 2016; Stein 2015) of digital fans and associated fan scholars will one day assume positions of academic power and thus assure that the struggle for fan studies' legitimation is definitively completed.

Legitimacy remains a key point of concern for fan studies; scholar-fandom may inhabit and represent a flourishing affective community of the kind (very rightly) called for by Paul Booth, open to engagements with different fan activities and different platforms of digital fandom. Yet by seeking to remain, at least in part, outsiders to disciplinary/disciplining norms—or even romanticizing this outsider status—there is a danger of fan studies embracing its position of relatively insecure legitimacy, rather than continuing to battle for greater scholarly acceptance.

Just as Tumblr may sometimes facilitate "filter bubbles," the same thing can happen in academic disciplines, and we need to follow Booth's call for a more interdisciplinary and open, hybridized fan studies in order to avoid the containment of disciplinary silos. But I think we also need to remain aware of ongoing struggles over academic legitimation at the same time, rather than presuming that fan studies has straightforwardly won this battle and can enter a post-legitimation phase. If striking such a note seems pessimistic, then it is not meant to be. Trading in affect, as do the cultures of fan studies *and* digital fandom, also means relating these emergent and "affective publics" (Papacharissi 2015) to unfinished cultural struggles over discourses of fan identity, value and recognition, whether these are articulated with issues of gender, sexuality, race and ethnicity (Carrington 2016), or with matters of disciplinarity and academic authority.

Perhaps I'm imagining another opponent here, one in academia that could be set against the rise and the recruiting power of fan studies. Academic disciplines such as English Literature—or even Media Studies in its guise of "hard news" analysis and public sphere preoccupations—may sometimes intersect with fan studies and the analysis of digital fandom, given the cultural presence of literary fans or fan-journalist bloggers. But by the same token, even if fan studies has moved into a position of relative academic normalization, it is true to say that we don't yet have lecturers or professors in fan studies, nor Departments of Fandom Studies (even if my new academic home, the University of Huddersfield, has a Research Centre for Participatory Culture, just as De Paul University has a degree program related to media fandom thanks to Paul Booth's own initiatives). Fan studies doesn't have a singular institutional academic home—it remains a hybrid stitched together out of scholars following very different cross-disciplinary trajectories into or toward work on fandom, I would say. If anything, fan studies has some of its "safe spaces" slightly off to one side of academic centrality and legitimacy—the Organization for Transformative Works, via the journal *Transformative Works and Cultures*, is undoubtedly the most established body here. Yet its fan-facing nature means that it arguably lacks the set-apartness and hierarchical elevation that can be claimed and performed by strictly academic organizations with more vociferous (as well as infinitely more problematic) gatekeeping functions and processes of audience/consumer/non-academic othering. The distinctiveness of fan studies that Booth rightly wants to argue for, defend, and even intensify, can sometimes lead to an imputed lack of academic distinction, as the subject's very open-ness to non-academic thinkers, newcomers and a diversity of voices concomitantly weakens its claims to old-school or established academic cultural capital. It may be that fan studies will develop different wings, factions, and versions (if it hasn't already)—some oriented toward multidisciplinarity and fandom's own organic intellectuals such as writers of "meta," and other versions oriented toward disciplinary academic traditions and the authorizations of sociology, philosophy, or literary theory. If so, one might hope that fan studies doesn't imagine or take rival images of itself as opponents in a zero-sum struggle over who can speak for "true" fandom or "authentic" digital fandom. Thankfully, *Digital Fandom 2.0*'s finale sets out an inspiring vision for the expansion, ambition and potential inclusiveness of fan studies. But struggles for, and over, academic legitimacy will no doubt continue at the same time.

This book consistently offers a skillful, impressive guide to thinking about contemporary fans. Booth's work also reminds us that digital fandoms are only one kind of participatory culture, and indeed one kind of community that can be "more or less" participatory (Jenkins, Ito, and boyd 2016). The perspectives and theories set out here, from "demediation" to the "Digi-gratis economy" to fan "playfulness," may also offer ways to tackle modes of participatory culture in the digital age that are not discursively positioned or labeled as "fandom" *per se*, whether these involve participatory politics (Highfield 2016), participatory memory cultures (Haskins 2015), participatory media heritage (Giaccardi 2012), or even participatory cultures surrounding architectural icons like the Sydney Opera House (Garduno Freeman 2016). Perhaps fandom is just one available discourse that we have for our digital and affective relations to media, culture, community, and self-identity, whilst the participations and affects that are often characteristic of fandom are already beginning to outrun and exceed this discursive classification thanks, in part, to affordances of social media. To tackle such additional reconfigurations will no doubt call for *Digital Fandom 3.0* in a few years' time. Academic work is nothing if not a process of sequelization, as authors respond to commentaries and critiques, incorporating new data and new phenomena. And if *Digital Fandom 2.0* is *The Empire Strikes Back* to *Digital Fandom*'s own *Star Wars*—or something academically akin to "NuWho" as the reimagining of classic *Doctor Who*—then *Digital Fandom*'s third edition might already be speculated about as a possible future regeneration, or anticipated as the completion of an intellectual trilogy.

By musing over the possible existence of "imaginary opponents" for digital fandom and fan studies—whether these are attempts at reinstalling binaries of online/offline or affirmational/transformational fandom; the emergence of neoliberal structures of feeling within individualizing, entrepreneurial digital fandom; or the splitting of fan studies itself into rival groups aimed at evading or courting disciplinary legitimation—I have risked imposing narrative structures, just like Corso in *The Club Dumas*. As such, these opening remarks may offer up the "wrong opponent." Yet such a possibility (of knowing who the "true" opponent is) also implies an omniscient narrator who can set out a last-gasp "reveal" or a shocking twist. In the genre of academic writing rather than the popular mystery, however, we are spared such final certainties. Instead, the dominant reading pleasures are more likely to be those of ongoing debate, influence or persuasion, and situated, intertextual originality. Luckily for you, the following pages—or ebook locations—have such qualities in spades … just like the best of digital fandom.

Bibliography

Barker, Martin, Kate Egan, Tom Phillips, and Sarah Ralph. *Alien Audiences*. Basingstoke and New York: Palgrave Pivot, 2016.

Bury, Rhiannon. "Technology, Fandom and Community in the Second Media Age. *Convergence* (published online before print, 2016): 1–16. http://con.sagepub.com/content/early/2016/05/24/1354856516648084.long

Carrington, André M. *Speculative Blackness: The Future of Race in Science Fiction*. Minneapolis: University of Minnesota Press, 2016.

De Kosnik, Abigail, Laurent El Ghaoui, Vera Cuntz-Leng, Andrew Godbehere, Andrea Horbinski, Adam Hutz, Renee Pastel, and Vu Pham. "Watching, Creating, and Archiving: Observations on the Quantity and Temporality of Fannish Productivity in Online Fan Fiction Archives." *Convergence* 21, no. 1 (2015): 145–64.

Derrida, Jacques. *The Truth in Painting*. Chicago: University of Chicago Press, 1987.

Galuszka, Patryk. "New Economy of Fandom." *Popular Music and Society* 38, no. 1 (2015): 25–43.

Garduno Freeman, Cristina. *Participatory Culture and the Social Value of an Architectural Icon: Sydney Opera House*. London and New York: Routledge, 2016.

Giaccardi, Elisa. *Heritage and Social Media: Understanding Heritage in a Participatory Culture*. London and New York: Routledge, 2012.

Haskins, Ekaterina V. *Popular Memories: Commemoration, Participatory Culture, and Democratic Citizenship*. Columbia: University of South Carolina Press, 2015.

Hellekson, Karen. "Making Use Of: The Gift, Commerce, and Fans." *Cinema Journal* 54, no. 3 (2015): 125–131

Highfield, Tim. *Social Media and Everyday Politics*. Cambridge, UK and Malden, MA: Polity Press, 2016.

Jenkins, Henry, Mizuko Ito, and danah boyd. *Participatory Culture in a Networked Era*. Cambridge, UK and Malden, MA: Polity Press, 2016.

Lobato, Ramon, and Julian Thomas. *The Informal Media Economy*. Cambridge, UK and Malden, MA: Polity Press, 2015.

Merrin, William. *Media Studies 2.0*. London: Routledge, 2014.

Newman, Michael Z., and Elana Levine. *Legitimating Television: Media Convergence and Cultural Status*. New York and London: Routledge, 2012.

Papacharissi, Zizi. *Affective Publics: Sentiment, Technology, and Politics*. Oxford: Oxford University Press, 2015.

Seltzer, Mark. *The Official World*. Durham, NC and London: Duke University Press, 2016.

Stein, Louisa Ellen. *Millennial Fandom: Television Audiences in the Transmedia Age*. Iowa City: University of Iowa Press, 2015.

ACKNOWLEDGMENTS

I am indebted to so many people, both for this second edition and for shaping my thoughts on fandom and fan studies over the past decade. I'm particularly grateful to the folks at Peter Lang—Mary Savigar in particular—who suggested I undertake this revision in the first place. It has been a real learning experience to go back to this, my first major work, and see how much has changed (and how much really hasn't). I'm also grateful to Steve Jones, the editor of the Digital Formations series, for his help in guiding the first edition into publication. I learned a lot from him over breakfast at Lou Mitchell's back in 2009.

The following permissions are gratefully acknowledged. I presented portions of the introduction at the 2008 meeting of the National Communication Association in San Diego, CA, and at Columbia College Chicago's Cultural Studies Department colloquium series (October 1, 2009), and am grateful to Ann Hertzel Gunkel and Jaafar Aksikas for feedback. Sections of chapter 3 were published in Winter 2009 in *Narrative Inquiry* 19, no. 2 as "Narractivity and the Narrative Database: Media-Based Wikis as Interactive Fan Fiction" (used with kind permission by John Benjamins Publishing Company, Amsterdam/Philadelphia, www.benjamins.com). Portions of chapter 4 were published in December 2008 in *Critical Studies in Media Communication* 25 no. 5 as "Re-Reading Fandom: MySpace Character Personas and Narrative

Identification" (used with kind permission by Taylor & Francis, New York, www.informaworld.com). Some new material in this edition has been published in earlier forms. Chapter 6 includes some previously published work from my chapter "Quest of the Magi: Playful Ideology and Demediation in MagiQuest," in *Produsing Theory: The Intersection of Audiences and Production in a Digital World*, edited by Rebecca Ann Lind (New York: Peter Lang, 2012). Chapter 7 includes material from "Crowdfunding: A Spimatic Application of Digital Fandom," *New Media and Society* 17, no. 2 (2015) and from my afterword to *Crowdfunding the Future*, edited by Lucy Bennett, Bertha Chin, and Bethan Jones (New York: Peter Lang, 2015). Thanks also to my dissertation committee from 2009—Michael Century, Ekaterina Haskins, and Jim Zappen—and my amazing advisor June Deery.

I also wish to thank the many fans I've spoken with or worked with over the years. When I started my dissertation I had a very different idea about researching fandom—I wish I had been more communicative with the fans whose work I analyzed in the first edition of *Digital Fandom*. I've since reached out and discussed the ethics of using fan work more concretely with as many as I could, and I apologize to those I haven't been able to reach. A lot of my thinking on fandom has come from the conversations I've had with my students, and I'm particularly grateful to those who taught me about being a fan. Teaching fandom and fan studies has been a highlight of my time in higher education, and I am impressed by the compassion and thoughtfulness of so many of my students. I can only hope to emulate them as I grow up too.

I'm grateful for the community of fan scholars who have influenced my thinking and writing. I hesitate to list, as there are far too many to name here. However, I would be remiss if I did not thank Kristina Busse for her discussions with me about Tumblr; Lucy Bennett, Bertha Chin, and Bethan Jones for their work on crowdfunding; Casey McCormick for her work with organizing fan studies groups; and Matt Hills for his insight, thoughtful conversations, and feedback on a variety of fan studies work, and for his foreword. Extra special thanks to Lucy Bennett for her thoughtful comments on chapter 8, and Amanda Brennan for the conversations about Tumblr. Thank you also to Mariah Cowan for her editing help, Jen Kelly for her thoughts on Tumblr, and Laura Bluett for the amazing design work.

Many thanks to my family: Colin and Deb Booth, Anna, Jason, and Henry Jackson Bird, Tom Vincent and Wendy Vincent. Thanks also to all the furry friends that make life worth it—Slinky and Rosie, Gizmo and Black Kitty, Sam and Spooky, Hurley, Oliver, and Maggie. Dusty will be forever missed.

I started *Digital Fandom* around the same time as I started dating Katie Vincent; I finished the manuscript a month before we got married. This second edition is being finished just a few days before our seventh anniversary. I couldn't have asked for a more supportive partner (and she for a more thankless job than listening to me pontificate—or worse, sing—while I wrote). I am grateful for her critical eye, her interest in fandom, and her forgiving nature. And while I may be irritating at times, at least I introduced her to *Doctor Who* (so I guess that makes up for it).

INTRODUCTION TO THE SECOND EDITION

Time moves differently in academic publishing and in digital culture. New technologies are designed and created, flourish, and wither, so quickly that publishing about them is guaranteed to be out-of-date; academic publishing is notoriously time-consuming and prone to obsolesce. Writing academically about new technology is therefore a seemingly futile gesture, as one can never hope to be caught up.

I originally wrote *Digital Fandom* in 2008–2009; it was based on my dissertation from Rensselaer Polytechnic Institute, *Fandom Studies: Fan Studies Re-Written, Re-Read, Re-Produced. Digital Fandom* was published in 2010, not even a decade ago. And yet in the time between the book's publication and today, enormous changes in the digital environment have signaled a need for a significant revision to the original. At the same time, my own views on fandom and fan scholarship have evolved since 2008. My writing has changed as well. Reading *Digital Fandom* again is like revisiting a childhood home. It's familiar and comforting but strangely uncanny as well, as if it was written by someone like but also unlike me. A specter of Paul Booth resides in these pages, but no longer fully encapsulates me or my thinking. Of course, no book can fully capture the entirety of one's mindset—it's a snapshot, designed for permanence only through elision—and as scholarship on fandom has grown over the past decade, *Digital Fandom* remains permanently fixed in time.

The book you are currently reading is my attempt to revise *Digital Fandom* to retain the unique aspects of the original that made it something (I hope) worth reading, but also to update it to changes that have happened in the past five years in the digital environment. It is an attempt to stay true to the original intention (examining fan work in digital technology as new forms of textuality and new media theory) while expanding from what were some limitations of the original. It is a more focused version of *Digital Fandom* with an expanded outlook. In this second edition of *Digital Fandom*, I want to do more than just fix some typos and add in some contemporary references. I want to explore new aspects of fandom in the digital age that emerged in the time since first publication, and I want to expand my discussion of three key topics: the Digi-gratis economy, the philosophy of playfulness, and demediation. And, as *Digital Fandom* had focused on particular technologies (LiveJournal blogs, Wikia wikis, and myspace [or, what used to be—and what I historically reference as—MySpace] social media sites), I want to retain this technological focus for the second edition. Therefore, this is more than just a reworking of the language of the original; it is an attempt to historicize some of the more dated elements and update with reference to new scholarship and new technologies.

At the same time, I also wanted to retain the original content as an historical marker of a particular time and way of thinking in fandom and fan studies. The first half of this second edition is thus a compression and revision of the first edition, and the entirety of the first edition of *Digital Fandom* has been condensed into the first half of this second edition. Chapter 1 introduces fandom and the digital environment. Chapter 2 looks at fandom and blogs, defining "Intra-textuality" as a concept that relies on the interaction between blog posts and comments. Intra-textuality is the flip side of transmediation; whereas transmedia finds unity between disparate texts, intra-textuality finds unity within disparate elements within a text. Chapter 3 examines wikis and the state of fan knowledge, defining the concept of "narractivity" as a fandom's communal construction of a narrative. Chapter 4 focuses on MySpace as a site of fan identity construction, defining "identity roleplay" and "interreality" as concepts that deepen our understanding of how fans "take on" alternate identities in their fan activities (note that although MySpace does not currently capitalize its name, I have retained the original capitalization as I am discussing MySpace from 2009, not myspace in 2016). The focus on MySpace is probably the most dated in the book, as MySpace was the biggest social network site for years until Facebook's domination started in 2007. Yet, as historical documentation of a particular time—and a particular technology—MySpace's character personas still have relevance in a digital fandom era. In fact, precisely because Facebook doesn't allow the type of identity construction that MySpace does (and, for the record, at least at the time of this writing in spring 2016, one can still log into myspace and find profiles of characters populating the landscape, surrounded by artists, musicians, and celebrities), it becomes even more important to record the presence of this type of identity roleplay in the digital age. Indeed, in many ways, MySpace prefigured Tumblr more so than it did Facebook.

However, there was also a much harder decision for me to make: Most significantly (and perhaps most controversially) I have condensed my writing about Alternate Reality Games within the manuscript. I have moved the majority of the discussion about Alternate Reality Games to chapter 5, and thus changed the overarching metaphor that I used throughout the book. While I still love researching and writing about ARGs, I wanted to make this book more focused onto *fandom* and a meditation on the field of *fan studies*.[1] It made sense to me to see ARGs as one more digital technology rather than a (limited) metaphor guiding fandom and fan cultures. Chapter 5 is thus a revisit to my original conclusion, a refocused look at ARGs within fan cultures, and a transition to the second half of the book, which opens up new forms of interaction between fans and technology.

The second half of this new edition contains three new chapters, each of which focuses on how fans have embraced a particular digital technology. I have used each of these new chapters to further explore what I see as the three concepts from *Digital Fandom* that could use more explanation: the Digi-gratis economy, demediation, and the philosophy of playfulness of fandom. In chapter 6 I open up my discussion of demediation to explore the interaction between mediated and face-to-face fandom through an analysis of the live-action playspace *MagiQuest*. In *MagiQuest*, players use both tangible and digital technology to explore mediated environments. By making obvious the mediation, and by immersing players within this ubiquity of media, *MagiQuest* creates a demediated environment. At the same time, the game has clear links to the ARG and to pervasive gaming, and serves as a mechanism by which fans can perform a particular type of fandom within a game environment. In chapter 7 I deepen my exploration of the Digi-gratis economy by focusing on crowdfunding. The digital environment has opened up new opportunities for fans to engage with the production process of their favored texts. The successful 2013 and 2014 Kickstarter campaigns for *Veronica Mars*, *Wish I Was Here*, and *Reading Rainbow*—and the failed 2013 Kickstarter campaign for Melissa Joan Hart's film *Darci's Walk of Shame*—illustrate how the entertainers used fandom to generate funds through this mash-up of gift and market economies. In this chapter I explore how crowdfunding fandom seemingly influences participatory culture by integrating the fan into the production process, although severely limiting and policing the fans' roles. The final chapter of *Digital Fandom 2.0* relates Tumblr and the philosophy of playfulness engendered by this social site to larger issues raised by the increasing disciplining of fan studies. Of all the social sites discussed in *Digital Fandom*, Tumblr appears to be the most attuned to fannishness, and in some ways reflects the visibility of fandom—it is populated by fans and non-fans, but it gives most of the content posted the feeling of fannishness. Tumblr has revolutionized fandom; fans on Tumblr feel "they were more 'themselves' on Tumblr than other social media sites."[2] Tumblr is a short-form blogging site, although unlike sites like Twitter there is no character limit to posts. Rather, Tumblr users tend to use images, animations, videos, and GIFs (Graphics Interchange Format) to reflect an emotional state. In chapter 8 I explore Tumblr *not* as a specific site of fan production, but rather as a metaphor for fan *studies* moving forward in the 21st century. The field of fan studies has significantly changed in the past decade. This chapter examines the consequences of the mainstreaming of fandom and the encroaching disciplining of fan studies as an academic field, using the polyvocality of Tumblr as a symbol and metaphor for these changes.

The second half of this book is thus less focused on specific *sites* and more focused on the way that fans are using technology in multiple ways in the digital environment. This, to me, is true to what is happening with today's transmediated and multi-mediated audience. As audiences become more and more used to second (and even third) screen viewing and scanning across multiple sites simultaneously, it no longer makes sense to view individual technologies out of their particular contexts. Just a year after the first edition of *Digital Fandom* came out, the *New York Times* released a study of technology in which they argued that "the blurring of [technology] lines is readily apparent among users of Tumblr."[3] Young people are moving from more static forms of social media (e.g., written blogs) to more dynamic (Tumblr and Facebook) even while the same types of fan activities are going on in multiple environments at once. I hope that *Digital Fandom 2.0* will provide a resource for these new types of digital interactions, although as with the first edition I'm sure that within a year a new technology will emerge that changes everything—fan creativity will always trump academic publishing. Time is funny that way.

Notes

1. Chess and Booth, "Lessons Down the Rabbit Hole"; Booth, *Game Play*.
2. Hillman, Procyk, and Neustaedter, "'alksjdf;lksfd,'" 6.
3. Kopytoff, "Blogs Wane."

Bibliography

Booth, Paul. *Game Play: Paratextuality in Contemporary Board Games*. New York: Bloomsbury, 2015.

Chess, Shira, and Paul Booth. "Lessons Down the Rabbit Hole: Alternate Reality Gaming in the Classroom." *New Media and Society* 16, no. 6 (2015): 1002–17.

Hillman, Serena, Jason Procyk, and Carman Neustaedter. "'alksjdf;lksfd': Tumblr and the Fandom User Experience." 2014. Accessed 23 March 2016. http://dl.acm.org/citation.cfm?id=2600887&dl=ACM&coll=DL&CFID=594181856&CFTOKEN=93137949.

Kopytoff, Verne G. "Blogs Wane as the Young Drift to Sites Like Twitter." *New York Times*, 20 February 2011. Accessed 29 December 2015. www.nytimes.com/2011/02/21/technology/internet/21blog.html.

INTRODUCTION TO THE FIRST EDITION

This is a book about fans, fan fiction (of a sort), and the fan communities that center on that fiction. But it is more than that, too: It's also a book about digital media and necessary changes in our contemporary study of media. Fans are a microcosm of this change, and the study of fans can become emblematic of studies in contemporary media.

I do not take this grandiose claim lightly, for the 21st century brings with it new forms of media beyond traditional models of broadcasting, beyond mass media, and beyond convergence. I believe that the contemporary media environment is more than the sum of its parts. This book is also about these changes to the media environment. And I believe that the best way of examining these claims is with the new tools and methodologies that take them into account. This book is about the future of media as it is lived, experienced, and loved.

What is it about media fans that continues to fascinate media scholars? For over thirty years, fan studies has been a way to explore audience participation with a media text. From the pioneering studies of John Fiske, through the germinal ethnographic research of Henry Jenkins, to more recent explorations of fans by Kristina Busse and Karen Hellekson; Jonathan Gray, Cornel Sandvoss, and C. Lee Harrington; Matt Hills, Cornel Sandvoss, Louisa Stein, Lynn

Zubernis, and Katherine Larsen (among many others), the productive work of media consumers has been at the forefront of audience analysis. The evolution of fan studies, however, has reached a critical moment: Traditional studies of media fandom in the digital age seem inadequately equipped to describe and analyze what I call the "philosophy of playfulness" we can observe in fans' use of today's digital technology.

What is a "philosophy of playfulness"? As I describe throughout the book, the contemporary media scene is complex, and rapidly becoming dependent on a culture of ludism: Today's media field is fun, playful, and exuberant. More so than at any other time, the media we use in our everyday lives have been personalized, individualized, and made pleasurable to use. We play with our media; it is malleable in our hands. The field of media studies needs to take into account this philosophy of playfulness in order to represent the media texts created by fans not just as individual fan fiction, fan videos, fan songs, or fan research, but rather as pieces of what fans use as a larger multivocal media "game."

The philosophy of playfulness represents a new way of looking at the practice of contemporary media studies that takes into account, and uses, the technologies that audiences are using to engage with media. In an age where digital media is becoming convergent and ubiquitous, it becomes important to analyze not just that media, but also the very discipline of media studies. This book is, therefore, a response to William Merrin, who writes, "ongoing changes in digital media needed to be placed at the core of the discipline; backward-looking perspectives needed to be left behind; and the historical basis of the discipline needed to be opened up to critical scrutiny."[1] The language we speak and the terminology with which we define our technology, will always delimit and define the ways in which we can understand our theoretical conceptions.[2] Fans offer a chance not only to explore the destabilization of concepts in mediation, but also to emphasize and paradoxically underplay New Media mediation at the same time.

But what do I mean by "New Media?" Quite plainly, I define New Media as those media forms that are digital, interactive, updatable, and ubiquitous: in this book represented by textual blogs, wikis, and social network sites like MySpace.[3] In the new section of the book for this revised edition, New Media are also represented by image-based blogs, sites to collate fan financing, and hybrid interactions between physical and digital media. Digital media are, at their most basic, defined by their constituent parts: the 1's and 0's of binary code. This has the effect of making all New Media texts boundless—or, rather, bounded in the same infinite mediation as all other New Media texts. Whereas once the technological device determined which media entertainment one might experience (TV programs were on TV, newspapers were in print, films were in the cinema), now the digitization of media means that the mediated entertainment does not depend on the technology of its viewership. Interactive media means not only that the media product can influence its viewership's identities, but also that viewers themselves can influence their interpretation of the media product.[4] The wiki I edit today may not be the wiki you view tomorrow, thanks to the interactive actions of a multitude of editors. Specific online interactive New Media "texts" include blogs, wikis, online comments, social network sites, crowdfunding sites, interactive technology, and all interactions between them.

New Media are updatable, which follows naturally from the interactivity of the mediation. Users of New Media can easily update the media "texts" as they use them. New Media are not static products, but instead are akin to what Raymond Williams describes as cultural processes.[5] A scholar cannot cite a wiki, for example, without acknowledging the date she downloaded that wiki, because the wiki exists just as much in time as it does in space. Indeed, looked at differently, wiki texts are ultimately time-less because they are also, ultimately, endless. Finally, New Media are ubiquitous, meaning not only that New Media surround and greet us on a daily basis, but also that they do so by attempting to disappear or naturalize within quotidian life. To *not* use new media technology today is to stand out in a crowd.

New Media and fandom are closely tied academic subjects; and indeed, it is important to look at fans in New Media because of the close ties between fandom and extant media objects.[6] The show *Babylon 5* is a good example of this, as creator J. Michael Straczynski made major changes to the show after fans offered input online about the pilot episode.[7] Deery has demonstrated how producers of the *X-Files*

> did read fan sites and did take on some suggestions occasionally. Writers used online fan names for bit parts and even dedicated one episode to a prominent online fan who had recently died. Viewer feedback also determined the prominence of character roles. The character of Skinner, for instance, was apparently expanded due to positive viewer reactions.[8]

In 2013 producers of Sy-Fy channel's *Defiance* integrated video games over the show's three seasons. Other fan-created content has made it into shows, albeit usually through officially sponsored channels. For example, the original run of the cult show *Heroes* featured a "Create Your Own Hero" fan-based promotion in which fans could go online and vote on various characteristics for two new heroes. The results of the voting determined the personality, appearance, and abilities of the heroes, among other attributes. After producers tallied the votes, the new heroes premiered in an online series.[9] The producers of *Heroes* maintained rights over the fan-created content, but the fact they sought out and utilized this content speaks to the ubiquity of fan culture in television production, and may have helped spark the reboot mini-series that premiered in 2015.

By enacting these changes, fans perform New Media; academics need to keep pace in order to keep current. In this book, I refer to this type of scholarship into fan performance specifically as *digital fandom*, a way of searching for new paradigms and new ways of seeing the media technology we use on a daily basis. Fans are only one way of looking at New Media, but fans' use of online, interactive technologies demonstrates an important step to an augmentation of scholarship in media studies. It is "digital" fandom not because it assumes that there is some inherent difference in the way digital technology affects fans, but rather because many creative fan practices rely on the characteristics of the digital. By integrating digital scholarship into fan studies, I hope to provide a unique view of contemporary audiences.

The trajectory of *Digital Fandom 2.0* thus follows two parallel paths. First, I augment traditional studies of media fandom with descriptions of the contemporary fan in a digital media environment. Second, I use this analysis of digital fandom to discuss media studies as a contemporary field of study. I undertake both a critical and a historical analysis of contemporary media by looking at contemporary uses of New Media. Importantly, I should note here that to make large generalizations of New Media based on a relatively small sample of texts, as I do, can be problematic: One is faced with offering the work of a few to represent larger trends. However, I use this form of textual analysis for a specific effect: Namely, I show under-examined and under-theorized aspects of New Media through an analysis of specific ways audiences are using New Media. I do not mean to indicate that this is the only way fans use New Media, nor do I argue that the fan texts at which I look in this book are universal. I simply highlight particular uses of New Media to indicate changes in our scholarly perception of this new, exciting, digital environment. The very fact they are happening is indicative of larger shifts in our use of media in general.

In order to examine this new toolbox and the contents within, I undertake a form of cultural criticism of fan-created, online interactive media texts. In this book I focus on fans of cult television programs, because their interaction with serial, on-going, extant media objects offers unique insights into fan communities and their use of online media. Cult texts have "vast, elaborate and densely populated fictional world[s] that [are] constructed episode-by-episode, extended and embellished by official secondary-level texts (episode guides, novelizations, comics, magazines) and fan-produced tertiary texts (fan fiction, cultural criticism essays, art, scratch videos)."[10] Fans "fill in the gaps" between episodic narratives: For example, a cult text asks the audience to answer questions like, who will win the *Game of Thrones*? Who is the Doctor in *Doctor Who*? In what ways will the future be different in *Star Trek*? Who are the hidden Cylons in *Battlestar Galactica*? Cult television's meaning exists not in any one place, but rather in the ethereal location in-between answer and question, in-between desire and the fulfillment of desire.[11] Jason Mittell argues that this form of "narrative complexity" complicates the cult world so much that we watch in order to "crack each program's central enigmas."[12]

My exploration of digital fandom and New Media Studies begins in chapter 2, with an examination of the blog as an intra-textual document. I show how the textual blog represents a form of interaction neither wholly intertextual nor individual. Specifically, a blog is made up of both the posting and the comments about that posting, and the "author" of a blog fan fiction is not a fan, per se, but rather a fandom. I use Bakhtinian theories of the carnivalesque to begin an *intra-textual* discussion of blogs. In contrast to intertextuality, which sees the ways texts work together and where meaning is uncovered between texts, intra-textuality examines the meaning that occurs inside the document text itself. If intertextuality connects disparate texts through implicit connection, intra-activity is the mechanism by which transmedia texts cohere. Through an examination of blog fan fiction written about the television show *Battlestar Galactica* (2003), I posit and discuss six factors that construct and determine intra-textuality.

The next chapter reveals the interactive potential of wikis as archives of fan narrative information by looking at the wikis for the cult TV shows *Lost* (2004) and *Heroes* (2006). I examine the implications of this narrative reconceptualization as an interactive narrative database. The narrative database forces a reappraisal of traditional narrative form, where narrative is split between what is told (the *story*) and the telling of that tale (the *discourse*). I explore the interactive narrative construction of *narractivity*, the process by which communal interactive action constructs and develops a narrative structure. Through an analysis of spoilers on these wikis, I describe three different ways fans can construe and construct the story.

In the fourth chapter, I show the blurring of boundaries between the real and the virtual in social network sites, specifically MySpace. Although MySpace is no longer one of the dominant SNSs (from 2005 until 2008, when the first version of this chapter was written, it was the largest social networking site in the world with 75.9 million users; in 2013 it was redesigned as a home for music networking and as of 2015 has 50.6 million users), it remains a historically relevant example of digital fandom at the end of the first decade of the 2000s.[13] In addition, the practices of fans on MySpace, which I call *identity roleplay*, have been taken up on more recent social media like Tumblr (see chapter 8).

My analysis of MySpace details how fans create a new form of textual "space" that is open to user interpretation and play, and normalizes the virtual and the real-world identities of their users. Taking off from the work of de Certeau, I reconceptualize the role of strategies and tactics in reading a cult television text, and demonstrate that MySpace offers an *interreality*, a third space of fan creation. I describe three ways this identity roleplay allows fans to rewrite a media text's characters on MySpace. Specifically, I show how other fans reread this roleplay as a narrativized sense of identity in online character personas, and then reproduce their fan community in this interreal space. For users of MySpace, creating a profile of a character is more than fan fiction, and more than textual poaching: It is a space where identities mingle.

In chapter 5 I interrogate the very methodology I have used in the book, by re-examining the concept of the Alternate Reality Game refracted through the lens of New Media Studies. An ARG is a game of hybrids. It is played in the physical world but utilizes digital technology to help players solve puzzles and decipher clues and puzzles. Although much of the game takes place online, some real world excursions blur contemporary distinctions of media and technology. Why do I link fans to ARGs in this manner?[14] I do not mean to suggest that only fans participate in ARGs, nor do I intend to imply that ARGs are created solely for fans' uses. However, I believe there is an important linkage between the two on the metaphorical level. I show that the ARG is the result of the outcome of the heavy mediation in a contemporary mediated society. Through an articulation of concepts described in Jay David Bolter and Richard Grusin's *Remediation*, I show how ARGs actually *demediate* real life by reversing and surpassing the polarity of hypermediacy and immediacy, a point emphasized in chapter 7 as well.[15] Hypermediacy represents the way media are made obvious: When we admire the special effects in a film, we are in awe of its obviousness. Immediacy, alternately, is the way a media can seem to disappear, and we can seem to forget that we are using or experiencing it. When we emerge bleary-eyed from a film, forgetting that we were even watching, we have been experiencing immediacy. I argue that ARGs represent a future of media entertainment that reverses these two conceptions through a demediated, playful hyper-immersion, and assert that just as useful a metaphor as the ARG makes, so too must it fall under the very re-examination it portends.

The new chapters added to the second edition extend the conversation about fans' use (and the use of fans within) of digital technology. Chapter 6 deepens my exploration of demediation by bringing the physical to bear in an analysis of role playing through gaming. I develop my analysis through an examination of *MagiQuest* as a site of interaction between the digital and the tangible. Chapter 7 explores crowdfunding as a way of harnessing fans' affect of a media product. By exploring both successful and unsuccessful crowdfunding campaigns, I investigate the influence of digital technology in teasing out elements of the Digi-gratis economy. Finally, chapter 8 uses Tumblr as a metaphor for understanding the impending disciplining of fan studies in the academy. I show that the polyvocality of Tumblr can serve as a powerful metaphor for negotiating and maintaining the multiple viewpoints within fan studies.

Digital Fandom 2.0 illustrates the connection between fans' use of online, interactive technologies and New Media in general. It also discusses the ways that digital fandom provides a lens through which we can focus on changes in our contemporary media landscape. We must examine the unquestioned assumptions that gird contemporary analysis of media. As Benkler indicates, the traditional theories of media and cultural studies cannot account for both potential and the actualization of online practices.[16] We must rethink traditional ways of dealing with issues such as ownership, originality, authorship, commercialization, and copyright in the digital world.

This book is about media. It is media. It relies on and examines media. But more so, it offers a glimpse of a mediated future as seen through contemporary fan practices. It should not be the end of the conversation, though; this should not be the last post in our conversation. Media, like fans, are continually developing, and it is only through a constant and vigilant observation of these changes that scholars, students, and practitioners of media can hope to stay current. If anything, therefore, this book is about a particular time and a particular mediated state.

This book is an attempt, therefore, to describe fandom in a digital culture, and to introduce, problematize, and explain new conceptions in media studies as demonstrated by fans. It is my hope that readers will come away not only with some new ideas about the contemporary media landscape, but also with new questions to ask, and new avenues of research to explore. To that end, I do not intend this book to be a conclusion; rather, it is what I hope to be the first line of dialogue in a conversation—the first Tumblr comment, perhaps—about digital media, about digital fandom, and about the place of both in digital scholarship.

Notes

1. Merrin, "Media Studies 2.0," 19.
2. Gunkel, *Hacking Cyberspace*, 1.
3. These are not the only flavors of New Media; but for the sake of brevity it is on these I focus.
4. Sandvoss, *Fans*, 101.
5. Williams, *Culture and Materialism*, 48.
6. I use the term "extant media object" in this book to refer to the text for which a fan shows appreciation, or the object of the fan's attention.
7. "What Effect Have Fans Had?"
8. Deery, "TV.com," 177.
9. "Create Your Own Hero."
10. Gwenllian-Jones, "Virtual Reality and Cult Television," 84.
11. Hills, "Defining Cult TV, 509–10: Hills shows how the definition of "cult" relies on the intersection of production, intertextuality, and fan appreciation; see also, Hills, *Fan Cultures*, 131–5.
12. Mittell, "Narrative Complexity," 38.
13. Smith, "By the Numbers."
14. In the first edition of the book, ARGs were a focus throughout. I've retained some of that language for this chapter in the second edition, but edited out other portions for the sake of space.
15. Bolter and Grusin, *Remediation.*
16. Benkler, "Coase's Penguin," 169.

Bibliography

"Create Your Own Hero" <http://heroeswiki.com/Create_Your_Hero>

"What Effect Have Fans Had?" The Lurkers Guide to Babylon 5, 19 March 2006. Accessed 21 March 2016. http://www.midwinter.com/lurk/resources/fans.html#effect.

Benkler, Yochai. "Coase's Penguin, or, Linux and the Nature of the Firm." In *CODE: Collaborative Ownership and the Digital Economy*, edited by Rishab Aiyer Ghosh, 169–206. Cambridge, MA: The MIT Press, 2005.

Bolter, Jay David, and Richard Grusin. *Remediation*. Cambridge, MA: The MIT Press, 1999.

Deery, June. "TV.com: Participatory Viewing on the Web." *The Journal of Popular Culture* 37, no. 2 (2003): 161–83.

Gunkel, David. *Hacking Cyberspace*. Boulder, CO: Westview Press, 2001.

Gwenllian-Jones, Sara. "Virtual Reality and Cult Television." In *Cult Television*, edited by Sara Gwenllian-Jones and Roberta E. Pearson, 83–98. Minneapolis: University of Minnesota Press, 2004.

Hills, Matt. "Defining Cult TV: Texts, Inter-texts, and Fan Audiences." In *The Television Studies Reader*, edited by Robert C. Allen and Annette Hill, 509–23. London: Routledge, 2004.

Hills, Matt. *Fan Cultures*. London: Routledge, 2002.

Merrin, William. "Media Studies 2.0: Upgrading and Open-Sourcing the Discipline." *Interactions: Studies in Communication and Culture* 1, no 1 (2009): 17–34.

Mittell, Jason. "Narrative Complexity in Contemporary American Television." *The Velvet Light Trap* 58 (2006): 29–40.

Sandvoss, Cornel. *Fans: The Mirror of Consumption*. Malden, MA: Polity Press, 2005.

Smith, Craig. "By the Numbers: 17 MySpace Stats and Facts Then and Now," 14 August 2015. Accessed 23 March 2016. http://expandedramblings.com/index.php/myspace-stats-then-now/.

Williams, Raymond. *Culture and Materialism*. London: Verso, 1980.

PART I

HISTORICAL DIGITAL FANDOM

· 1 ·

FANDOM IN THE DIGITAL ENVIRONMENT

> *[Texts] are mirrors in the image of those who wrote them. They reflect their concerns, questions, desires, life, death. … They're living beings: you have to know how to feed them, protect them …*
>
> —Pérez-Reverte, *The Club Dumas* (p. 60)

> *[I]n literature there are never any clear boundaries. Everything is dependent on everything else, and one thing is superimposed on top of another. It all ends up as a complicated intertextual game …*
>
> —Pérez-Reverte, *The Club Dumas* (p. 95)

In traditional parlance, a fan is a person who invests time and energy into thinking about, or interacting with, a media text: in other words, one who is enraptured by a particular media object. As the quotations from Arturo Pérez-Reverte's novel *The Club Dumas* indicate, fandom can even extend towards antiquarian books. Corso, the hero of Pérez-Reverte's novel, deals in centuries-old occult manuscripts that their collectors religiously revere and for whom these texts lead to new worlds: worlds inhabited by demons and devils, angels and authors. Indeed, in their quasi-religious devotion to these ancient texts, the book collectors are also, in a sense, fans, although perhaps of a more negative tenor. *The Club Dumas*, while ostensibly a novel about book collecting and the occult, also provides us with a traditional popular cultural reading of fandom, and through it we can glean a number of salient details about popular conceptions of the fan. As something no "respectable" book collector would consider himself, fans are "vulgar," are "miserable wretch[s]," cannot distinguish "the line between fantasy and reality," and are akin to "innocents and children."[1]

However, the study of fans over the past three decades has found that more complex associations can be made about fans than these crude generalizations indicate. The very interaction with the text that these antiquarian bibliophiles crave mirrors, in many ways, the interactions of fans with cult television texts. For Corso texts are more than just the words on a page. Corso discovers that texts can also grow and change with every reader and every reading, and can adopt new meanings as new readers reread them throughout time. At first, the connection between the text and its reader is tactile and sensual: a collector/fan holds a book, strokes the crisp leather binding, and smells the ink and printing. Further in the story, however, Corso learns that the connections between texts and their readers can be more intimate than the five senses allow. Texts enter a reader's soul, and become part of that reader: Readers imagine themselves to be in the story, a part of the literature.

For fan scholars, fans are representative of audiences as those book collectors are for Pérez-Reverte. Fans make explicit what we all do implicitly: That is, we actively read and engage with media texts on a daily basis. In this book, I add to this body of fan scholarship by positing that, in the digital age, fans are an exemplar of a change in media studies itself. This is a change William Merrin refers to as "Media Studies 2.0," a form of scholarship that uses the digital technology of today's audiences to examine media as it is currently being consumed and produced. Merrin argues that "today, only a post-broadcasting, digital paradigm can explain our contemporary media experience," and this is available to us by exploring not just new types of media, but also new users of and practices in that media environment.[2] I show that what William Merrin calls Media Studies 2.0 can be tangibly witnessed and enacted in the study of fans as "digital fandom."[3]

One key characteristic of digital fandom is how fans' uses of technologies bring a sense of playfulness to the work of active reading. The work that fans put into creating fan fiction, fan blogs, fan videos, fan wikis, or other fan works reveals the ludic nature of digital technology and communication. In this book I illustrate a trend in contemporary media to ludicize texts, or how audiences create a *philosophy of playfulness*. This is not to indicate that fans don't take themselves—or their work—seriously. Far from it. Rather, I mean to show that there would be no fan fiction, no fan work, if fans did not derive some playful pleasure from the act or existence of their fandom. The type of playfulness fans engage with in their interaction with each other and with extant media texts undergirds fan communication in the digital age. Quite simply, being a fan, engaging with texts, is *fun*; and as fandom becomes a more mainstream and visible identity, this ludism—this philosophy of playfulness—has permeated all levels of fan and audience interaction. Participating in philosophy of playfulness means engaging in the media landscape as if each element were a game; as if fans were players; and as if each text were open to all the imaginative possibilities of free play.

In *Digital Fandom 2.0*, using a methodology from Matt Hills's *Fan Cultures*, I critically examine eight dialectics at play within media studies.[4] By looking at how fans, as explicitly pro-active media audiences, use key digital environments, I show that New Media destabilizes these traditional dialectics, and that destabilization symbolically functions to represent larger issues at stake in media studies. The dialectics I examine in this book are:

- Work/Play (Chapter 1)
- Text/Intertext (Chapter 2)
- Narrative/Interactivity (Chapter 3)
- Real/Virtual (Chapter 4)
- Immediate/Hypermediate (Chapter 5)
- Tangible/Digital (Chapter 6)
- Fan/Producer (Chapter 7)
- Fandom/Fan Studies (Chapter 8)

Re-Writing Media Studies

Fandom has become a ubiquitous frame of reference for media professionals. As more and more television shows develop serialized plots that extend beyond the boundaries of the television set, producers court fannish sensibility. Fans engage in television shows in a way that makes explicit the activity that many contemporary viewers perform implicitly. Further, fans offer us something we can all relate to. Whether cult television, sports, food, opera, or books—we are all fans of something, and we all become emotionally involved over something. Studying fans, in effect, is a way of studying ourselves.

Fandom is a heavily researched area of scholarship. The start of popular media fandom in the early 1930s coincides with an important point in the history of popular media technology.[5] Within approximately a decade of each other, *The Jazz Singer*, the first motion picture with sound, debuted (1927), Philo Taylor Farnsworth invented the first electric television (1927), FM radio was invented (1933), and regular electronic television service began in the United States (1939). The tumultuous decade of the Great Depression may have seen the economic downfall of much of the world community, but it also saw the invention and spread of some of its most technologically innovative media devices. As these devices became popular, so too did the media entertainment displayed on them. From the 1930s onward, groups of people who all enjoyed the same media entertainment were referred to as "fans," from the pejorative meaning of the term "fanatic."[6]

The importance of fans and fan studies cannot be overestimated. Fan studies invokes and participates in an ongoing discussion with, among others, popular culture studies, audience analysis, communication studies, psychology, media studies, and technology studies. Fandom can be experienced individually or in a community, and as such, fan studies has recently been situated next to, and contributes to, ongoing academic dialogues about the nature of identity, the concept of the text, and issues of community. Different types of fan studies examine different types of fans: some, like Henry Jenkins's *Textual Poachers*, and John Tulloch and Henry Jenkins's *Science Fiction Audiences*, have specifically examined fans' experiences with individual media texts. Nicholas Abercrombie and Brian Longhurst examine fans as one important component on a spectrum of audience types. Matt Hills examines the cultural theory behind analyses of fans and the development of a "fan studies" in the academy, while fan studies authors like Camile Bacon-Smith, Constance Penley, and Rhiannon Bury all look specifically at different elements of fandom, e.g., conventions, the writing of fan fiction, and communities. Collections like Kristina Busse and Karen Hellekson's *Fan Fiction and Fan Communities in the Age of the Internet* and Jonathan Gray, Cornel Sandvoss, and C. Lee Harrington's *Fandom: Identities and Communities in a Mediated World* examine fans in online discourses, and monographs like Louisa Stein's *Millennial Fandom* highlight fans in the digital era.[7] Since the publication of the first edition of *Digital Fandom*, even more writers, thinkers, and scholars have turned fan studies into an even more developed discipline. Fan studies shows the *productive consumption* at the heart of fan affect.[8]

Fans are also well-integrated into the media landscape. As Jenkins shows, television media consumers can be broken into three general categories: Zappers, Casuals, and Loyals. Zappers flit around the television channels, flipping between shows but watching no program in particular. Casuals enjoy watching different television shows when they are on the air, but rarely schedule time to watch. Loyals, as fans,

> actually watch fewer hours of television each week than the general population: they cherry pick those shows that best satisfy their interests, they give themselves over fully to them, they tape them and may watch them more than one time; they spend more of their social time talking about them; and they are more likely to pursue content across media channels.[9]

Although broadcast-era advertising tended to target Zappers, recent media advertisers have started to target Loyals. In fact, according to Lawrence Lessig, media corporations are even creating new types of content to appeal directly to these fans.[10] Loyals are fans, and although they may be in the minority of viewers, "advertisers are increasingly realizing that they may be better advised investing their dollars behind shows, that have a high favorability rather than shows that have high ratings."[11] Indeed, Loyals tend to purchase more products, percentage-wise, than do other types of media audiences. When these producers court fans and fan communities, the resulting increase in critical praise and profits can also help the media company. For example, according to Elana Shefrin, one of the main factors behind the success of Peter Jackson's *The Lord of the Rings* film trilogy was the fact that the producers actively sought the advice and support of the *The Lord of the Rings* fan community.[12] Fans seek out shows, interact with them, and influence the type of shows that are made. As advertisers target fans, media corporations create new types of content that appeals to fans. We can see this in advertising campaigns for films such as *District 9*, which blanketed internet sites with what Jenkins, Ford, and Green call "spreadable media content," and in the mainstream popularity of transmediated cult texts like the Marvel Cinematic Universe.[13]

Thus, it becomes important to look at fans as a target audience for advertisers; for, to have a critical understanding of public relations and modern media political economy means we must not only understand fandom, but we must also examine the extant media objects of fandom. The study of fandom is becoming a crucial analytic tool in our changing, digital culture. Fans typically utilize their technological capacities, their communal intelligence, their individual knowledge base, and their social interaction skills to investigate and explore media. Fans lie at the forefront of our rapidly changing media environment. As Jenkins has written, the future of media entertainment belongs to the fan; at the same time, he also notes how media commentators avoid using the term "fan," both to avoid the same tired stereotypes, and also because the practices of the fan have become so common that to differentiate fans from consumers seems futile.[14]

Finally, whether or not we are fans of cult television, of the type I describe in this book, chances are that everyone is a fan of *something*. We may not be the fans of *Doctor Who* or *Star Trek* that Tulloch and Jenkins describe, but we might be fans of music, of sports, or even food. An analysis of cult media fans reveals as much about the analysis as it does about the fan, and this analysis can be used in other contexts, in terms of other types of fans or audiences. As Sandvoss shows, a media object "is part of the fan's (sense of) self," as some fans use their identity *as* fans as a way to differentiate themselves from other media audiences.[15] Being a fan means identifying with a media text, and studying fans allows us to "explore some of the key mechanisms through which we interact with the mediated world at the heart of our social, political, and cultural realities and identities."[16] Whatever we are fans of, we base part of our identity on our appreciation of that fandom.

As I explore throughout *Digital Fandom 2.0*, the more fans become of interest to those that produce and provide media, the more the study of fans becomes integral to understanding media production and scholarship. Media scholar Hector Postigo demonstrates that "practices of one community (fandom) are bleeding into the other (mass culture); ... changes in technologies that mediate cultural products have made more porous the boundaries to practicing the mores of fan culture."[17] In an era when the mass media saturate our lives, and the boundaries between cult and mass cultures have blurred, fandom becomes an important mechanism to understand contemporary digital culture. For, as Sandvoss goes on to show, "it has become impossible to discuss popular consumption without reference to fandom and fan theory, just as it has become next to impossible to find realms of public life which are unaffected by fandom."[18] Fans are becoming ubiquitous audience members, and fans' interactions with and within online technologies provides a rich area of media scholarship.

Throughout this book, I follow two seemingly contradictory claims. The first is determinist in nature and argues that *new technology develops New Media, which calls for new scholarship*. The second is constructivist in nature and mirrors Robin Williams and David Edge's assertion that "technology does not develop according to an inner technical logic but is instead a social product, pattered by the conditions of its creation and use": In other words, *new technology cannot be understood without a reading of the cultural theory that has come before.*[19] Both claims, by themselves, are limiting and self-contained. But the contradiction between determinism and constructivism is itself a useful heuristic for the examination and analysis of the media, which are, after all, both constructed and determined. In reality, both arguments are, by themselves, reductive: To assert that cultural effect or change has only one cause is to discount the multiple forces at work in society.

I believe that neither determinism nor constructivism is enough to explain the complicated relationship between media, fandom, and technology. Neither factor can be entirely responsible for the other. I agree with Potts, who calls for a more sophisticated form of analysis that integrates both theories: a bridge between determinism and constructivism.[20] Thus, by seeing *beyond* the paradox—by using the contradiction as a way to break out of a determinist dialectic—we can effectively witness the inherent tensions at the heart of media studies; not to resolve them, but to embrace them.[21] In this way, and as I describe in detail later, I do not attempt either a purely determinist argument, nor an entirely constructivist one, but use the inherent tension both as, and to open up new avenues of, digital scholarship.

As Raymond Williams cautions us, however, any new discussion of scholarship must be sure to start at the level of language: In intellectual discourse, scholars must be careful to use the same vocabulary with the same meanings, and to articulate exactly what they mean in their discourse.[22] My investigation of digital fandom is no exception. I present three concepts in this chapter to lay the groundwork for the rest of my argument; however, each is discussed in greater depth throughout the book.

Digital Fandom

At the heart of my argument, *digital fandom* implies an augmentation to scholarly research about fans, a further stepping-stone in scholarship about fandom. Digital fandom examines fandom as the work of a collective community, and not fans as individual audience members. By "community," I refer to the social grouping of individuals with shared interests, joined together through some form of mechanism of membership—the self-selected organization of a group of fans who both enjoy an extant media object, and who create additional content about that extant media object. I refer to the conception of the community as written by Kollock and Smith, who describe communities as "groups of people who meet to share information, discuss mutual interest, play games, and carry out business."[23] Erickson brings up the topic of membership.

Williams asserts that term "community" is problematic because "unlike all other terms of social organization (state, nation, society, etc.) it seems never to be used unfavourably, and never to be given any positive opposing or distinguishing term."[24] Studies of fandom have concentrated on the communalization of fans, as *New York Times* columnist Rob Walker points out in his description of *Star Trek* fandom.[25] Fan communities have always been the object of study in fan studies, but usually from the point of view of the individual fan's contribution to the whole. I ask, instead, what of the contribution of *the community* to the whole? How does a community contribute collectively?

A fan's work tends to be described on a production/consumption dialectic. For example, the work first undertaken by Jenkins places fandom on a continuum between producer and consumer—a metaphor based in market or commodity economics.[26] The fan "consumes" the text that media conglomerates "produce" in order to sell advertising time. These fan studies use the metaphor of the market economy to describe the work of fans.[27] Academic studies of fandom have described fans as a combination of both producer and consumer—with such neologisms as "produser," "prosumer," and "prosumption"—but still remain tethered to a binary that views fan productivity in relation to the fan's consumptive attitudes.[28] This commodity metaphor, however, has limitations. In order for the production/consumption metaphor to function, we must assume that *consumption also implies destruction*—the destruction of possession, or the destruction of demand. As I describe the Digi-gratis economy, this market economy model assumes a limited text and a limited market for the text.

Yet, traditional studies of fandom discuss the fans' productive abilities; these abilities resemble what de Certeau terms "perruque": labor undertaken and disguised as work for an employer. With "perruque," for example, a worker might sneak into his employer's carpentry shop after hours with his own tools and materials to make his own chair. This worker steals nothing tangible (for nothing leaves the place of work), but instead uses the tools of others to complete work for him-/herself. He steals only from his own leisure time.[29] Axel Bruns extended the idea of digital labor to examine the "produser," who:

> highlights that within the communities which engage in the collaborative creation and extension of information and knowledge … the role of 'consumer' and even that of 'end user' have long disappeared, and the distinctions between producers and users of content have faded into comparative insignificance.

In other words, the produser is a digital perruque, producing digital goods with a near-professional quality. Defined as both a producer and a user of content, the produser has several characteristics similar to those that fans possess: The produser exists in a community of other produsers, is part of a participatory culture, and collaborates through "intercreativity."[30] In many ways, produsage is integral to digital fandom.

Digital fandom provides a balance to this "productive" dichotomy metaphor of internet communication by revealing non-market economic antecedents to fans' use of media. Key to both Bruns's and my argument is the shift in the nature of the digital product. Whereas once works of fan fiction, fan art, or any fan-created imaginative work inspired by an extant media object would have been *discrete* and *whole* works of original production akin to the materials used by de Certeau's perruque, the ubiquity of a digital-based creation and reception of media products has led to more fluid and distributed works of creative enterprise. Indeed, as I go on to argue throughout the book, this one key shift—from finished product to unfinished process—affects all aspects of the way scholars study fans, audiences, and the media.

Fans are exemplars of a New Media Studies, as they rewrite versions of extant media objects not to revise them, but to reimagine them. Fans move and shape source objects, appropriating and re-appropriating material in what Lev Manovich refers to as a remix culture.[31] Fans take advantage of what Jenkins notes about media characters: "once television characters enter into a broader circulation, intrude into our living rooms, pervade the fabric of our society, they belong to their audience and not simply to the artists who originated them."[32] But fans do not write in a vacuum. As Bacon-Smith has noted, fans write in order to be *read*, to be interpreted by a community.[33] By examining fan-created texts as key aspects in the formulation of active communities, digital fandom articulates a new way of conceiving of textual creation. Fan community is playful and central.

In the past, fan fiction has been described as the writing of stories using characters and/or plot lines that have been created either by professional media producers or by other fans. In this conception of fan fiction, the fan, as a consumer of a media product, produces a new strand of a story that may or may not fit into the canon of the source text. A story canon is the complete fictional universe deemed (either by the fans or by the media creators) "authentic," or an accurate history of that story world. Cult television shows, which have an extensive hyperdiegesis, often complicate issues of canonicity, because specific units of knowledge about a show might contradict other units of information (e.g., Spock is an only child in the original *Star Trek* series, but has a half-brother in the fifth film, *The Final Frontier*: The "canonicity" of

Spock's sibling is thus in doubt for many fans).[34] Fan fiction writers produce strands of the story that are not officially considered "canon," but that other fans can read as adjacent to canon; as "a work in progress," to use Busse and Hellekson's phrase.[35]

The open-ended enigmas of shows like *Lost* (2004), *Battlestar Galactica* (2003), and *Doctor Who* (1963/2005) encourage fandom. The meaning of that show for the audience exists between what is revealed and what is eternally hidden, and exists not in any one place, but rather in the ethereal location in-between answer and question, in-between desire and the fulfillment of desire. This cultish desire, in a Lacanian sense, can never be entirely fulfilled in a cult narrative, for the cult show rarely resolves.[36] Fans attempt to fill in this ethereal space through the fulfillment of the "hyperdiegesis," an immense and immersive narrative structure.[37] Fan fiction becomes a way to complete a narrative, through a rewriting of the boundaries of the cult text canon.

Digital fandom builds on and contributes to this historical development of fan studies.[38] Fan studies looks at the socio-political dynamics of fandom, how fans create and re-create "social and cultural hierarchies … as a reflection and further manifestation of our social, cultural, and economic capital."[39] Taking off from work like Bourdieu's *Distinction*, fan studies also looks at the fan community's hierarchical societies. Indeed, for this type of fan studies, scholars analyze the fan not merely as a "viewer," but as an active participant in a complicated social structure. Hierarchy in fan communities, however, is not always clear-cut. Often, those at the "top" of the fan hierarchy are there because they either have "an active and knowledgeable" comprehension of the extant media object, or because they have a radically different view of the extant media object, one so distinctive and individual it reveals itself as an equal companion to the original.[40] Key here is the fan's reproduction, or representation, of the source text's canon. To know the most arcane details from the extant media object, even if one changes those details in the fan fiction, is to be intimately tied with the object of devotion. Fans that can "quote whole sections" of a narrative are thus positioned higher on that hierarchy, in a culturally determined manner.[41]

Digital fandom is perhaps most closely aligned with the fan studies that examine the relationship between fan communities and the internet. For example, Baym focused on the interaction between technologies and practices of fan communities.[42] Jenkins and Hills, among others, have also taken this stance, and both illustrate how fans use the internet and the World Wide Web to create new texts. For example, in *Convergence Culture*, Jenkins illustrates how message boards and forums become knowledge communities that structure and organize the fans' knowledge about a particular text.[43] Hills describes how, for instance, *X-Files* fans "have become increasingly enmeshed within the rhythms and temporalities of broadcasting, so that fans now go online to discuss new episodes immediately after the episode's transmission time—or even during ad-breaks" for a "just in time" fandom.[44] (Hills's quotation is relevant even as I rewrite this chapter: The 2016 mini-series of the *X-Files* has been released, spurring hundreds of thousands of fan works online.) Dialogue through forums and listservs creates meaning for the members of the community separated from the meaning contained (or read) within the extant media object.

By examining how fans cooperate with each other, digital fandom helps us explore the philosophy of playfulness that fans utilize in their interactions. This philosophy of playfulness lies at the heart of any fan interaction: Examining fan-created texts as "imaginary entertainment" online is not enough. Digital fandom not only looks at how fans produce new texts, but also how, in that process, they playfully reproduce old ones. This production-through-reproduction is perhaps best demonstrated by *Star Wars Uncut*, a remarkable online recreation of *Star Wars: A New Hope*). Users of the site choose a 15-second clip from George Lucas's film, and then remake that 15-second clip in whichever style they would like. Some animate their clip using Lego toys. Others re-create the scene as closely as possible. At least one user has subtitled his one-year-old's babbling with dialogue from the Han Solo/Greedo Mos Eisley Cantina scene (and yes, Han shoots first). As the collection of user-generated clips grows, the creator of the site, Casey Pugh, reassembles them in their original filmic order, inserts traditional *Star Wars* music, and creates anew a fan-made version of their favorite film.[45]

What is remarkable about the *Star Wars Uncut* project is not that *individual* fans are creating their own versions of *Star Wars*, but that a *group* of fans is. The fannish rewriting of *Star Wars* here takes on a completely different tenor from that originally described by Jenkins; individual fans have been rewriting filmic texts for well over half a century, in the forms of fanzines originally, and then in more and more complexity as the technology has improved and aided the fan work. *Star Wars* has been a perennial fan favorite, and fans embrace the web to take their fan productions public.[46] The fan process with *Star Wars Uncut*, however, is different. It's not an individual fan re-making a scene or a film for others to watch, but rather the entire community remaking the complete film for each other to watch.

What the existence of *Star Wars Uncut* reveals is that a re-evaluation of the way fan studies functions is necessary in order to fully understand the dramatic and dynamic process of fannish (re)production. This process is also complicated in this case by the fact that the entire film is available in multiple versions because different people can all remake different scenes:

> Multiple people are allowed to have a scene so that everyone can be a part of Star Wars: Uncut! A rating system is in the works to choose everyone's favorite version of each scene. Don't worry though- All scenes will be kept forever so that we can watch Star Wars a completely different way every time![47]

Rereading the text is never a singular event, but a continual process or rearticulation and reevaluation.

Digital fandom becomes a useful tool for examining *Star Wars Uncut* not because the fan community is creating something new, but rather because they're reproducing something they've already known: *Star Wars*. The ultimate goal with *Star Wars Uncut* is not to produce a new take on *Star Wars*, or to showcase individual talents at digital filmmaking (although both of those are appreciated within the individual 15-second clips). Rather, the ultimate goal is to recreate *Star Wars* as filtered through the fan community; a film seen through the eyes of its fans.[48]

As I show throughout this book, digital fandom does not look at these fans as doing something entirely new, but rather looks at these fans in new ways. Fans both influence and are influenced by technology not just as tools but also as necessary and catalytic mechanisms to alter their subjective experiences of cultural life. Fans are techno-literate, liberating the experience of viewership through self-conscious knowledge about the technology of mediation. Fans use digital technology not only to create, to change, to appropriate, to poach, or to write, but also to share, to experience together, to become alive with community. Fans rewrite not just the extant media object, but also the state of media studies itself.

Web Commons

Digital fandom thus requires a re-evaluation of how we view the web, as fans form a type of Web Commons. The term *Web Commons* characterizes not a shift in the technological development of the web, but rather a conceptual change in the ways the web is used. By using the term Web Commons, I indicate a connection between the web and the feudal commons: "a resource shared by a group of people that is subject to social dilemmas."[49] I derive the term "commons" from Lessig's description of the "creative commons."[50] Customarily, we understand the commons metaphor in one of two ways. In "common-pool" commons, anyone can use the resources of the commons independently of external control: For example, air is common-pool, in that anyone can breathe it without paying for it. In a "common-property" commons, an external party (or the community within the commons) sets particular rules in order to establish order and stability within the commons. For example, anyone in an office might use the communal coffeemaker, but everyone must follow the communally established rules and contribute supplies, funds, or other necessities for this common-property to function.

The web is a unique entity that straddles the commons property/pool divide. On the one hand, the web is open and free. Anyone can use it and there are few rules to follow. The components needed to access the web may be necessary—a computer, a connection—but the actual web itself could be considered common-pool.[51] My use of the web does not impinge on your use of the web: We cannot "use up" the web, a concept on which the 2014 debates on Net Neutrality hinge. On the other hand, there are a great many socially constructed and culturally supported norms for online communication that make the web seem more like common-property. There is a vast infrastructure that prohibits much amateur usage of the net, and economic class affects availability of the materials. Internet protocol helps define and delimit the standards by which people write and publish online. Social norms prescribe ways of acting on the web and in different technologies through the web. Thus, while the actual *use* of the web may be common-pool, the structure of the web can be seen as common-property.

Fan works also straddle this commons property/pool-divide. Fans who write fan fiction do so in a perilous position in between subversion and support. In many ways, many fan fiction authors see the extant media object as common-pool, while some media producers see it as common-property—or as non-commons and proprietary. This struggle over textual property is what Jenkins describes as "one which has had to be actively fought or at least negotiated between fans and producers in almost every media fandom."[52] For example, while fans are aware of their infringement of copyright, some openly mock the copyright system through ludic comments and playful banter, and others take it more seriously. Yet, there is a contemporary recognition on the part of some media producers that the fans' use of the extant media object as common-pool might benefit the media company. As Anelli has shown in the cases of *Harry Potter* fan fiction, Warner Brothers tends to turn a blind eye to most forms of non-pornographic fan fiction, as they see fan fiction working as free marketing for *Harry Potter*.[53]

I am not the first to articulate a connection between the web and the commons: Benkler also points out the close connection between the two.[54] However, in contrast to these other descriptions, I highlight the commons in its relation to a different conception of the web: that of an informational portal.[55] Two non-mutually exclusive practices exist online, not as different technologies or tools, but rather as different conceptions of the web's usage.[56] The Information Web focuses on ways the web remediates print. In this conception, the web is like a vast library of books, hypertextually linked to one another. Although it has a global reach and a ubiquitous nature, the "Information Web" is ultimately static and non-interactive. In the Information Web, the web is conceptualized as a vast, hypertextually connected remediation of print and broadcasting.[57] Users look at web pages and browse through links in order to find information or answer questions. In his transmission view of communication, James Carey argues that the goal is the movement of information: the transportation of a message across space.

The Web Commons marks an alternate conception of the web, one that Carey might describe as not "the extension of messages in space but … the maintenance of society in time; not the act of imparting information but the representation of shared beliefs." This "ritual" view of communication represents the sociality of the web, the way in which users of the web log on to chat, to play games with others, or to communicate to other people about what Carey calls "a particular view of the world."[58] Communication as a form of dialogue or shared meaning is key here, as the use of the Web Commons informs social practices.

Importantly, people use the web—just as they have used any communication technology—for both transmission and ritual communicative purposes at the same time. The Information Web and the Web Commons work together—they have always worked together. To picture this, think of the web as a library. One can walk through a library and see scores of individual cubicles. You can think of these cubicles as static, Information Web portals. Walking past, you can see what the person is doing, but unless you are very rude (a troll?) or want to copy off people's work (a hacker?) you cannot actually change anything. Now walk past the cubicles and see the large table in the center of the library—around the table sit dozens of people all working on one project. Input is communal and interactive. The community may input your opinions into that project as you walk by. This is the Web Commons, and it exists, must exist, with the Information Web.

In contrast, other recent conceptions of web communality have focused on a technological change in the web. The most salient example is the term "Web 2.0" (a problematic term for some media scholars).[59] For publisher Tim O'Reilly who coined the term, "Web 2.0" defines the way that the web has developed from a technologically static medium to a dynamic one. In the "Web 1.0" world, he notes, the key to accessing the internet was the browser—Netscape, Internet Explorer, Safari, Opera, etc.—that was contained on the user's desktop.[60] This browser could access the internet, but by virtue of the fact it was proprietary software, users were beholden to the design of the product. The browser window, quite literally akin to Burke's terministic screens, "necessarily directs the attention" of the user "into some channels rather than others."[61] What this means is that the way we use technology colors our interpretation of what we see in that technology. For O'Reilly, when browsers are given "control over standards for displaying content and applications in the browser," we would eventually only see what Netscape wanted us to see.[62]

For the "Web 2.0" world, however, O'Reilly presents a key piece of software: Google, the online application/web tool that operates as a web portal/search engine to the rest of the web. For O'Reilly, Google is a technological innovation, located "in the space between browser and search engine and destination content server, as an enabler or middleman between the user and his or her online experience."[63] Google, a free, easy-to-use, and constantly updatable application that anyone online can access and use, is, for O'Reilly, the shepherd of a new technological revolution, one in which the key commodity is not a *product* but *access to a product.*

O'Reilly falls prey to a key fallacy here: That is, Google is still viewed in a browser (the Google Chrome browser seems to hybridize O'Reilly and my arguments). O'Reilly creates false dichotomy: Google is not really equivalent to Netscape, any more than *The Tempest* is equivalent to The Globe Theatre. One can be a part of the other, but either can exist without the other as well. Ultimately, "Web 2.0," describes a paradigm shift in technology via technology. While useful for describing a change in the way the web works, what the conceptual difference between "Web 1.0" and "Web 2.0" lacks is a description of how we work in the web.

The Web Commons is thus not a "new web," but rather a new way of *conceptualizing* how people use the internet. Like recent technological developments in the web the Web Commons indicates an evolution in our understanding of the web.[64] The Web Commons is a *mind-set*, not a specific form of technology. It is more than just social media. Although it exists online, it extends beyond the internet and represents a substantive alteration in how we conceive of the socialization of information. Lawrence Lessig suggests that the interactions and precedents we set online spill over into our offline culture: It matters what happens online, because it influences our offline lives.[65] Just as there is a change in the interconnected nature of contemporary media, so too is there a revolution in the way media technologies are used. We can see it in the collective, communal nature of the web, in the self-conscious nature of the use of the web, and in the assertiveness of fans. According to W. Birdsall, these changes "are congruent with and central to those articulated as constituting a human right to communicate. Thus, recent developments in Web development are a significant phase in … a component in a larger social movement."[66]

In sum, we "are seeing a shift from individuals who depend on social relations that are dominated by locally embedded, thick, unmediated, given, and stable relations, into networked individuals—who … weave their own web of more or less instrumental, relatively fluid relationships."[67] What helps to illustrate this shift from an online to an offline world is an economic bridge, a new form of economy posited in the Web Commons by the fan: the Digi-gratis economy.

Digi-Gratis Economy

The presence of digital fandom in the Web Commons exists as a mash-up of two different economic forces that, while always having been present in contemporary culture, are reaching a climatic and dramatic juncture. By "mash-up," I reference a bastard art form for which artists would illegally appropriate and fuse together two different copyrighted audio recordings; however, I use the term in a more general sense, to refer to the combination of two (or more) disparate elements that forms a unique object or concept that retains characteristics of its constituent components while simultaneously presenting something new.[68] For my purposes, I argue that the existence of a "Digi-gratis" economy is based upon a mash-up of the market economy and the gift economy. The Digi-gratis, being a mash-up, retains key elements of both economies, but exists as unique unto itself. Importantly, it is *not* a convergence or a hybrid of the two. It is the simultaneous existence of both economies as both separated and conjoined, existing in various guises through various mechanisms.

The term *Digi-gratis* indicates an economic structure where money is not (necessarily) exchanged, but that retains elements of a market structure. In a "gratis" economy, people create and share content without charge or recompense, or, at least, by charging a variable, user-determined fee.[69] In many ways, this exchange is related to the gift economy described by Marcel Mauss as one where "exchanges and contracts take place in the form of presents ..."[70] The gift economy builds social bonds—for example, we bring wine or appetizers to a dinner party not because we necessarily think our hosts *need* them, but because we create the social environment for our relationship to develop. Fandom and fan communities have long been tied to gift economies.[71] This economic structure, in many respects, differs from our contemporary idea of a market economy, in which goods and services are exchanged for monetary recompense, and in which the purpose is to build wealth.

However, as Mauss's quotation goes on to say, "... in reality [exchanges and contracts] are given and reciprocated *obligatorily*."[72] To give a gift necessarily implies reciprocity, a return of a gift in kind. This gift reciprocity falls under certain social rules: If you recently brought a bottle of wine to the last dinner party, and this time it's your turn to host and no one brings anything, you feel insulted, cheated, deprived even. The reason for this giving/receiving /reciprocation triumvirate has to do with the loss of the tangible object: Jacques Derrida shows us that "in general, it is thought that one can give only what one has, what one possesses as one's own, and give it to the other who, in his or her turn, can thus have it, come into possession."[73] To give thus implies a loss, a consumption. Yet, in the digital, there *is* no loss of the tangible, no deficit of object during transfer. There is, instead, a *reproduction* of the object, a copy/exchange. The "gift" of digital objects comes with infinite reproducibility—my giving an MP3 file to one friend does not limit me from giving it to a second, a third, a fourth, as well as then keeping it for myself.

The Digi-gratis, therefore, is not quite a gift economy, and not quite a market economy. Linguistically, the term also resonates with another portmanteau, "digirati," as coined by John Brockman, for whom the digirati are the "digital elite," the people at the forefront of the computer and internet industries.[74] Fans become these elites not through sociological means, but through specialized knowledge. Fans exist within this mash-up economy, as "the 'Digi-gratis' economy thrives because *neither* the gift *nor* the commodity economy outweighs the other. Instead, through mutual reciprocity, their mash-up forms a third type of encounter" where there is "a mutual relationship between the two economies, and [which] places no emphasis between production and consumption: both are weighted equally."[75]

The Digi-gratis economy applies to fandom in the way that media and fandom coexist. In regards to the market economy, fans routinely purchase merchandise based around the object of their fandom. When producers release DVDs for a fan's favorite TV show, many fans buy them even if they have already seen them when the show aired, and sales can be staggering.[76] Fandom, however, can also be perceived in a non-monetary environment online, highlighting instead the social standing of the individual. As Benkler states,

> For any given culture, there will be some acts that a person would prefer to perform not for money, but for social standing, recognition, and probably, ultimately, instrumental value obtainable only if that person has performed the action through a social rather than a market, transaction.[77]

Fans participate in this gift economy in multiple ways. Carolyn Stevens, looking here at music fans, shows how music can be seen as a type of "gift" wherein "the 'gifts' [fans] believe they receive are those of pleasure, inspiration, and even nostalgia. Receiving this gift prompts some fans to return the favor by sending objects or letters to the star … [which] has created a circular movement of capitol, objects, and affect between fans and stars."[78] For Karen Hellekson, fandom's gift economy does not exist to exchange fan works for money, but rather "to engage … to click, read, comment, write, make up a song and sing it, to hotlink, to create a video, to be invited to move on, to come over here or go over there—to become part of a larger metatext."[79] The way fans buy, sell, trade, relate, and exchange on the internet retains individual characteristics of both the market and the gift, but, importantly, also diverges from these economic systems to form something new. Suzanne Scott argues, in addition, that the media industries use models of the gift economy—what she calls the "regifting economy"—to "regift a narrowly defined and contained version of fandom to a general audience."[80] In other words, media industries use the trappings of the gift economy—free giveaways as advertising and fan work as incentivizing consumption—to their own financial advantage. The creation of fan communities and content all exist in this mash-up economy, as cult fans make use of products that have been designed in a market economy, but use them in order to promote social construction in a manner reminiscent of a gift economy.

These economic forces—the market economy of industrial production and the gift economy of shared exchange—mark two separate, but unified, areas of economic research. As economist Alan Schrift shows, "where commodity exchange is focused on a transfer in which objects of equivalent exchange value are reciprocally transacted, gift exchange seeks to establish a relationship between subjects in which the actual objects transferred are incidental to the value of the relationship established."[81] In other words, while the market economy describes the buying and selling of goods, the gift economy articulates the establishment of social relationships among its participants, the formation of a community.

The Digi-gratis economy resonates with Lawrence Lessig's conception of the hybrid economy. The main difference between Lessig's hybrid economy and the Digi-gratis economy lies in the way the two requisite economies interact within the new economic system. Initially, Lessig argues that both requisite economies have to work *with* each other: "no distinction between [the] 'sharing' and 'commercial' economies can be assumed to survive forever, or even for long." In other words, the hybrid economy converges the two distinct economies into a single, unified economic structure. He later states, however, that the link between the two, "simpler" economies "is sustained, however, only if the *distinction* between the two economies is preserved."[82] In other words, Lessig's contradictory stance shows the hybrid economy as both a *convergence* of economies and in an irreconcilable *tension*. Lessig's solution to this contradiction is to assume that one economy alternatively dominates the other, depending on the situational transaction. For example, the video hosting site YouTube offers a space for users to upload their own videos for free; however, it also provides the mechanics for video creators to link to online stores as an overlay on the videos and a pay service (YouTube Red). Both economies are working, but the overlay offers a visual confirmation that one always dominates the other.

In contrast, the Digi-gratis economy assumes that both economies are crucial for the functioning of a complete economic system, and work together, always. Fans create "gifts" out of the products they purchase in the market economy, and the market caters to the fan culture by offering free services that fans can interpret as gifts. The Digi-gratis economy is not *new*, in as much as both economies, the market and the gift, have been working in tandem for hundreds of years. Digital technology simply highlights this economic interaction.

The coexistence of fan texts as both commodities and gifts forms through the way fans view the inherent worth or value of a television text. An audience can inscribe this value into the television, or producers can design a cult show with the intention of audiences inscribing value to it. For example, Gwenllian-Jones describes textual devices put into *Xena: Warrior Princess* by media producers to attract a cult fan audience: "intertextuality, metatextuality, self-referentiality, story-arc and stand-alone episodes within the same series, an exaggerated play of fracture and textual excess and generic interconnections with wider subcultures."[83] Additionally, genre specialty channels like the Sy-Fy Channel create and target cult audiences by designing media products like *The Expanse* or *12 Monkeys* as long-running serial narratives.[84] This duality assumes that cult fandom cannot be associated with *just* the market economy or the gift economy. In other words, value comes not just from market production, or from fan consumption, but through the complex interaction of both. To describe more fully this Digi-gratis economy, we must understand that first, while market economies underscore the *production/consumption* dichotomy relationship, digital technologies alter this relationship; and second, that the nature of the media text itself as multi-modal process (and not static object) results in an Lacanian-esque pursuit of desire, one that cultivates fans as both consumers and as gift-givers.

Traditional media studies concentrates on the fan as part of an economic binary between *producer* of media content and *consumer* of media content in a market economy. The "media text" itself is seen as an object received and consumed by the fan. The fan may poach the text, may reinterpret it, or may tactically respond to it; but importantly, they will first consume the text, and only then will interpret it. This the gastronomic metaphor of fandom has provided a sense of the media text as a concrete "whole" upon which the fan devours. The metaphor only becomes complete when we realize that profit in a market only comes about because *consumption also implies destruction*—the destruction of possession and the destruction of demand. (Taken to its logical conclusion, of course, this metaphor also leads one to see of what caliber most people consider fan fiction: the product of extreme gastronomic excess.)

In the digital environment, however, commodities/gifts are no longer tangible, but instead are infinitely reproducible. "Consumption" as a concept no longer applies. The *modus operandi* of the Web Commons is not the production of material goods, but rather the production of knowledge.[85] As information becomes the commodity traded online, digital technology, the Web Commons, and the activity of fandom has questioned this gastronomic assertion of the market economy.[86] Unlike physical goods, there is no "loss" from the production of information.[87] Indeed, there can be no true "ownership" of information, as Ghosh points out: To keep information to yourself is to make sure that no one knows that you have it to begin with.[88] We can copy documents, email and transfer them, with no appreciable loss of quality or value. Even when media goods became infinitely reproducible, when DVDs, VCRs, and the like were invented and widespread, media producers worked from this assumption. For example, every television commercial that airs that urges you to "tune in at 9 pm" for the next episode of *Doctor Who* is operating under the assumption that you can't or won't just go to your computer to download or stream the episode immediately afterward. The advertisement attempts to create a demand, to pretend that the media object of *Doctor Who* will only be available at 9 pm. Because the television rating system still operates under the top-down measurement of *how many eyes are watching when a show airs*, media producers hope to create a desire—a need—for audiences to tune in.

This form of information ownership helps make visible the Digi-gratis economy, a gift economy that deals less with monetary exchanges and more with the socialization of information. Specifically, as Jenkins et al. show, aspects of the gift economy have been tied into the web from its inception: "[V]alues were built into the infrastructure of the web which was designed to facilitate the collaboration of scientists and researchers rather than to enable the metered access expected within a commodity culture."[89] This is one reason why major media corporations, whose economic foundation has roots in the market economies of the late 19th and early 20th centuries, are having trouble containing the illegal downloading of consumers in the 21st. Piracy appeals to fans not just because it is free, but also because it becomes a form of subcultural capital, a practice that takes advantage of the gift mentality fostered by the web.

As opposed to the market economy, gift-giving does not depend on monetary value: Gift exchange is defined by the absence of financial calculation (this is why we remove the price tag before giving a gift). Gifts also differ from commodities in that they depend on some sort of interpersonal dependence: "[T]he giver of a gift remains an element of the good or service and does not alienate himself from it."[90] Ritualized gifting—the structured and ordered exchange of items of significant symbolic value—depends more on the community in which the giving occur than it does on the actual items given.[91] Giving creates social life, as William Merrin shows in his description of gift-giving cultures:

> Giving [brings] **social rank**, with the capacity and willingness to give bringing honor and respect; it [creates] **strong social relations**, as a positive act of communication forming affective ties and alliances between families or tribes; and it [brings] **social power**, creating a power relation of indebtedness, obligation and loss of face, with honor only being regained through a counter-gift.[92]

Although we are more familiar with our contemporary market economy, relatives of gift-giving economies still exist today: an example would be the giving of rings as a ceremonial gesture in marriage, one that symbolizes, as Baudrillard cites, a "social logic" of consumption.[93] We give rings to symbolize union, companionship, and completeness.[94] Further, to bring a bottle of wine or an appetizer to a social gathering is not just to provide libation or sustenance, but to bring camaraderie and friendship as well.

Giving gifts, as described by Mauss, is a three part system of ritualized exchange: The giving of the gift leads to an obligation to reciprocate, that is, to give a gift back at a later date, which in turn leads to the obligation to receive the gift.[95] In fact, to reciprocate a gift means not just that one *should* return a similar gift, but also that one *must*, by all rights, give a gift. Once a gift is given, one is obligated to receive it—we cannot refuse an appetizer brought to a dinner party (no matter how slimy) without risking insulting our guests, damaging our social reputation, and complicating our relationships within the community. Once received, we are then obligated to reciprocate that gift—to bring our own (less slimy) appetizer when we are invited over for dinner.[96] If we do not reciprocate, we have committed a faux pas and damaged social relations. These three obligations—to give, to receive, and to reciprocate at a later date—describe a social situation in which the actual gifts given matter less than the fact that they were given and received.

Ultimately, the relation between gift and market both emerge from a central concept. Both economies rely on the notion that there is or will be a limited supply of goods: Once they are sold or gone, there are no more. In a market economy, this becomes apparent because, at some point, the consumer will not have something he/she needs or wants, and will be forced to buy it from a supplier. The consumption of the goods implies the destruction of the goods, as the term "consumption" implies the utilization of perishable goods. Similarly, a gift economy also relies on the notion that there are no more objects, but conversely: A gift is meaningless unless the giver no longer has it. Indeed, having an object represents the possession of power. Giving a gift implies loss and sacrifice. The gifting of an item implies the destruction of the possession of that item: By giving it, I destroy my possession of it but create my own loss.[97]

Thus, if digital technology no longer requires the destruction of the original, and key to gift-giving was the sense of obligation created from the loss of the gift, how can we call the sharing of information "gift-giving"? If, as Derrida shows, we can only give knowing that we will no longer have it—but we still retain what we have given, digitally—does what we transfer over the network constitute a gift?[98] In fact, it does, but with a caveat: The new gift, the digital gift, is a gift without an obligation to reciprocate. Instead of reciprocity, what the gift in the digital age requires for "membership" into the fan community is merely an obligation to reply.

The difference between reciprocity and reply lies in the status of the object given. To reciprocate, one must give back or return mutually, in kind. There must be an equal amount given as taken. However, by "reply" one simply needs to respond, that is, to give an answer in receipt. "Reply" does not imply a quantitative degree of response, but rather an acknowledgment. To obligate a reciprocation of a digital gift is to necessitate a measurement either in bits or in length. To obligate a reply to a digital gift, however, is only to ask for a moment for communion. To institute a communicative reply is to draw "persons together in fellowship and community," as Carey writes, and is thus to strengthen the communal ties amongst members of a group.[99] The more communication undertaken, the stronger the group becomes. Thus, while giving a digital gift might differ from the terms set out in traditional gift economies, it is similar in its social bonding experiences.

In the Web Commons, the market and the gift economies exist side-by-side. The commercial aspect of the web—the buying and selling of goods—exists in full force (as the popularity of amazon.com and other e-sellers continues to eclipse traditional brick-and-mortar stores), but adjacent to it is the sense of the web as a vast repository of digital gifts (free downloads, free streaming video, free offers abound). It's not that one dominates, or will dominate depending on the situation, but that both exist *simultaneously*. In both the market and the gift economies, the sense of consumption/destruction is strong; but the existence of the *other* economy reverses that sense for the first. The consumptive attitude in the market economy is negated by the digital reproducibility of the gift; the sense of loss in gift-giving is annulled by the sense of possession taht emerges from the market economy. In the next section, I'll give a specific example of how fandom in the digital cannot be seen either as a market or as a gift economy, but rather as a mash-up of the two.

Digital Fandom, the Web Commons, and the Digi-Gratis Economy in Practice

We can see how these three concepts function through an extended example of contemporary fandom at work. The BBC documentary *Love Off-Air* describes a community of fans of the cult science-fiction show *Doctor Who* and how their fandom supported and completed re-releases of missing titles from the series.[100] Through the practice of digital fandom, these *Doctor Who* fans helped produce—and reproduce—key *Doctor Who* texts, using the Web Commons in a Digi-gratis style economy.

During the early years of the series (e.g., before the advent of the VCR), many fans would record *Doctor Who* episodes on audio cassette by holding the microphone of the recorder up to the speaker on the television set. According to the documentary, fans often played these fan "off-air" recordings as they effectively re-played the series (sometimes with action figures[101]), re-enacting the visuals of the episodes while using recovered audio. Off-Air recording fans would often also meet with other fans to play back the recording, "performing" the television show together in a communal atmosphere. As fan and *Doctor Who* author Justin Richards put it, the "sound archiving of *Doctor Who* was a very serious business" for the community of young fans.

This form of *Doctor Who* fandom came to a head in the early 2000s, when fans and media producers worked together to reconstruct missing serials from the long run of the show. Tragically, many of the early *Doctor Who* episodes are missing. The show has been on the air (with a few hiatuses) since 1963. From the start of the series in 1963 to the current run beginning in 2005, the British Broadcasting Corporation has broadcast over 800 episodes of the series. During that 50+ year-run, the BBC has lost or destroyed over a hundred episodes, due to policy or to negligence.[102] Some have been found; as of January 2015, there are only 97 missing. One serial that remains incomplete is the now-classic 1968 "The Invasion." Representing a pivotal moment in the program's history, "The Invasion" only existed in fragments: Of the original eight parts, two were completely missing.

Thanks to digital recording technology, however, the BBC used the off-air recordings of those missing episodes made by fans in 1968 in a re-play of the serial. By digitally combining and cleaning up the recordings of a number of fans, audio engineer Mark Ayres was able to reproduce the broadcast audio from the missing two episodes. It's not that the BBC cleaned up one recording to make the new soundtrack; rather, they used the collaborate unity of *many* to create a master-track. Using this new master soundtrack, BBC then commissioned film studio Cosgrove Hall to animate the missing episodes in order to produce the entire serial. The fandom of *Doctor Who* saved the audio—and the episode—from complete destruction. More episodes are being saved, as fan Michael Stevens puts it: "That hobby I used to have, when I was eight or nine years old, of recording *Doctor Who* off the telly, we now use people's similar recordings and polish them up and put them on CDs." Since "The Invasion," other *Doctor Who* missing episodes have been released, including "The Reign of Terror," "The Tenth Planet," "The Moonbase," and "The Ice Warriors."

In one view, then, it appears that the BBC used the work of fans to produce another product for those self-same fans to purchase: a DVD. This market view is undoubted, and the BBC profited off of the work of the fan community. Yet, looked at differently, the fans that worked on this project helped to shape and to contribute to the continued success and promotion of the object of their fandom. Their work—both as audio recorders in 1968 and as a fandom in the 21st century—could be interpreted as gifts to *Doctor Who*. Indeed, fan's work in *Doctor Who* helped bring the show back on air in 2005.[103] By saving and contributing their work, these fans became active participants in the creation and preservation of the object of their fan affection. Only through the Digi-gratis economy, an economy based on a mash-up of market economics and gift-giving, could such transactions have happened.

Throughout the rest of this book, I use examples from online fan creations to augment the theories of New Media I propose. However, this warrants a caveat: obviously, no scientific conclusions can be drawn from small samples like the ones I investigate in this book. This is not intended to be a definitive account of all fan practices in the digital, nor is it intended to elucidate every use of the web. Instead, I examine specific instances of digital fandom and describe how they illustrate the necessity for a change in scholarship of New Media. Although the samples are small, the work these fans create is immense, both in terms of its meaning for the individual fan (and the fan community), and in terms of its larger cultural context. It is significant in the most important way: to those who create it.

Notes

1. Pérez-Reverte, *The Club Dumas*, 246, 156, 260, 322.
2. Merrin, "Media Studies 2.0," 32.
3. Merrin, *Media Studies 2.0*.
4. See Hills, *Fan Cultures*, for other dialectics.
5. Coppa, "A Brief History of Media Fandom," 42.
6. See Jenkins, *Textual Poachers*, 12.
7. See about fans: Jenkins, *Textual Poachers*; Tulloch and Jenkins, *Science Fiction Audiences*. About types of fans: Abercrombie and Longhurst, *Audiences*, 138–42. Theory of fans: Hills, *Fan Cultures*; Stein, *Millennial Fandom*. About fan fiction: Bacon-Smith, *Enterprising Women*; Penley, *Nasa/Trek*; Bury, *Cyberspaces of Their Own*. Recent edited

collections: Hellekson and Busse, ed., *Fan Fiction and Fan Communities*; Gray, Sandvoss, and Harrington, ed., *Fandom*.

8. de Certeau, *The Practice of Everyday Life*, 29–42. See also, for a reconsideration of this metaphor, Sandvoss, *Fans*.
9. Jenkins, *Convergence Culture*, 74–9, echoing Abercrombie and Longhurst.
10. Lessig, *Remix*, 221.
11. Jenkins, *Convergence Culture*, 76.
12. Shefrin, "*Lord of the Rings*, *Star Wars*, and Participatory Fandom," 262.
13. Jenkins, Ford, and Green, *Spreadable Media*.
14. Jenkins, "Afterward," 361.
15. Sandvoss, *Fans*, 101.
16. Gray, Sandvoss, and Harrington, "Introduction," 10.
17. Postigo, "Video Game Appropriation through Modifications," 71.
18. Sandvoss, *Fans*, 3.
19. Williams and Edge, "The Social Shaping of Technology," 2.
20. Potts, "Who's Afraid of Technological Determinism?" 3. My argument in this book tends to fall more on the deterministic side, if only because fans generally cannot change the technology and must work with the tools they are given: Thus, the technology does affect the practice of creating fan fiction in the guises I discuss in this book. However, in the larger realm of New Media Studies, I do believe that society has an effect in the development of technology.
21. Gunkel, *Thinking Otherwise*.
22. Williams, *Keywords*, 9.
23. Kollock and Smith, Introduction, 16.
24. Erickson, "Social Interaction on the Net," 2.1; Williams, *Keywords*, 66.
25. Walker, "Enterprising."
26. Jenkins, *Textual Poachers*.
27. As we will see, recent studies of fans use the metaphor of the gift to explain fan transactions. However, I show that both gift and market are limited in themselves, and only through a mash-up of the two do we understand contemporary fan consumptive practices.
28. Bruns, *Blogs, Wikipedia, Second Life, and Beyond*, coined the term "produser"; Toffler, *The Third Wave* coined "prosumer"; Tapscott and Williams, *Wikinomics*, use "prosumption."
29. de Certeau, *The Practice of Everyday Life*, 25–6.
30. Bruns, *Blogs, Wikipedia, Second Life, and Beyond*, 2, 16. "Intercreativity" is a term coined by Tim Berners-Lee, the inventor of the World Wide Web, which means networked creativity: the way information communities collaborate to construct creative projects, like wikis (see http://intercreativity.com/?p=1).
31. Manovich, "What Comes After Remix?" ¶1.
32. Jenkins, *Textual Poachers*, 279.
33. Bacon-Smith, *Enterprising Women*, 57.
34. Hills, *Fan Cultures*, 137.
35. Busse and Hellekson, "Introduction," 6.
36. Žižek, *Looking Awry*.
37. Hills, *Fan Cultures*, 137.

38. Gray, Sandvoss, and Harrington, "Introduction," 3–9. The authors describe three "waves" of fan studies as three epistemological breaks in the discipline. See also Jensen, "Fandom as Pathology," 27 and Fiske, *Reading the Popular*, 121. See also, Fiske, *Understanding Popular Culture*, 99; Fiske, "Cultural Economy of Fandom"; Amesley, "How to Watch *Star Trek*," 335.
39. Gray, Sandvoss, and Harrington, "Introduction," 6.
40. Soukup, "Hitching a Ride on a Star," 323, 327.
41. Eco, "*Casablanca*," 3; Gray, Sandvoss, and Harrington, "Introduction," 9; Lancaster, *Interacting with Babylon 5*, xxvii–xxviii.
42. Baym, *Tune In, Log On*, 21.
43. Jenkins, *Convergence Culture*, 32–4.
44. Hills, *Fan Cultures*, 178.
45. http://www.starwarsuncut.com.
46. Jenkins, *Convergence Culture*, 135–6.
47. http://www.starwarsuncut.com.
48. Brooker, *Using the Force*, 173–98.
49. Hess and Ostrom, "Introduction," 3–6; see also, Ostrom, *Governing the Commons*, 1.
50. Lessig, "The Creative Commons"; Lessig, *The Future of Ideas*, 9–20.
51. With the exception of the fact that one must pay for internet service; although free Wi-Fi and Wi-Max have made "paying" superfluous for many users of the web. Plus, even access to the web can be free, if it is provided by a major metropolitan area, or used at a library or other provider.
52. Jenkins, *Textual Poachers*, 32.
53. Anelli, *Harry, a History*, 96–100.
54. Benkler, "Coase's Penguin, 171; Benkler, *Wealth of Networks*.
55. By "informational portal," I refer to the conception of the web as a "realm of pure information," a vast digital library containing all of human knowledge; see, Benedikt, "Cyberspace," 20.
56. Carey, "A Cultural Approach to Communication," 16–20.
57. Bolter, *Writing Space*, 27–46.
58. Carey, "A Cultural Approach to Communication," 18.
59. For instance, the creator of the "The Web Is Us/Ing Us" video about Web 2.0, Michael Wesch, describes the problem as one of strategic non-description, as the term can mean whatever any practioner or scholar wants: see, Wesch, "What Is Web 2.0?" 30.
60. O'Reilly, "What Is Web 2.0?" ¶12–15.
61. Burke, "Terministic Screens," 45.
62. O'Reilly, "What is Web 2.0?" ¶13.
63. O'Reilly, "What is Web 2.0?" ¶1.
64. The term "commons" has been used to describe the web before, notably in Benkler's "commons-based peer production" and Lessig's "creative commons." New web spaces use the term commons as well: many amateur photographers use a Creative Commons license for the pictures on the web, while Wikipedia houses a Commons-based encyclopedia. Benkler, "Coase's Penguin," 169; Benkler, *Wealth of Networks*; Lessig, "The Creative Commons."

65. Lessig, *Free Culture*," xiii–xvi.
66. Birdsall, "Web 2.0 as Social Movement," ¶34.
67. Benkler, *Wealth of Networks*, 362.
68. Kendall and Schmidt, "Mash-ups," 15; Shiga, "Copy-and-Persist"; Gunkel, "Rethinking the Digital Remix."
69. Baym, "The Lost Librarians of National Defense."
70. Mauss, *The Gift*, 3.
71. Hellekson, "Fannish Field of Value," 114.
72. Mauss, *The Gift*, emphasis mine.
73. Derrida, *Given Time*, 48.
74. Brockman, *Digerati*.
75. Booth, "ARGS, Fandom, and the Digi-gratis Economy."
76. Arnold, "'Lost 2' Finds Way to Top of DVD Sales," ¶1–2.
77. Benkler, *Wealth of Networks*, 96.
78. Stevens, "Buying Intimacy: Proximity and Exchange at a Japanese Rock Concert," 80, quoted in Hills, "Michael Jackson Fans on Trial?" 469.
79. Hellekson, "Fannish Field of Value," 113.
80. Scott, "Repackaging Fan Culture," ¶1.6.
81. Schrift, "Introduction," 2.
82. Lessig, *Remix*, 150, 77, my emphasis.
83. Gwenllian-Jones, "Web Wars," 166.
84. Gwenllian-Jones and Pearson, Introduction, xii.
85. Benkler, *Wealth of Networks*, 41: "The widely held intuition that markets are more or less the best way to produce goods, that property rights and contract are efficient ways of organizing production decisions, and that subsidies distort production decisions, is only very ambiguously applicable to information."
86. See the essays in Ghosh, ed., *CODE*; also, Bruns, *Blogs, Wikipedia, Second Life, and Beyond*, 9–36. For an early view of gift economy in cyberspace, see Kollock, "The Economics of Online Cooperation," 221–2.
87. This is not a new idea: Thomas Jefferson once wrote: "If nature has made any one thing less susceptible than all others of exclusive property, it is the action of the thinking power called an idea … [I]ts peculiar character, too, is that no one possesses the less, because every other possesses the whole of it" (quoted in Lessig, *Remix*, 290).
88. Ghosh, "Why Collaboration Is Important (Again)," 2.
89. Jenkins, Li, Krauskopf, with Green, "If It Doesn't Spread, It's Dead (Part Three)," ¶17; see also Rheingold, *The Virtual Community*, 49; boyd, "Facebook's Little Gifts," ¶8–9; Hellekson, "Fannish Field of Value."
90. Bell, "Modes of Exchange," 156.
91. Sahlins, "The Spirit of the Gift," 84.
92. Merrin, *Baudrillard and the Media*, 13, emphasis his.
93. Baudrillard, *The Consumer Society*, 60–1.
94. And, perhaps in a holdover from less enlightened days, possession and ownership.
95. Mauss, *The Gift*, 39–43.
96. The dinner situation itself is based in gift-giving—we "gift" our company, our food, and our shelter, which in turn requires a reciprocal invitation to share in the guests' food, company, and shelter at a later date.

97. See, particularly, Derrida, *Given Time*, 12; Mauss, *The Gift*, 15–18.
98. Derrida, *Given Time*, 12.
99. Carey, "A Cultural Approach to Communication," 18.
100. Goss and Francis, *Love Off-Air*.
101. Geraghty, "Aging Toys and Collectors."
102. Molesworth, *Wiped!*
103. Perryman, "*Doctor Who* and the Convergence of Media"; Hills, *Triumph of a Time Lord*.

Bibliography

Abercrombie, Nicholas, and Brian Longhurst. *Audiences: A Sociological Theory of Performance and Imagination*. London: Sage Publications, 1997.

Amesley, Cassandra. "How to Watch *Star Trek*." *Cultural Studies* 3, no. 3 (1989): 323–39.

Anelli, Melissa. *Harry, a History*. New York: Pocket Books, 2008.

Arnold, Thomas K. "'Lost 2' Finds Way to Top of DVD Sales." *Hollywood Reporter*, 14 September 2006. Accessed 01 November 2008. http://www.hollywoodreporter.com/hr/search/article_ display.jsp?vnu_content_id=1003121876. Link no longer available.

Bacon-Smith, Camile. *Enterprising Women: Television, Fandom and the Creation of Popular Myth*. Philadelphia: University of Pennsylvania Press, 1992.

Baudrillard, Jean. *The Consumer Society*, translated by Chris Turner. London: Sage, 1998.

Baym, Nancy. "The Lost Librarians of National Defense." *Online Fandom: News and Perspectives on Fan Communication and Online Life*, 30 April 2008. Accessed 22 March 2016. http://www.onlinefandom.com/archives/the-lost-librarians-of-national-defense.

Baym, Nancy. *Tune In, Log On: Soaps, Fandom, and Online Community*. London: Sage, 2000.

Bell, Duran. "Modes of Exchange: Gift and Economy." *Journal of Socio-Economics* 20, no. 2 (1991): 155–67.

Benedikt, Michael. "Cyberspace: First Steps." In *Cybercultures Reader*, edited by David Bell and Barbara M. Kennedy, 19–33. London: Routledge, 2000.

Benkler, Yochai. "Coase's Penguin, or, Linux and the Nature of the Firm." In *CODE: Collaborative Ownership and the Digital Economy*, edited by Rishab Aiyer Ghosh, 169–206. Cambridge, MA: The MIT Press, 2005.

Benkler, Yochai. *The Wealth of Networks: How Social Production Transforms Markets and Freedom*. New Haven, CT: Yale University Press, 2006.

Birdsall, W. "Web 2.0 as Social Movement." *Webology* 4, no. 2 (2007). Accessed 22 March 2016. http://www.webology.org/2007/v4n2/a40.html.

Bolter, Jay David. *Writing Space: Computers, Hypertext, and the Remediation of Print*, 2nd ed. Mahwah, NJ: Lawrence Erlbaum Associates, 2001.

Booth, Paul. "ARGS, Fandom, and the Digi-gratis Economy." Interview with Henry Jenkins. *Confessions of an Aca-Fan*, 13–18 August 2010. Accessed 22 March 2016. http://henryjenkins.org/2010/08/args_fandom_and_the_digi-grati.html.

boyd, danah. "Facebook's Little Gifts." *Apophenia*, 13 February 2007. Accessed 21 March 2016. http://www.zephoria.org/thoughts/archives/2007/02/13/facebooks_littl.html.

Brockman, John. *Digerati: Encounters with the Cyber Elite*. San Francisco, CA: Hardwired, 1996.

Brooker, Will. *Using the Force: Creativity, Community, and Star Wars Fans*. London: Continuum Press, 2002.

Bruns, Axel. *Blogs, Wikipedia, Second Life, and Beyond*. New York: Peter Lang, 2008.

Burke, Kenneth. "Terministic Screens." In *Language as Symbolic Action: Essays on Life, Literature and Method*, 44–62. Berkeley: University of California Press, 1966.

Bury, Rhiannon. *Cyberspaces of Their Own: Female Fandoms Online*. New York: Peter Lang, 2005.

Busse, Kristina, and Karen Hellekson. "Introduction: Work in Progress." In *Fan Fiction and Fan Communities in the Age of the Internet*, edited by Karen Hellekson and Kristina Busse, 5–32. Jefferson, NC: McFarland and Co., 2006.

Carey, James. "A Cultural Approach to Communication." In *Communication and Culture: Essays on Media and Society*, 13–36. New York: Routledge, 1992.

Coppa, Francesca. "A Brief History of Media Fandom." In *Fan Fiction and Fan Communities in the Age of the Internet*, edited by Karen Hellekson and Kristina Busse, 41–59. Jefferson, NC: McFarland, 2006.

de Certeau, Michel. *The Practice of Everyday Life*, translated by Steven Randall. Berkeley: University of California Press, 1984.

Derrida, Jacques. *Given Time: I. Counterfeit Money*, translated by Peggy Kamuf. Chicago: University of Chicago Press, 1992.

Eco, Umberto. "*Casablanca:* Cult Movies and Intertextual Collage." *SubStance* 47 (1984): 3–12.

Erickson, Thomas. "Social Interaction on the Net: Virtual Community as Participatory Genre." Proceedings of the Thirtieth Annual Hawaii International Conference on System Sciences, *IEEE* (1997).

Fiske, John. "The Cultural Economy of Fandom." In *The Adoring Audience: Fan Culture and Popular Media*, edited by Lisa A. Lewis, 30–49. London: Routledge, 1992.

Fiske, John. *Reading the Popular*. New York: Routledge, 1989.

Fiske, John. *Understanding Popular Culture*. New York: Routledge, 1989.

Geraghty, Lincoln. "Aging Toys and Players: Fan Identity and Cultural Capital." In *Finding the Force of the Star Wars Franchise: Fans, Merchandise, and Critics*, edited by Matthew Wilhelm Kapell and John Shelton Lawrence, 209–23. New York: Peter Lang, 2006.

Ghosh, Rishab Aiyer. "Why Collaboration Is Important (Again)." In *CODE: Collaborative Ownership and the Digital Economy*, edited by Rishab Aiyer Ghosh, 1–6. Cambridge, MA: The MIT Press, 2005.

Ghosh, Rishab Aiyer, ed. *CODE: Collaborative Ownership in the Digital Economy*. Cambridge, MA: The MIT Press, 2005.

Goss, James, and Rob Francis (producers). *Love Off-Air*. Documentary feature on *Doctor Who: The Invasion*, DVD. London: British Broadcasting Corporation, 2006.

Gray, Jonathan, Cornel Sandvoss, and C. Lee Harrington, eds. *Fandom: Identities and Communities in a Mediated World*. New York: New York University Press, 2007.

Gray, Jonathan, Cornel Sandvoss, and C. Lee Harrington. "Introduction: Why Study Fans?" In *Fandom: Identities and Communities in a Mediated World*, edited by Jonathan Gray, Cornel Sandvoss, and C. Lee Harrington, 1–16. New York: New York University Press, 2007.

Gunkel, David. "Rethinking the Digital Remix: Mash-Ups and the Metaphysics of Sound Recording." *Popular Music and Society* 31, no. 4 (2008): 489–510.

Gunkel, David. *Thinking Otherwise*. West Lafayette, IN: Purdue University Press, 2007.

Gwenllian-Jones, Sara. "Web Wars: Resistance, Online Fandom and Studio Censorship." In *Quality Popular Television*, edited by Mark Jancovich and James Lyons, 163–77. London: BFI, 2003.

Gwenllian-Jones, Sara, and Roberta E. Pearson. Introduction to *Cult Television*, ix–xx. Minneapolis: University of Minnesota Press, 2004.

Hellekson, Karen. "A Fannish Field of Value: Online Fan Gift Culture." *Cinema Journal* 48, no. 4 (2009): 113–8. http://dx.doi.org/10.1353/cj.0.0140.

Hellekson, Karen, and Kristina Busse, eds. *Fan Fiction and Fan Communities in the Age of the Internet*. Jefferson, NC: McFarland, 2006.

Hess, Charlotte, and Elinor Ostrom. "Introduction: An Overview of the Knowledge Commons." In *Understanding Knowledge as Commons: From Theory to Practice*, edited by Charlotte Hess and Elinor Ostrom, 3–26. Cambridge, MA: The MIT Press, 2006.

Hills, Matt. *Fan Cultures*. London: Routledge, 2002.

Hills, Matt. "Michael Jackson Fans on Trial? 'Documenting' Emotivism and Fandom in 'Wacko about Jacko.'" *Social Semiotics* 17, no. 4 (2007): 459–77.

Jenkins, Henry. "Afterward: The Future of Fandom." In *Fandom: Identities and Communities in a Mediated World*, edited by Jonathan Gray, Cornel Sandvoss, and C. Lee Harrington, 357–64. New York: New York University Press, 2007.

Jenkins, Henry. *Convergence Culture: Where Old and New Media Collide*. New York: New York University Press, 2006.

Jenkins, Henry. *Textual Poachers: Television Fans and Participatory Culture*. New York: Routledge, 1992.

Jenkins, Henry, Sam Ford, and Joshua Green. *Spreadable Media: Creating Value and Meaning in a Networked Culture*. New York: New York University Press, 2013.

Jenkins, Henry, Xiaochang Li, Ana Domb Krauskopf, with Joshua Green. "If It Doesn't Spread, It's Dead (Part Three): The Gift Economy and Commodity Culture." *Confessions of an Aca-Fan*, 16 February 2009. Accessed 23 March 2016. http://henryjenkins.org/2009/02/if_it_doesnt_spread_its_dead_p_2.html.

Jensen, Joli. "Fandom as Pathology: The Consequences of Characterization." In *The Adoring Audience*, edited by Lisa Lewis, 9–29. London: Routledge, 1992.

Kendall, Kenneth E., and Allen Schmidt. "Mash-Ups: The Art of Creating New Applications by Combining Two or More Web Sites." *Decision Line*, 15–16 March 2007. Accessed 23 March 2016. https://www.researchgate.net/publication/228400062_Mashups_The_Art_of_Creating_New_Applications_by_Combining_Two_or_More_Web_Sites.

Kollock, Peter. "The Economics of Online Cooperation: Gifts and Public Goods in Cyberspace." In *Communities in Cyberspace*, edited by Marc A. Smith and Peter Kollock, 220–39. London: Routledge, 1999.

Kollock, Peter, and Marc A. Smith. Introduction to *Communities in Cyberspace*, edited by Marc A. Smith and Peter Kollock, 3–27. London: Routledge, 1999.

Lancaster, Kurt. *Interacting with Babylon 5*. Austin: University of Texas Press, 2001.

Lessig, Lawrence. "The Creative Commons." *Florida Law Review* 55 (1994): 763–77.

Lessig, Lawrence. *Free Culture: The Nature and Future of Creativity*. New York: Penguin, 2004.

Lessig, Lawrence. *The Future of Ideas: The Fate of the Commons in a Connected World*. New York: Random House, 2001.

Lessig, Lawrence. *Remix: Making Art and Commerce Thrive in the Hybrid Economy*. New York: Penguin, 2008.

Manovich, Lev. "What Comes after Remix?" 2007. Accessed 23 March 2016. http://remixtheory.net/?p=169..

Mauss, Marcel. *The Gift: The Form and Reason for Exchange in Archaic Societies*, translated by W. D. Halls. New York: Norton, 1990.

Merrin, William. *Baudrillard and the Media*. Cambridge, UK: Polity Press, 2005.

Merrin, William. *Media Studies 2.0*. London: Routledge, 2014.

Merrin, William. "Media Studies 2.0: Upgrading and Open-Sourcing the Discipline." *Interactions: Studies in Communication and Culture* 1, no 1 (2009): 17–34.

Molesworth, Richard. *Wiped!* Doctor Who's *Missing Episodes*. London: Telos Press, 2010.

O'Reilly, Tim. "What Is Web 2.0: Design Patterns and Business Models for the Next Generation of Software." *O'Reilly*, 30 September 2005. Accessed 23 March 2016. http://www.oreillynet.com/pub/a/oreilly/tim/news/2005/09/30/what-is-web-20.html.

Ostrom, Elinor. *Governing the Commons: The Evolution of Institutions for Collective Action*. Cambridge, UK: Cambridge University Press, 1990.

Penley, Constance. *Nasa/Trek: Popular Science and Sex in America*. London: Verso, 1997.

Pérez-Reverte, Arturo. *The Club Dumas*, translated by Sonia Soto. New York: Vintage, 1997.

Perryman, Neil. "*Doctor Who* and the Convergence of Media: A Case Study in Transmedia Storytelling." *Convergence: The International Journal of Research into New Media Technologies* 14, no. 1 (2008): 21–39.

Postigo, Hector. "Video Game Appropriation through Modifications." *Convergence: The International Journal of Research into New Media Technologies* 14, no. 1 (2008): 59–74.

Potts, John. "Who's Afraid of Technological Determinism?: Another Look at Medium Theory." *Fibreculture* 12 (2008). Accessed 23 March 2016. http://twelve.fibreculturejournal.org/fcj-084-who%E2%80%99s-afraid-of-technological-determinism-another-look-at-mediumtheory/.

Rheingold, Howard. *The Virtual Community: Homesteading on the Electronic Frontier*. Reading, MA: Addison-Wesley, 1993.

Sahlins, Marshall. "The Spirit of the Gift." In *The Logic of the Gift: Toward an Ethic of Generosity*, edited by Alan D. Schrift, 70–99. New York: Routledge, 1997.

Sandvoss, Cornel. *Fans: The Mirror of Consumption*. Malden, MA: Polity Press, 2005.

Schrift, Alan D. "Introduction: Why Gift?" In *The Logic of the Gift: Toward an Ethic of Generosity*, edited by Alan D. Schrift, 1–22. New York: Routledge, 1997.

Scott, Suzanne. "Repackaging Fan Culture: The Regifting Economy of Ancillary Content Models." *Transformative Works and Cultures* 3 (2014). http://doi:10.3983/twc.2009.0150.

Shefrin, Elana. "*Lord of the Rings*, *Star Wars*, and Participatory Fandom: Mapping New Congruencies between the Internet and Media Entertainment Culture." *Critical Studies in Media Communication* 21, no. 3 (2004): 261–81.

Shiga, John. "Copy-and-Persist: The Logic of Mash-Up Culture." *Critical Studies in Media Communication* 24, no. 2 (2007): 93–114.

Soukup, Charles. "Hitching a Ride on a Star: Celebrity, Fandom, and Identification on the World Wide Web." *Southern Communication Journal* 71, no. 4 (2006): 319–37.

Stein, Louisa Ellen. *Millennial Fandom: Television Audiences in the Transmedia Age*. Iowa City: University of Iowa Press, 2015.

Stevens, Carolyn. "Buying Intimacy: Proximity and Exchange at a Japanese Rock Concert." In *Fanning the Flames: Fans and Consumer Culture in Japan*, edited by William W. Kelly, 59–77. New York: SUNY Press, 2004.

Tapscott, Don, and Anthony Williams. *Wikinomics: How Mass Collaboration Changes Everything*. New York: Penguin, 2006.

Toffler, Alvin. *The Third Wave*. New York: Bantam Books, 1980.

Tulloch, John, and Henry Jenkins. *Science Fiction Audiences: Watching* Doctor Who *and* Star Trek. London: Routledge, 1995.

Walker, Rob. "Enterprising: Her *Star Trek* Conventions Harnessed the Power of Media Fandom Long before the Barons of Content Did." *New York Times Magazine*, 23 December 2008. Accessed 23 March 2016. http://www.nytimes.com/2008/12/28/magazine/28wwln-consumed-t.html.

Wesch, Michael. "What Is Web 2.0? What Does It Mean for Anthropology?" *Anthropology News* (2007): 30–1.

Williams, Raymond. *Keywords: A Vocabulary of Culture and Society*. Oxford, UK: Oxford University Press, 1976.

Williams, Robin, and David Edge. "The Social Shaping of Technology." Accessed 23 March 2016. https://www.researchgate.net/publication/222482133_The_Social_Shaping_of_Technology.

Žižek, Slavoj. *Looking Awry: An Introduction to Jacques Lacan through Popular Culture*. Cambridge, MA: The MIT Press: 1991.

· 2 ·

DIGITAL FANDOM BETWEEN TEXT AND INTERTEXT

> *Place some [books] over others? … I simply couldn't do it. They all have the same immortal soul.*
>
> —Pérez-Reverte, *The Club Dumas* (p. 150)

> *Wow, it was really brilliant! I have totally enjoyed it, the whole story, the whole interactions between the characters and the plot! You fabulously mixed all the little things from the canon and made your own story. … Your characters are very canon.*
>
> —pellamerethiel, comment on "Hubris"

In a converged culture, transmediation is central to analyses of contemporary cult media, as it foregrounds the connections between disparate media texts. Jenkins's classic example of *The Matrix* functions as an exemplar for "the age of media convergence, [by] integrating multiple texts within a single medium." *The Matrix* franchise consists of three blockbuster motion pictures, a straight-to-DVD animated volume, a number of Manga tomes, comics, a video game, a phenomenally large web presence, and an MMORPG (Massively Multiplayer Online Role Playing Game). The connections between these separate texts become a transmediated story, which "unfolds across multiple media platforms, with each new text making a distinctive and valuable contribution to the whole."[1] The whole becomes greater than the sum of its parts through the web of connections woven between texts and media objects.

For fans, of the type described by Pérez-Reverte in the above quotation, the extant media objects enjoyed can hold deep and meaningful truths. Yet, it's not just individual texts that hold meaning, but also vast intertextual networks of connected texts—some of which can be fan created in themselves. For example, as representative of a new form of interaction with texts, the second quotation inscribed at the head of this chapter comes from a comment on a piece of blog fan fiction. For readers of pellamerethiel's blog comment, many of these seemingly transmediated connections already do exist, thanks to the relationship between the blog fan fiction and the extant media object on which it is based. Traditional fan fiction is intertextual by its very nature: It exists as part of a larger corpus of works. Textual blog fan fiction, alternatively, offers a unique space to comment on, interpret, and acknowledge the object of their devotion. Blogs exist as ontologically whole but characteristically divided: They are constructed not just of the post but of the comments as well. Pellamerethiel's blog comment is situated in a singular location: both *part of* the fan fiction on the blog and *apart from* the fiction, it is both distinctive and elemental to the whole.

In this chapter, I examine the nature of fan rewriting not as a revision, but as communal reimagining. I use a reading of cultural critic Roland Barthes's essay "From Work to Text" to chart this rewriting as *intra-textuality*. In this piece, Barthes describes two different conceptions of textuality. On the one hand, he calls a "work" that which is "a fragment of substance, occupying a part of the space" of physical objects. The work is tangible, is able to be seen, and is complete and whole in and of itself. On the other hand, he describes the "text" as a "methodological field," an "activity of production" that "cannot stop … cutting across" works.[2] Put more concretely, the "work" is a tangible object, a physical entity, while a "text" is a series of relationships, an intertextual web of meanings and connections *between* works.

I use Barthes to define intra-textuality as the tension that coheres elements within a whole. As contrasted to intertextuality, which connects elements within a network, intra-textuality functions within a single narrative to position and define each element. While we might be able to describe *The Matrix* as "intertextual," in that each of the media objects are themselves discrete, I believe an intra-textual definition helps to understand the transmediated narrative *as a discrete whole in and of itself*. Ultimately, intra- or intertextuality are not opposites, but merely two different ways of conceptualizing the relationship between media objects within a transmediated whole. To give another example: each of the separate texts of Marvel's Cinematic Universe

(*The Avengers*, *Iron Man*, *Agents of S.H.I.E.L.D.*, *Daredevil*, etc.) could be seen as cohesive in and of itself; or, they could be described as components within a larger, discrete corpus, held together through intra-textual connections that continually reference and augment the MCU as an entity.

A piece of blog fan fiction, as an example of an online "text" and as representative of larger issues in media studies, illustrates intra-textuality. The blog most specifically represents a new form, one not described as either a "work" or a "text." It is not enough to go "From Work to Text," but rather we must now look *between* work and text to an intra-textual view of textuality. In popular discussion of blogs, the blog post itself is valued more highly than the comments. References to blogs tend to refer to the post by default (i.e., if I talk about a particular "blog," what goes unstated is the fact I'm referring to the post of that blog, not the blog as a whole). Yet, such determination slights the important function of the comments on the blog "text" itself.[3] (The fact that there is even a common mantra on the web to "ignore the comments" speaks to their importance.) In digital environments, "texts" can be made up of many different elements, each one wholly discrete and complete unto itself. The blog is not just a post, but is rather the combination of the post plus the comments (plus the multitudinous blog entries written over time). The "writer" of the blog as a *whole* is ultimately a group, not an individual (even as posts/comments are individually contributed). Importantly, for every digital document that the creator imbues with their own personal, unique stamp, there will always be contributions over which the owner has little to no control. Blogs, consequently, are never finished. They are in a constant state of revision and rewriting, as additional comments contribute more information to the blog. For media fans, blog fan fiction represents a new style of communication; for media studies scholars, blogs are a new type of digital document.

In this chapter, I examine blog fan fiction based around the 2003 television remake of the show *Battlestar Galactica* to demonstrate this new intra-textuality. I use the Bakhtinian concept of the carnivalesque to explore how intra-textuality permeates web interactions. While cultural scholars have used the concept of the carnivalesque before as a theoretical methodology, only rarely have they done so to describe digital technology and fandom.[4] One notable exception is fan scholar Rachel Shave, who uses Bakhtin's work with carnival to describe slash fan fiction.[5] Taking off from her work, I describe blog writing in relation to the carnivalesque, and aim in this chapter not only to articulate a new methodology for examining New Media texts, but also to re-examine the carnivalesque as a useful tool for media research.

The Blog: A Critical Understanding

The text-based blog (against which I compare the image-based blog, in chapter 8) functions as a repository for continually updated writing.[6] Importantly, the physical proximity between the blog post and the blog comment allow us to see how both exist on the same site, spatially connected to each other. Lessig shows how "comments [are] an integral part of blogging," and thus by adding a way to talk back, blogs "changed how they were read."[7] Comments become a way for the fan blogger to elicit feedback and participation by members of a fan community. One can imagine a blog post and comment connection: For example, a comment might correct a misspelling or suggest a change in phrasing, uniquely tying the comment to the post. In effect, the "blog," as amalgam of post and comment, is authored by both poster and commenter.

In traditional media studies, the blog would primarily be seen through the text and images in the post. The comments are generally considered paratext, extra-text, or superfluous. As spaces for a writer to promote his/her own agenda, as a publicly visible text for ego casting, blogs have traditionally been seen as a space for one person to voice an opinion.[8] In this traditional view, the blog is like a diary, a filtering service, a commonplace book, or a media monitoring service.[9] To describe the blog in this way, however, is to limit it: other uses of blogs exist. For example, few scholars have discussed fiction writing on blogs, and even fewer blog researchers have discussed fan fiction writing.[10] Yet, to integrate the comments into our notion of the blog is to allow a new reading of ritual communication as it establishes a community.

For fans, writing on blogs might be aligned with writing on forums, especially when it comes to fan fiction.[11] Forums are online bulletin-board-type systems that allow users to post an item to which other users can then comment. The key difference between the blog and the forum, however, is that any comments on a forum exist separate from the post itself (i.e., a reader must hyperlink between the post and the comment), while comments on a blog post are part of the same document.

Such a difference may seem to be small, but ultimately has major consequences. The blog creates a document that continually expands with each additional comment; a fan fiction forum, however, remains static, and only the comments' webpage changes. On the forum website, the posts and the comments remain separated, and thus, the "text" of a forum is always the product of one particular author. For example, on fanfiction.net, 205 static fiction texts about *Battlestar Galactica* (the original series) have been posted by hundreds of individuals (as of 2009; in 2015 it is 330). One such story, "Battlestar Galactica 1980/1985 Finding Earth," details the exploration and eventual location of the mythic Planet Earth. However, in contrast to the blog, where the comments lay directly beneath the post, on this forum post there is only a hyperlink to the comment: It reads "Submit a Review" as part of a drop-down menu, to direct readers to post comments on a separate page.[12] Forums, by virtue of the different format, are not constructed in the same way as blogs. On a blog, although only one author may write the post, many additional voices can comment on the same webpage, creating a multi-layered, multi-authored document. Some comments can create their own extensions to the story, or even their own original stories called CommentFic—however, CommentFic represents not "mass authorship" but as Turk describes, the coexistence of multiple voices as a "system (or series of systems) within which all fans participate in various ways."[13] Multivocality is not the same thing as mass authorship.

For example, livejournal.com/storyY might direct to a document with both fiction and comment. While the fiction post itself may remain static, the comments may increase in number continually. Because blogs are one document, any change to the comments necessarily changes the blog.[14] Thus, by juxtaposing a blog post with comments about that post, blogs spatially unite two disparate elements: the text and the discussion/criticism of that text. Fan blogs, in this view, are thus dependent upon the community of fans that form in and around them.[15] To post a blog online is to be aware that others might read your writing—and is to join in a community of other blog readers.

The blog is a Barthesian multi-worked text: a mash-up, an object with more than one component that retains the unique character of each. Tellingly, Barthes himself indicates as such in "From Work to Text": On one page he alternately describes the text as "a new object," and yet also declares, "The Text is *not to be thought of as an object* that can be computed."[16] In this contradiction, Barthes calls the "text" both something that is tangible and specific, as well as something that cannot be pinned down or defined, something that cannot be "computed" or figured out.

There are, of course, antecedents to this form of textuality: Footnotes become similarly spatially connected to their requisite text.[17] Footnotes comment or extend the conversation, but do so in a separate location. The blog's timeliness, along with the spatial connection between the post and the comment, produces a document that is different in nature from a footnoted work. In the past, critiques would have been diegetically separated from the object of that critique: The author/critic relationship existed in time as well as space.[18] Now, however, critiques are part of, not apart from, the blog post itself.

First, the temporal swiftness of the comment on the post allows the blog writer almost instantaneous feedback and critique on her work, and the blog readers a view of a rapidly and visibly growing connection among many fans. In the blog story "Seelix/Cally, Lubing the Viper," for example, e-commenters life_on_queen and lyssie leave three comments back and forth on the space of the blog: The rapid pace of the commenting occurs within an hour of one another. This interaction is telling: It's not just that life_on_queen debates the fiction post, but also that she seems to assert a communalization with her fellow blogger. "Dude," she writes, of the slash fiction, "I'm torn between applauding and tearing my frickin' eyes out." She enacts a playfulness with this comment, reaching out across the internet to assert a value of connection, but this assertion is phrased playfully, as if to indicate a more lighthearted sentiment.

To comment on a blog is to assert not only that you have read the post, but also that you care enough about the post to *act* in some manner, highlighting the Digi-gratis economy at work on the blog. The commenters cannot alter the text of the fiction post, nor can the writers of the fiction change a comment once it has been posted (unless it is their comment).[19] The rapidity of the comments indicates the fact that both commenters cared enough about each other and the blog itself to contribute to it. Both commenters and writers of fiction can change that blog document as a whole. The rewriting that occurs on a blog happens not because the fiction post is revised, but because the blog itself changes, is rewritten with each comment posting.

Second, the function of the blog comment qualitatively differs from the function of the annotated footnote. Footnotes, especially in works of literary criticism, tend to elucidate or explain the text to a reader.[20] Authors might read those critical footnotes later, and even respond in a future edition or reprint, but the conversation would be slow and ultimately the interaction stilted. Comments on a blog, however, are tangible links between the author of the blog post and the author of the blog comment. Regardless of what is written, the mere act of comment asserts a connective relationship: life_on_queen and lyssie engage in a playful conversation about the blog post rather than a textual critique. Although it would not be difficult to imagine a comment that critiqued a work of blog fiction, can you imagine Barthes engaging in a discussion of *Sarrasine*, "Dude, I'm torn between applauding and tearing my frickin' eyes out. LOL"?

If annotations and critiques could originally have been analyzed as intertextually connected to the object of that critique, blog fan fiction thus invites a rereading of the theory of intertextuality because of its incongruous existence as both fictional and meta-critical work. The confluence of fiction and critique on one document indicates a reevaluation of the way we examine traditional textuality and intertextuality. The "meaning" of the blog post—of any media text—is traditionally seen between it and the comments that are separated from it. Kristeva illustrates this separation through her language: "*inter*textuality and *trans*position" indicate "the text's constitutive state [as] one of motion."[21] Through a rewriting of how media scholars examine textuality, a revision in media studies can be undertaken.

Text, Intertext, Intra-Text

Throughout literary theory, the term "text" has been used to describe different things. The most commonly used definition of text, at least in popular parlance, is that of something *read*, that is, a written document utilizing language and symbolic understanding with the intent of transmitting information from one place or source ("the author") to another ("the reader"). We can see this in a quotation from Russian literary scholar Sergei Gindin: "A text is a 'quantum' of communicative activity possessing relative autonomy (distinctness) and unity (integrity)."[22] For years, literary studies have had "a long history of fetishizing the text as a solitary, pristinely autonomous object."[23] Traditional media theory, which examines the blog post and the blog comments as separate components, uses the language of intertextuality.

Intertextuality, as a theoretical concept, emerges from the post-structuralist semiotic analyses of both Kristeva and Barthes in the late 1960s and early 1970s.[24] Julia Kristeva coined the term "intertextual" to describe the process of subconscious textual relationships: Every text that is written is written with the knowledge and covert textuality of every other text written before. As she describes, "any text is constructed as a mosaic of quotations; any text is the absorption and transformation of another."[25] Further, Barthes's description of intertextuality examines texts not as discrete units, "not [as] a line of words releasing a single 'theological' meaning," but rather as links in a chain of interrelated and interconnected meanings, "a multi-dimensional space in which a variety of writings," becomes a "tissue of quotations drawn from the innumerable centers of culture."[26] Intertextuality is a thread interwoven into the tapestry of all "texts," all other references and works, part of an intertextual web spanning the eons.

A new intertextual theory of authorship emerges, one where "the logic regulating the [t]ext is not comprehensive … but metonymic; the activity of associations, contiguities, carryings-over coincides with a liberation of symbolic energy."[27] Media critic Jonathan Gray in his intertextual analysis of the television show *The Simpsons* echoes Barthes when he says, "even a simple sentence requires a reading history to understand."[28] Any text is constructed not through any particular writing down or authorship, but rather in the space *between* the reader and the written.[29] The viewer moves across media to construct the narrative intertextually. In fact, we must look at a transmediated environment not just as a way of presenting narrative story, but also as a way of perceiving it as well. The reader moves between discursive texts, actively constructing the story.

Just as for broadcast-era media we needed to look at entertainment through intertextuality rather than mere textuality, for digital era media we need to look at entertainment through an intra-textual analysis. As fan fiction writers write on a blog, they not only intertextually contribute to the ongoing discussion about the extant media object, but they also intra-textually construct that text itself through their own playful contributions. New Media Studies provides us with an understanding of new methods of analysis. I posit "intra-textuality" as one such method. In contrast to the traditional intertextuality, where meaning is uncovered in a transmediated sense *between* texts and "any text is the absorption and transformation of another," intra-textuality examines the meaning that occurs *inside* the transmediated text itself.[30] For digital fandom, this has enormous consequences for the construction of blog fan fiction, as the fan fiction author writes on the blog not just to reference an original text, but also to playfully build a fan community, a fandom, of other blog writers through comments about a mutually shared media text.

Readers exist at the nexus of an intertextual web of the already-read and the stories as they are being-read. Media studies posit that such interactions occur every time someone picks up a book or turns on the television. To create or inscribe meaning is a major component of how we decode texts. Every time we sit down to experience some form of mediated entertainment or engage with media in some way, we bring to that experience the "already-read," the experience of a "text-between of another text."[31] Every text has a plurality of voices within it, interacting and speaking to each other. In this way, no two readings of any work can *ever* be the same, for each reader automatically brings to the reading his/her own background and experiences.

Fans are no different from any other reader in the respect that they bring to the extant media object ideas, notions, thoughts, and experiences that shape their understanding of the narrative. However, many fans also participate viscerally in this intertextuality through the writing of fan fiction. Fan fiction is, by definition, peripheral and ancillary to the primary extant media object. It functions as a text outside and apart from the extant text.[32]

I argue a new theoretical construction helps us to explain this complicated fannish interaction: intra-textuality is a Barthesian tissue of connections that propels the reader deeper into a media object, as it connects aspects of the text to other aspects of that text.[33] Complementary to this, intra-textuality describes the relationship between aspects in any transmediated narrative, like ARGs or other distributed narratives. Whereas intertextuality exists *in-between* texts, intra-textuality implies *movement within a whole*. The US highway system is a good example of this linguistic linkage: The interstates travel between states, while an intrastate highway connects towns within a single state.[34]

Intertextuality is central to the web, as it emerges from the connections between fragments of narratives.[35] A fragmented narrative, like a distributed narrative, is a conception of serialized narrative form experienced in discrete units separated by time and space.[36] What holds a distributed narrative "together," what unites the disparate elements, is the desire for narrative unity or completeness. Yet, an intra-textual sensibility comes from the interplay of elements within a larger document, self-referentiality instead of *exter*-referentiality. For example, the Whedonverse—a common view of all of Joss Whedon's media texts from *Buffy the Vampire Slayer* to *Dollhouse*—could be viewed as intertextual (all texts are discrete) or intra-textual (each is merely part of the larger whole). By calling it the "Whedonverse," one implies each text contributes to a whole. Viewing one narrative amongst multiple outlets means that the connections themselves cohere in a more unified sense; rather than intertextuality, which would see the narrative splintered amongst texts, intra-textuality relies on the links as stronger than the texts themselves.

Intertextuality and intra-textuality both hinge on reconceptualizations of the definition of text. Intertextuality sees texts as spread out, open, and defined by the connections between discrete units. Intra-textuality sees documents as complete, whole, and unified entities, defined by the connections within internal elements. Intertextuality is a useful tool for investigating narratives that cross technologies of distribution channels. For example, Bennett and Woollacott's famous analysis of the James Bond meta-narrative considers the intertextual relationship between components within the Bond textual corpus. For Bennett and Woollacott, any reading of a James Bond work—be it movie, book, comic, toy, or review—is a process by which the "inter-textually organized reader meets the inter-textually organized text ..." as the work becomes "not the place where the business of culture is conducted," but rather the site "around which the pre-eminently social affair of the struggle for the production of meaning" occurs.[37] This study of James Bond, like the study of *Batman* conducted 15 years later by Brooker, cannot, by definition, be a textual analysis of any one source, but must exist as an intertextual analysis that spans multiple works, readerships, and cultures.[38]

But this definition of intertextuality necessarily sees the narrative as fragmented. Why is this the case? Why is a gap across technologies necessarily a fragmentation? If narratives are whole and complete, then can't we say that they are all intra-textually connected? Ultimately, these questions depend largely on how one defines "text." For example, Tulloch and Jenkins's analysis of science-fiction audiences reveals loose definitions of text and intertext. In the book, Tulloch uses the term "intertextuality" to refer to times when fans of *Doctor Who* comment not on the particular episode they are watching, but on other episodes in the series.[39] He discusses the episode "The Monster of Peladon" in detail with the fans, but in reference to it, they often cite other episodes in the *Doctor Who* canon as comparison. If one considers each episode of *Doctor Who* a separate text, then Tulloch would be correct in calling this intertextual. If one looks at the *Doctor Who* corpus as a single complete narrative, however—as, indeed, many fans do[40]—then this becomes not an intertextual cite, but rather an intra-textual one.[41] In this conception of the larger "text," each *Doctor Who* episode is one in a serial narrative, contributing pieces to a complete, existent cult story. Indeed, intertextual references specifically to *Doctor Who* are problematic in themselves, although Tulloch doesn't mention it: for most of the show, *Doctor Who* "episodes" were multi-part, broadcast over a period of weeks. Thus, to refer to "The Monster of Peladon" as "an episode" is actually a misnomer—it is rather a serial composed of six episodes, broadcast over the course of a month and a half.[42] As Fairclough states, "The interpretation of inter-textual context is a matter of deciding which series a text belongs to."[43] *Doctor Who* "texts," therefore, can be seen as enacting both intertextuality and intra-textuality, depending on the way the viewer interprets the media content and the term "text." If a body of media work can have both inter- and intra-textuality, then the two terms are not mutually exclusive, are not dialectically defined, but are instead linked by a common approach to textuality as a fluid concept. We can read something through inter- or intra-textual moments, and in doing so, we define the parameters of our analysis.

As we've seen, there is both an inward (intra-) and an outward (inter-) movement when reading between texts. Similarly, Russian literary scholar Mikhail Bakhtin sees an outward force to language, the centrifugal, which describes the social and historical forces at play within each individual speech act; and an inward force, the centripetal, which unifies the way we speak.[44] A language references, consciously or not, any and all languages that have come before, as speakers interpret from the range of their experiences. Yet, the forces that bind a language together, that cohere it, work through self-referentiality, through the internal construction of that language itself. And as it is for language, so too is it for media: the web is rife with both inter- and intra-textual media. For example, one common feature on blogs is the "trackback" function, which allows one post to refer, intertextually, to another post of which it makes reference. By clicking on the "trackback" link (also called the "linkback" or "pingback"), users can follow a network of referencing links outward and beyond the confines of the blogosphere. But many blogs also feature links to other entries by the same author on the same blog: Such "internal links" reference inward, toward more of the same blog. The "permalink" link on many blogs is a tangible representation of this: Clicking on that link will take you to the original page of the blog featuring that self-same permalink. Aggregate sites that compile information from other websites similarly function in an intra-textual manner.

For fan studies, knowing inter- and intra-textuality can bolster analyses of fan fiction, whether on blogs, forums, or in print. Every piece of fan fiction on a blog exists between these two pulls—one connecting it to a larger corpus of work and the other building a more cohesive document. If Bakhtin's concept of heteroglossia may be a theoretical antecedent to intertextuality, then, a different Bakhtinian concept—that of the carnivalesque—leads to a productive discussion of intra-textuality. If we think of heteroglossia as one voice revealed as many, then the notion of the carnivalesque becomes many voices spoken as one.

The New Carnival

The carnivalesque most saliently deals with issues of collectivity and unification. A "carnival," in the sense that Bakhtin uses it, is a rambunctious festival, a Mardi Gras, in which the individuality of each participant becomes subsumed by the festival itself: for "Carnival is not a spectacle seen by the people;

they live in it, and everyone participates because its very idea embraces all of the people."[45] The theory of the carnivalesque becomes a useful tool for New Media scholars. The carnivalesque represents a state of consciousness—a theoretical state problematically and ultimately unachievable by humankind—where the underprivileged and repressed subvert and overturn the socioeconomic (and religious) hierarchies at the core of human society.[46] In this way, the carnivalesque encourages collectivity, although that collectivity must always be tempered by the concept of anonymity. In fact, as Kristeva says, any subject in a carnival is reduced to "nothingness," as the concept of hiding the self, of shrouding the face in a mask and embracing anonymity amongst the crowd, is enacted by all participants of carnival.[47] Similarly, for a LiveJournal blog, each commenter can pick a unique name and a unique icon to represent themselves in the online community.[48] The mask is deliberately ambiguous and, for Clark and Holquist, represents "the variety and flux of identities that otherwise, unmasked, are conceived as single and fixed." One consequence of this anonymity is that no one person dominates the carnivalesque, as "everybody makes carnival, everyone is carnival." This collectivity implies an equal and non-hierarchical structure to the carnivalesque. Clark and Holquist thus describe the carnivalesque as "fundamentally opposed to all hierarchies in epistemology, all canons and dogmas."[49] Bakhtin scholars Morson and Emerson describe carnival as "a social ritual of pure antinomianism" and as the "joyful negation of everything."[50] Bakhtin himself describes it as "the suspension of all hierarchical rank, privileges, norms, and prohibitions," and "of all hierarchic differences, of all ranks and status." In this sense, carnival is a hierarchy-free culture, which creates "a special type of communication ... permitting no distance between those who came in contact with each other."[51]

Yet, carnival does have an inherent hierarchy: specifically, it has a reverse hierarchy that subverts—but does not break—the official order through the inversion of social roles.[52] A reversal is, of course, not the same thing as a negation. Bakhtin's critics also recognize this, although they rarely mention the contradiction between non-hierarchy and reversal-hierarchy. Clark and Holquist describe carnival as a "reverse hierarchy" in which there exists "a humbling, debunking, or debasing of whatever is lofty by the lowly."[53] Bakhtin further describes the carnival as a "travesty" that underscores "a reversal of the hierarchic levels: the jester was proclaimed king, a clownish abbot, bishop, or archbishop was elected at the 'feast of fools,' and ... a mock pontiff was even chosen."[54] Dentith describes carnival as "the *inversion* and undoing of hierarchical or centripetal discourses."[55] In this description, it is not that social hierarchies do not exist, but rather it is more important to examine the inversion of these traditional hierarchies.

In examining blog fan fiction as a contemporary representation of the carnivalesque, therefore, we see a new form of quasi-hierarchical writing. Blog comments lie beneath the post, and, as we have seen, readers generally consider the post more important. But blog fan fiction comments often reverse this spatial hierarchy by refocusing the attention of the community onto the comments. The fiction post becomes an object upon which to comment, a reason for writers to post comments and for a community to form. Just like in the carnivalesque, where social hierarchies exist to be overturned, so too do the hierarchies that structure the blog fan fiction overturn traditional hierarchies of criticism and critique through intra-textual writing. In the following analysis I continue this line of thought by illustrating six tenets of intra-textuality to highlight the carnivalesque aspects of the blog.

One of the few scholarly articles to focus on intra-textuality as a theoretical mode of analysis is Tom Brown's analysis of DVD menu features. For Brown, the DVD has heralded a new interactive cinema, where the many extra features on the DVD make it difficult to "extricate the film from the context in which it is presented."[56] Brown mirrors the description of extra-text as described by Brookey and Westerfelhaus: The intertextual relationship between primary and secondary text on a DVD refocuses into an intra-textual relationship, one in which the intra-text serves the purposes of the producers of the DVD to rhetorically influence the viewer.[57] As Brown describes, the DVD is a New Media text that contains a number of distinctive features, all of which require a new type of analysis. Drawing on his article, I postulate the following six characteristics that scholars can use to analyze a text for its intra-textuality: Self-Reflexivity, Organization, Direct Address, Meta-Knowledge, Ludicity, and Recursive Expansion. I use examples from blog fan fiction written about *Battlestar Galactica* (2003) on the website LiveJournal.

Self-Reflexivity

Self-reflexivity defines the specific, textual connection between one element and another.[58] By this, I refer to the way an individual element in a blog can textually reference another: for instance, when a blog comment directly references the post, either by quoting it, referring to it via pronouns or other grammatical means, or identifying it by name or reference to content. By defining and then commenting on key elements in a blog, a comment about a post is, in fact, also a comment about the document as a whole. Posters may often comment on the fact of their posting by writing something like, "Sorry I haven't posted for awhile," self-reflexively creating a blog by describing the creation of that blog. Or, the comments of a blog might refer back to sections of that same blog. For example, a blog commenter might refer to the writing of the comment as content in the comment.

We can see this self-reflexivity in the interaction between cynthia_arrow and lyssie on lyssie's blog fan fiction story based on *Battlestar Galactica*, "Kara/Anders: Uninterrupted Routine."[59] On December 28, cynthia_arrow writes,

> I really like the mood of this. Of course, I'm always a fan of the playful vibe you keep with them [Kara/Anders], but always with this undertone of seriousness, sometimes with something lyrical threatening (just enough) to poke through, like that lovely note you ended on.

Cynthia_arrow begins by directly commenting on a part of the blog post ("… of this"). The "this" grammatically links back to the post, focusing the reader's attention away from the comment and reflecting back to the post itself. At the same time, cynthia_arrow also offers her own version of the events in that post, effectively rewriting the content of the post through a particular prescribed and determined fan-reading of the story. By directly stating how to read the story ("playful vibe"; "undertone of seriousness"; "something lyrical threatening"), or at least how to read particular elements within the story, cynthia_arrow creates a unique view of the post. But, importantly, she is not critical, but rather voices her comments with a ludic tone using a philosophy of playfulness. Because the post and the comment are attached to the same document, the intra-textual connection between both coheres these disparate elements together: It strengthens the fan fiction through self-reference.

Lyssie, author of the fiction post, responds to cynthia_arrow's comment a day later:

> Thank you =) Occasionally, I'm really sad that we didn't get to see Sam and Kara adjusting to married life. I like to think I'm pretty close to accurate in what we've seen—although I'm sure there was a lot more booze involved.

In their own way, both cynthia_arrow and lyssie rewrite lyssie's post, through what I call "conditional fictionality," a potentiality within the narrative as of yet unattained. This conditional fictionality becomes a rewriting that would take place, were she to change the post: "I'm sure," she says, "there was a lot more booze involved." This self-reflexivity—both commenting on and contributing to the blog document—conditionally rewrites the post without changing it. The reader must keep two versions in her head at all times: the blog post and the blog post-with-comments. The existence of comments on fiction blogs highlights this sense of all-encompassing collectivity, as blog posts and commenters come together to jointly construct the blog.

Bakhtin's notion of the carnivalesque offers a unique insight into this self-reflexivity. The carnivalesque expresses a construction wherein everyone participates: a carnivalesque culture exists only through the relationships *intra* the carnivalesque culture itself.[60] Any personal input into carnival, any participation with carnival, is a construction of that carnival. Self-reflexivity is thus the most basic function of carnival, or of the blog: Any contribution of individuals feeds into the collective whole. The blog actualizes this dichotomy.

Organization

A second characteristic of intra-textuality, organization, refers to the writer's proclamation of various components' positioning on a particular document, and the location can function to concretize the genre of the piece. Organization thus describes both a reference as to where the document is located, and a discussion of the location of aspects internal within the document. Many contemporary fan fiction blogs not only overtly state the *intended* genre before the post (most posts on LiveJournal are organized by the genre), but also state the genre within the comments of the blog itself, directly citing the genre as part of the content of the blog.

For example, in a comment to runawaynun, author of "Cylons, Shoes and Squirrels," projectjulie writes, "[I'm] putting this on the masterlist despite its gen-ness."[61] What projectjulie means is that she is going to upload this piece of fiction and save it to a different site, focused on a different genre. In this, projectjulie illustrates an *organization* of the fan text: It is worthy of being "on the masterlist," even though it is considered "gen," or "general" fic. In this, projectjulie organizes and classifies the blog while she contributes to it. The classification becomes the content of the blog.

We can see another example of organization in the story "Hubris, or the Downfall of Helena Cain, part 4 of 4."[62] Wyrdwritere, commenting on her own story, describes how she

> started writing this story for an AU [alternate universe, a genre of fan fic] community that wanted to use Helena Cain on occupied New Caprica. The community folded, but I finished the story anyway.

For wyrdwritere, the choice of where to place her story is important—how she categorizes it determines how the LiveJournal community will read it. She posts a comment not to explain where it "fits" in the canon, but rather where it "fits" in the organization of genres of fan fiction. By classifying the blog, wyrdwritere also builds the blog through an intra-textual connection between the blog content and the blog organization.

Organization is historically related to blogs. As pointed out by Blood in her excellent history of blogs, the original use of a blog was to keep an updated and updatable listing of interesting websites.[63] This list, the blog roll, was often organized thematically, with different genres of websites placed next to each other to create a blog intra-textually through its placement within a particular genre. We can see a similar type of emphasis on organization in Bakhtin's conception of the carnivalesque. As Bakhtin makes clear early in *Rabelais and His World*, the "carnival spirit offers the chance to have a new outlook on the world, to realize the *relative* nature of all that exists, and to enter a completely new *order* of things." The carnival allows the participants to explore their culture from the outside looking in, to construct a new social organization by surveying and leveling the old. Similarly, there is no organization outside of carnival, and people in a carnival are "a whole … organized in their own way."[64] The collectivity of the blog displaces the traditional hierarchies of the fan author/fan audience dichotomy, by normalizing all authors as blog authors. Although fans have always commented on each other's stories, now these comments can become part of the fan document itself. Through the intra-textual joining of the blog post with the blog comment, both fan writers not only rewrite the document, but also in that same act, reread the community to which that text belongs and their own position on it.

Direct Address

A third category of intra-textuality details moments in a document when the author of the work speaks to the reader/viewer via a direct address.[65] Literary scholars have discussed such meta-moments before, most notably in discussions of postmodern narratives. Meta-narratives highlight the constructed quality of a story, the artificiality of the text, by directly addressing the reader. By communicating "the *re*-presentative nature of narrative," meta-narratives make visible and obvious the means of the narration.[66] By intruding in this way, the narrator makes obvious the narration, and the fact that what the reader is reading is itself a construction. In a blog, direct address is most easily visible through the authorial comments: an author of a fiction post can comment on her own fiction by adding to, commenting on, conditionally rewriting elements of, or even critiquing it.[67]

To see an example of some of these direct addresses, on the blog story "Four Times Lee Adama Almost Died (4/4)," the author of the blog entry, helen_c, directly addresses each of the comments she receives on the post. For example, she often comments on the comment itself: to suffolkgirl, who writes "The Apollo-Starbuck banter is pitch perfect …" helen_c writes, "Thank you! Glad you liked the banter." Yet, some of her comments also directly address the story: She finishes her comment, "most of the fic wrote itself."[68] But in directly addressing the audience for whom she writes, helen_c enacts a unique characteristic of New Media technologies—the ability of authors and readers to communicate nearly instantaneously with comments, criticisms, or compliments. This instantaneousness is what makes such a difference from traditional footnoting or annotations: The speed of the response can be influential.

Further, we can see this direct address most evidently in the blog when an author comments on a piece she has written. For example, on "Cylons, Shoes and Squirrels," runawaynun adds her own personal statement, "I <3 Tory shopping for a dress for Laura. Of course, there'd be lots of undressing and trying on and. …"[69] With this comment, runawaynun provides a direct address to the reader, and through building the text with her own comment, also helps to develop the community of fan readers through conditional fictionality. She intra-textually constructs the story both in her *fiction* text and also in her direct address comments, breaking through the discursive fourth wall to address the fan community.

In terms of Bakhtin, this direct address has antecedents as well in the carnival. The most important moment of carnival is the direct address of the mock king. The metaphorical crowning and decrowning of the mock king expresses for the people the transitory nature of hierarchy and power and represents "the joyful relativity of all structure and order, of all authority and all (hierarchical) position." The decrowning of the king functions precisely because it is a direct address to the carnival participants. By directly addressing the participants in his symbolic crowning/decrowning, the king becomes the focus of the carnival—far from existing "without footlights and without division into performers and spectators," the king *does* become a performer.[70] His very presence, the symbolic gesture of crowning/decrowning, focuses carnival on his role. The king's decrowning is a direct address of the most spectacular kind.

Meta-Knowledge

A fourth characteristic of intra-textuality occurs when an author references the source canon of the fan's extant media object. The canon becomes a universal referent, antecedent to the piece of fan fiction that all fan readers understand. One cannot understand the blog fan fiction without at least some familiarity, some reference, to the extant media object's canon. To comment on the canonical authenticity of a blog fan fiction is to describe an intertextual relationship between the extant media object and elements in the blog document; however, to comment about canon in the blog itself is to construct intra-textually the blog document. Referencing canon works both centripetally, referring inward to the constituent understanding of the blog document, and centrifugally, referring outward to the external media object. We can see this reference to canonical authenticity in a comment by pellamerethiel on "Four Times Lee Adama Almost Died (4/4)."[71] Pellamerethiel begins her comment by referencing outward, to aspects of the extant media object:

> … I'm in the biggest love with Lee and the whole Adamas story—the scenes with Apollo and his father always make me happy, cause they have so great and strong chemistry, cause they're having this true, yet very complicated relationship.

She continues her description and comment by then referencing *inward*, to the blog document itself:

> And all of these I've found in your fic—all these tiny pieces that make me love Adamas even more. I'm speechless, so let me quote my most favourite scenes [from the fiction post].

Pellamerethiel shows how blog fan fiction exists in constant tension between what is *outward*, the extant media object canon, and what is *inward*, the fiction in the document. The implicit dialogue between pellamerethiel and helen_c here points in two directions: It intertextually comments on the relationship between the fan fiction and the extant media object, and also intra-textually comments on the content of the blog document.

In the same way, participants in carnival cannot act without an inherent understanding of the official culture that is effaced. Bakhtin describes the carnival as "dual-faced"; that is, it has what could be considered the "official" face, which "was turned to the past and sanctioned the existing order," but it also had the "face of the people of the market place [who] looked into the future and laughed." One face lies as part of the "protective, timeless stability," of tradition, while the other face challenges "the unchanging established order and ideology, and stresse[s] the element of change and renewal."[72] In the same way, Jenkins has described the "moral authority" of fans who feel that they know the extant media object just as well—or better—than the media producers. The effect of this is that "the ideology of fandom involves both a commitment to some degree of conformity to the original program materials, as well as a perceived right to evaluate the legitimacy of any use of those materials," including fan use.[73] Fan fiction thus comes from a history of balancing tradition with innovation, and blog fan fiction intra-textually recognizes through comments that reflexively establish canon while simultaneously freeing the fan-created fiction of canonicity.

Ludicity

A fifth element of intra-textuality Brown describes as a ludicity of content repetition.[74] Ludicity describes the playfulness of a text; the many ways that a Barthesian *jouissance* erupts from the text. By interacting and experiencing a text, readers help produce their own meanings within the prescribed structure of the text. Fan fiction is itself a way of "playing" at professional writing, or playfully subverting the copyright of a commercial product for noncommercial fun. There is a playful irony—a wink wink/nudge nudge to the reader—that blog fan fiction post authors utilize in their work. Blog fan fiction allows its authors to present their own work in a playful and self-deprecating way. Shave notes fan practices as playful, especially in relation to "story disclaimers [which] … acknowledge that copyright of the characters and universe belong elsewhere. The fans usually admit to 'playing' with the characters."[75]

We can see an example of this form of play in the way rowanjade playfully disclaims her own infringement upon *Battlestar Galactica*'s copyright with her story "They Took My Towel! And Other Furry Tails."[76] She writes:

> This is a Battlestar Galactica/Star Wars cross over. … Yes, Tails is spelled correctly. Yes, I blatantly stole ideas from both Battlestar Galactica and Return of the Jedi and the Wraith Squadron [a series of *Star Wars* novels], please don't sue me for doing it. This is for amusement and nothing more.

Rowanjade understands copyright here ("I blatantly stole ideas") and the supposed necessity for acknowledgment ("This is for amusement and nothing more"), but playfully skirts the issue of legality/illegality ("please don't sue me"). In another example, stargazercmc writes in response to lyssie's slash story "Seelix/Cally: Lubing the Viper," "Mmmmmm. … Cally porn. Tasty." The playfulness with which stargazercmc professes her pleasure of the fan fiction reciprocally feeds back into the fiction itself, and subverts the "official" reading of *Battlestar Galactica*. This type of comment demonstrates Carey's ritual communication: The community is built through experiencing and commenting on similar feelings about information and texts.[77] The playfulness of the language and the knowing comments function to bond members of the LiveJournal community through the shared experience of writing/reading fan fiction and the shared, secret subversion of the copyright. Despite the fact these community authors write works for free that deliberately break the law, they are willing to do so through both Jenkins's moral economy and the willful look-the-other-way practices of the extant media object producers.

This ludicity mirrors one of the key components of the carnivalesque, the "ancient ritualistic practice of directing laughter."[78] A carnivalesque culture exists in opposition to more serious, traditional modes of hegemony. The laughter of the carnivalesque is a way of cohering a group: Laughter attracts and presents the participants of the carnivalesque with a community. Indeed, Dentith shows that Bakhtin uses the carnivalesque as "a more socially unspecific notion of 'play' … [it] become[s] merely 'fun.'"[79] Importantly, this laughter is both subversive and supported. The carnivalesque is subversive by existing "outside of and contrary to all existing forms of the coercive socioeconomic and political organization."[80] At the same time, however, the carnivalesque festival is allowed to occur by this socioeconomic and political organization. The notion of the non-hierarchy simply reinforces the notion of the hierarchical outside of carnival. Eagleton describes carnival as

> so vivaciously celebrated that the necessary political criticism is almost too obvious to make. Carnival, after all, is a licensed affair in every sense, a permissible rupture of hegemony, a contained popular blow-off as disturbing and relatively ineffectual as any revolutionary work of art.[81]

Stallybrass and White follow suit: "Most politically thoughtful commentators wonder, like Eagleton, whether the 'licensed release' of carnival is not simply a form of social control of the low by the high and therefore serves the interests of that very official culture which it apparently opposes."[82] Other scholars describe the hegemonic control over carnival: Both Lachmann and Averintsev show that one can only laugh in carnival when one has permission from officially supported culture to do so.[83] In all these views, carnival becomes a force of regeneration for official culture, not the dynamic break with it hypothesized by Bakhtin.

Fan fiction acts similarly: Both slash fiction and carnival are subversive acts that are, or can be, supported by the same culture that is being subverted. Fan fiction subverts traditional or dominant readings of media products. For example, fans may subvert the intended relationship between Kirk and Spock in *Star Trek* and reread it as a homosexual bond, subverting cultural expectations. Or, fan fiction authors may use copyrighted materials for their own creative work, deliberately subverting the political power of the established media oligarchy. Of course, media producers may also support this behavior, as it encourages fans to buy more products and spread knowledge about the media product by word-of-mouth advertising. As Pesce puts it, "there's a strong sense that [the] act of piracy … is unofficially encouraged," in order that media products get free and effective advertising.[84] As Shave shows, fan fiction ends up supporting hegemony "by providing these fans an outlet, which thereby prevents insurrection at a societal level."[85] Indeed, fan fiction can never truly be a threat to large media conglomerates—few fans read fan fiction *instead* of watching a TV show—and media producers could be said to passively support the minor copyright infringement. Indeed, this passive support of copyright infringement has, in at least two cases, benefited the corporations: with the TV shows *Battlestar Galactica* and *Doctor Who*, according to Pesce, the illegal downloading of the premiere episodes actually increased viewership, as the "piracy made it possible for 'word-of-mouth' to spread."[86]

Fan fiction needs an extant media object just as carnival needs an official culture for self-comparison. The "official culture" is necessary for fan culture: Fans, as pointed out by Jenkins, need their fan-created texts to echo with and to support the extant media object.[87] Without that connection, they lose their "moral economy" over the text. This support becomes a part of the fiction, just as it is a part of the carnival.

Recursive Expansion

A sixth feature of intra-textuality is a recursive expansion of the document. By recursive expansion, I refer to the capacity of the digital document to feed back onto itself in a dynamic reciprocity of information exchange. In this way, the document becomes a self-aggregating experience. For example, an email message can grow longer with each reply, as the text of the previous messages remains below the new text. The email grows through repetition of content, and as such, each email recursively expands. Similarly, on a blog, writers of comments will repeat text from the post as reference in their comment. In this way, elements of the document are repeated to build the document. As a document recursively expands, it creates a feedback loop, a citation system, where the repetition builds upon itself.

This recursive expansion mirrors another aspect of Bakhtin's notion of the carnivalesque: the grotesque, the "very act of becoming and growth, the eternal incomplete unfinished nature of being." Carnival relishes the grotesque, takes pleasure in ever-expanding body that is "unfinished, outgrows itself, transgresses its own limits."[88] Just as a grotesque body continues to grow past its normal limits, to fatten with excess, so too does the carnival expand exponentially. The carnival grows to incorporate all.

Recursive expansion is, by definition, incomplete: It is a cycle with no end, a reference with no antecedent. Every time a document recursively expands, it expands but does not develop new content. There is no forward momentum to the text, but rather a turn inward, a reflexivity of the document back onto itself. As the blog grows and intra-textually connects to itself, it supports itself through its continually expanding comments. Theoretically, there is no limit to how large a blog could grow.

This capacity of the digital document to feed back onto itself is displayed in comments that quote and track back to the original post. For example, in daybreak777's comment on helen_c's story "Four Times Lee Adama Almost Died (4/4)," we can see how the helen_c's story becomes fodder for the comment.[89] She writes, italicizing quotations from the post, "*How the hell would I know? I'm in your head, remember? Besides, you died once, so you should know.* Even in his head, Kara gives it to him straight. And helps him stay alive. Just like always." Further, pellamerethiel comments on "Four Times Lee Adama Almost Died (4/4)" by quoting directly (and liberally) from the blog post. She writes, "let me quote my most favourite scenes," and then uses three separate sections from the post to build her comment, finally concluding from all three that she has "saved it into my memories." Or in the aforementioned comment on "Cylons, Shoes and Squirrels," runawaynun describes of her own story, "I <3 Tory shopping for a dress for Laura," summarizing exactly what happened in the post.

By rewriting the fan text in the comment, both daybreak777 and pellamerethiel recursively expand the blog by including wordings and statements remixed from the original. Recursive expansion demonstrates the eternally unfinished nature of the blog: One can never finish a blog that continues to grow and expand inwardly. By including works from the blog fan fiction in their comments, these fan commenters build the blog as they comment upon it.

Intra-Textuality and the Blog Document

Fan fiction is a form of rewriting of the extant media object; not a revision, but literally a "writing-around" of different aspects of the cult fiction. Yet, in the process of rewriting, there is more than just a fan's input or work. Instead, it takes a community of fans all contributing communally to one project. By rewriting together, fan communities scribe their fandom into existence. But rewriting is not the end of fans' work. There is a deeper process at work in a New Media environment: one that involves not just the creation of new stories, but the rereading of these stories in new contexts and with new insights. Rewriting is not the end result of digital fandom, but rather the beginning.

Notes

1. Jenkins, *Convergence Culture*, 97–8.
2. Barthes, "From Work to Text," 156–7.
3. Sullivan, "Comments," ¶2–3 writes about comments: "Most other blogs have them; they give readers a place to write and vent and discuss. No one has to read them." Sullivan notably does not allow comments on his blog, which forces a rhetorical question, is it a blog if there are no comments? In common parlance, the "blog" is a space for writing, not the way that space functions. However, for this chapter, I am defining the "blog" as a way of writing: The blog is text written for an online community to read. In this, even without allowing his readers to comment, Sullivan understands the fact the blog must deal with communities: "This blog has become as much about you as me over the years." Even without comments, Sullivan intertextually (and passively) references the community of readers he has.
4. Shepherd, ed., *Bakhtin*.
5. Shave, "Slash Fandom on the Internet," ¶3.
6. Blood, "Weblogs."
7. Lessig, *Remix*, 59.
8. Miller and Shepherd, "Blogging as Social Action," ¶42; for more on blogs as texts, see Hevern, "Threaded Identity in Cyberspace"; Huffaker and Calvert, "Gender, Identity, and Language Use in Teenage Blogs"; and Huffaker, "Teen Blogs Exposed"; Foucault, "Self-Writing," 216; Keen, *The Cult of the Amateur*, 34.
9. See Miller and Shepherd, "Blogging as Social Action"; Serfaty, *The Mirror and the Veil*.
10. With the notable exception of Thomas, "Fictional Blogs."
11. See Bury, *Cyberspaces of Their Own*.

12. Firewolfe, "Battlestar Galactica 1980/1985 Finding Earth"; WHCnelson, Review of "Battlestar Galactica 1980/1985 Finding Earth."
13. Turk and Johnson, "Toward an Ecology of Vidding," ¶2.2.
14. Cooper, "This Is Not an Isolated Incident," 12; Shave, "Slash Fandom on the Internet," ¶2.
15. Rettberg, *Blogging*, 57.
16. Barthes, "From Work to Text," emphasis mine.
17. Grafton, *The Footnote*.
18. Moores, *Media/Theory*, part I.
19. Authors of posts can delete comments.
20. Grafton, *The Footnote*, 6.
21. Kristeva, "Word," 66; quoted in Gray, *Watching with The Simpsons*, 27.
22. Gindin, "Contributions to Textlinguistics in the Soviet Union," 261.
23. Gray, *Watching with The Simpsons*, 19; Harty, "Text, Context, Intertext," 4.
24. For texts that introduce intertextuality, see Kristeva, "The Bounded Text"; Kristeva, "Word"; Barthes, "Death of the Author"; Barthes, "From Work to Text"; Barthes, "Theory of the Text." For texts that describe intertextuality, see Morgan, "Is There an Intertext in This Text?"; Allen, *Intertextuality*; Ott and Walter, "Intertextuality"; Orr, *Intertextuality*.
25. Kristeva, "Word," 66.
26. Barthes, "Death of the Author," 146.
27. Barthes, "From Work to Text," 158; Foucault, "What Is an Author?"
28. Gray, *Watching with The Simpsons*, 21.
29. Johnson, *Everything Bad Is Good for You*; Mittell, "Narrative Complexity"; Gray, *Television Entertainment*; Jenkins, *Convergence Culture*; see also Booth, "Intermediality in Film and Internet," and what I call "intermedia" narratives.
30. Kristeva, "Word," 66.
31. Barthes, "From Work to Text," 160.
32. Kaplan, "Construction of Character through Narrative," 136.
33. For various uses of "intra-text," see Otto, White, and Camperell, "Text Comprehension Research to Classroom Application"; Schork, "Acoustic Intratexts"; Shen and Xu, "Intratextuality, Extratextuality, Intertextuality"; Cerdán and Vidal-Abarca, "The Effects of Tasks."
34. Why Hawaii would have interstates, however, remains a mystery.
35. Warnick, *Rhetoric Online*, 119.
36. Walker, "Distributed Narratives," ¶1.
37. Bennett and Woollacott, *Bond and Beyond*, 56, 59–60.
38. Brooker, *Batman Unmasked*.
39. Tulloch and Jenkins, *Science Fiction Audiences*, 136.
40. See Britton, *TARDISBound*; Parkin, *AHistory*.
41. Tulloch and Jenkins, *Science Fiction Audiences*, 134, my emphasis: "The discursive negotiation of the text of 'The Monster of Peladon' in this general studies interview occurs then via the ascription of its meaning to the *series of histories*: [specifically] the history of the *Doctor Who* series itself."
42. Cartmel, *Through Time*, 208.
43. Fairclough, *Language and Power*, 152.

44. Bakhtin, "Discourse in the Novel," 272. I do not mean to tie intertextuality or intratextuality to centripetal or the centrifugal forces; merely to indicate that a similar relationship exists for both.
45. Bakhtin, *Rabelais*, 7. Carnival is useful for cultural studies: See, Miller, "The Otherness of History in Rabelais' Carnival"; Webb, "Bakhtin at the Seaside"; Eagleton, "Bakhtin, Schopenhauer, Kundera," 178. Eagleton asserts, "Few modern critical concepts have proved more fertile and suggestive, more productively polymorphous, than the Bakhtinian notion of carnival."
46. This notion of carnival has been debated by critics such as Eagleton, *Walter Benjamin*; and Stallybrass and White, *The Politics and Poetics of Transgression*. They show that carnival is supported by the powerful in order to offer a "release" to the populous.
47. Kristeva, "Word," 78.
48. rowanjade, raincitygirl, comments on "They Took My Towel …"
49. Clark and Holquist, *Mikhail Bakhtin*, 302–4, 300, 310.
50. Morson and Emerson, *Mikhail Bakhtin*, 92, 94.
51. Bakhtin, *Rabelais*, 246, 10.
52. Dentith, *Bakhtinian Thought*, 73.
53. Clark and Holquist, *Mikhail Bakhtin*, 309.
54. Bakhtin, *Rabelais*, 81.
55. Dentith, *Bakhtinian Thought*, 85.
56. Brown, "DVD of Attractions," 170. Later, rewritten as a book chapter in Bennett and Brown, ed. *Film and Television after DVD*.
57. Brookey and Westerfelhaus, "Hiding Homoeroticism in Plain View," 23–4. See also, Brookey and Booth, "Restricted Play," for a similar argument about video games and interactivity, and Booth, "Mediating New Technology," 32–3.
58. Brown, "DVD of Attractions," 171, 173.
59. cynthia_arrow, comment on "Kara/Anders, Uninterrupted Routine"; lyssie, comment on "Kara/Anders, Uninterrupted Routine."
60. Bakhtin, *Rabelais*, 246; Kristeva, "Word," 78.
61. projectjulie, comment on "Cylons, Shoes and Squirrels"; runawaynun, "Cylons, Shoes and Squirrels."
62. wyrdwritere, comment on "Hubris, or the Downfall of Helena Cain, part 4 of 4."
63. Blood, "Weblogs."
64. Bakhtin, *Rabelais*, 34, my emphasis, 255.
65. Brown, "DVD of Attractions," 175–6.
66. Cobley, *Narrative*, 175.
67. In the blogs analyzed for this book, almost half of the comments (33 out of 70) were written by the author of the post.
68. helen_c, comment on "Four Times Lee Adama Almost Died (4/4)."
69. In "netspeak," <3 symbolically represents a heart, which translates as "love," as in "I love the thought of Tory shopping for a dress for Laura."
70. Bakhtin, *Problems of Dostoevsky's Poetics*, 124, 122.
71. pellamerethiel, comment on "Four Times Lee Adama Almost Died (4/4)."
72. Bakhtin, *Rabelais*, 81.
73. Jenkins, "*Star Trek* Rerun, Reread, Rewritten," 55.

74. Brown, "DVD of Attractions," 174.
75. Shave, "Slash Fandom on the Internet," ¶18.
76. rowanjade, "They Took My Towel! And Other Furry Tails."
77. Carey, "A Cultural Approach to Communication," 18–19.
78. Bakhtin, *Problems of Dostoevsky's Poetics*, 129.
79. Dentith, *Bakhtinian Thought*, 84–5.
80. Bakhtin, *Rabelais*, 255.
81. Eagleton, *Walter Benjamin*, 148.
82. Stallybrass and White, *The Politics and Poetics of Transgression*, 13.
83. Lachmann, "Bakhtin and Carnival"; Averintsev, "Bakhtin and the Russian Attitude to Laughter."
84. Pesce, "Piracy Is *Good?*" 3.
85. Shave, "Slash Fandom on the Internet," ¶22.
86. Pesce, "Piracy Is *Good?*" ¶2–3.
87. Jenkins, "*Star Trek* Rerun, Reread, Rewritten," 54–7.
88. Bakhtin, *Rabelais*, 52, 26.
89. daybreak777, comment on "Four Times Lee Adama Almost Died (4/4)."

Bibliography

Allen, Graham. *Intertextuality*. London: Routledge, 2000.

Averintsev, Sergei S. "Bakhtin and the Russian Attitude to Laughter." In *Bakhtin: Carnival and Other Subjects: Selected Papers from the Fifth International Bakhtin Conference University of Manchester, July 1991*, edited by David Shepherd, 13–19. Atlanta, GA: Rodopi, 1993.

Bakhtin, Mikhail M. "Discourse in the Novel." In *The Dialogic Imagination: Four Essays*, edited by Michael Holquist, translated by Caryl Emerson and Michael Holquist, 259–422. Austin: University of Texas Press, 1981.

Bakhtin, Mikhail M. *Problems of Dostoevsky's Poetics*, edited and translated by Caryl Emerson. Minneapolis: University of Minnesota Press, 1984.

Bakhtin, Mikhail M. *Rabelais and His World*, translated by Hélène Iswolsky. Bloomington: Indiana University Press, 1984.

Barthes, Roland. "The Death of the Author." In *Image-Music-Text*, translated by Stephen Heath, 142–8. New York: Hill and Wang, 1977.

Barthes, Roland. "From Work to Text." In *Image-Music-Text*, translated by Stephen Heath, 155–64. New York: Hill and Wang, 1977.

Barthes, Roland. "Theory of the Text." 1973. In *Untying the Text: A Post-Structuralist Reader*, edited by Robert Young, translated by Ian McLeod, 31–47. London: Routledge, 1981.

Bennett, James, and Tom Brown, ed. *Film and Television after DVD*. London: Routledge, 2008.

Bennett, Tony, and Janet Woollacott. *Bond and Beyond: The Political Career of a Popular Hero*. London: MacMillan, 1987.

Blood, Rebecca. "Weblogs: A History and Perspective." *Rebecca's Pocket*, 07 September 2000. Accessed 22 March 2016. http://www.rebeccablood.net/essays/weblog_history.html.

Booth, Paul. "Intermediality in Film and Internet: *Donnie Darko* and Issues of Narrative Substantiality." *Journal of Narrative Theory* 38, no. 3 (2008): 398–415.
Booth, Paul. "Mediating New Technology: The Realization of a Digital Intellect." *Nebula* 5, no. 1–2 (2008): 28–43.
Britton, Piers. *TARDISBound: Navigating the Universes of* Doctor Who. London: I.B. Tauris, 2011.
Brooker, Will. *Batman Unmasked: Analyzing a Cultural Icon*. New York: Continuum, 2001.
Brookey, Robert A. *Hollywood Gamers: Digital Convergence in the Film and Video Game Industries*. Bloomington: Indiana University Press, 2010.
Brookey, Robert, and Paul Booth. "Restricted Play: Synergy and the Limits of Interactivity in *The Lord of the Rings: The Return of the King* Video Game." *Games and Culture* 1, no. 3 (2006): 214–30.
Brookey, Robert, and Robert Westerfelhaus. "Hiding Homoeroticism in Plain View: The *Fight Club* DVD as Digital Closet." *Critical Studies in Media Communication* 19, no. 1 (2002): 21–43.
Brown, Tom. 'The DVD of Attractions'? *The Lion King* and the Digital Theme Park." *Convergence: The International Journal of Research into New Media Technologies* 13, no. 2 (2007): 169–83.
Bury, Rhiannon. *Cyberspaces of Their Own: Female Fandoms Online*. New York: Peter Lang, 2005.
Carey, James. "A Cultural Approach to Communication." In *Communication and Culture: Essays on Media and Society*, 13–36. New York: Routledge, 1992.
Cartmel, Andrew. *Through Time: An Unauthorized and Unofficial History of* Doctor Who. New York: Continuum, 2005.
Cerdán, Raquel, and Eduardo Vidal-Abarca. "The Effects of Tasks on Integrating Information from Multiple Documents." *Journal of Educational Psychology* 100, no. 1 (2008): 209–22.
Clark, Katerina, and Michael Holquist. *Mikhail Bakhtin*. Cambridge, MA: The Belknap Press of Harvard University Press, 1984.
Cobley, Paul. *Narrative*. London: Routledge, 2001.
Cooper, Dennis. "This Is Not an Isolated Incident: An Introduction." In *Userlands: New Fiction from the Blogging Underground*, edited by Dennis Cooper, 11–13. New York: Akashic Books, 2007.
cynthia_arrow, comment on "Kara/Anders, Uninterrupted Routine."
daybreak777, comment on "Four Times Lee Adama Almost Died (4/4)."
Dentith, Simon. *Bakhtinian Thought: An Introductory Reader*. London: Routledge, 1995.
Eagleton, Terry. "Bakhtin, Schopenhauer, Kundera." In *Bakhtin and Cultural Theory*, edited by Ken Hirschkop and David Shepherd, 178–88. Manchester, UK: Manchester University Press, 1989.
Eagleton, Terry. *Walter Benjamin, or Towards a Revolutionary Criticism*. New York: Schocken Books, 1981.
Fairclough, Norman. *Language and Power*. London: Longman, 1989.
Firewolfe. "Battlestar Galactica 1980/1985 Finding Earth." *Fanfiction.net*, 20 November 2007. Accessed 23 March 2016. http://www.fanfiction.net/s/3777645/1/Battle_Star_Galactica_1980_1985 Finding Earth.

Foucault, Michel. "Self-Writing." In *Ethics: Subjectivity and Truth Vol 1*, edited by Paul Rabinow, translated by Robert Hurley and others, 207–22. New York: The New Press, 1997.

Foucault, Michel. "What Is an Author?" In *Language, Counter-memory, Practice*, edited and translated by Donald Bouchard and Sherry Simon, 113–38. Ithaca, NY: Cornell University Press, 1977.

Gindin, Sergei I. "Contributions to Textlinguistics in the Soviet Union." In *Current Trends in Textlinguistics*, edited by Wolfgang U. Dressler. Berlin: de Gruyter, 1978.

Grafton, Anthony. *The Footnote: A Curious History*. Cambridge, MA: Harvard University Press, 1999.

Gray, Jonathan. *Television Entertainment*. New York: Routledge, 2008.

Gray, Jonathan. *Watching with The Simpsons: Television, Parody, and Intertextuality*. New York: Routledge, 2006.

Gray, Jonathan, and Jason Mittell. "Speculation on Spoilers: *Lost* Fandom, Narrative Consumption and Rethinking Textuality." *Particip@tions* 4, no. 1 (2007). http://www.participations.org/Volume%204/Issue%201/4_01_graymittell.htm.

Harty, E. R. "Text, Context, Intertext." *Journal of Literary Studies* 1, no. 2 (1985): 1–14.

helen_c, comment on "Four Times Lee Adama Almost Died (4/4)."

Hevern, Vincent W. "Threaded Identity in Cyberspace: Weblogs & Positioning in the Dialogical Self." *Identity: An International Journal of Theory and Research* 4, no. 4 (2004): 321–35.

Huffaker, David. "Teen Blogs Exposed: The Private Lives of Teens Made Public." *American Association for Advancement of Science*. St. Louis, MO, 2006. Accessed 23 March 2016. https://www.researchgate.net/publication/241644648_Teen_Blogs_Exposed_The_Private_Lives_of_Teens_Made_Public.

Huffaker, David A., and Sandra L. Calvert. "Gender, Identity, and Language Use in Teenage Blogs." *Journal of Computer-Mediated Communication* 10, no. 2 (2005). http://onlinelibrary.wiley.com/doi/10.1111/j.1083-6101.2005.tb00238.x/abstract.

Jenkins, Henry. *Convergence Culture: Where Old and New Media Collide*. New York: New York University Press, 2006.

Jenkins, Henry. "Star Trek Rerun, Reread, Rewritten: Fan Writing as Textual Poaching." In *Fans, Bloggers, and Gamers: Exploring Participatory Culture*, 37–60. New York: New York University Press, 2006.

Johnson, Steven. *Everything Bad Is Good for You*. New York: Riverside, 2005.

Kaplan, Deborah. "Construction of Character through Narrative." In *Fan Fiction and Fan Communities in the Age of the Internet*, edited by Karen Hellekson and Kristina Busse, 134–52. Jefferson, NC: McFarland, 2006.

Keen, Andrew. *The Cult of the Amateur*. New York: Random House, 2007.

Kristeva, Julia. "The Bounded Text." In *Desire in Language: A Semiotic Approach to Literature and Art*, edited by Leon S. Roudiez, translated by Thomas Gora, Alice Jardine, and Leon S. Roudiez, 36–63. New York: Columbia University Press, 1980.

Kristeva, Julia. "Word, Dialogue, Novel." In *Desire in Language: A Semiotic Approach to Literature and Art*, edited by Leon S. Roudiez, translated by Thomas Gora, Alice Jardine, and Leon S. Roudiez, 64–91. New York: Columbia University Press, 1980.

Lachmann, Renate. "Bakhtin and Carnival: Culture as Counter-Culture." *Critical Critique* 11 (1988): 115–52.

Lessig, Lawrence. *Remix: Making Art and Commerce Thrive in the Hybrid Economy*. New York: Penguin, 2008.

lyssie, comment on "Kara/Anders, Uninterrupted Routine."

Miller, Carolyn R., and Dawn Shepherd. "Blogging as Social Action: A Genre Analysis of the Weblog." In *Into the Blogosphere: Rhetoric, Community, and Culture of Weblogs*, edited by Laura Gurak, Smiljana Antonijevic, Laurie Johnson, Clancy Ratliff, and Jessica Reyman. Minneapolis: University of Minnesota Press, 2004. Accessed 23 March 2016. http://blog.lib.umn.edu/blogosphere/.

Miller, Paul Allen. "The Otherness of History in Rabelais' Carnival and Juvenal's Satire, or Why Bakhtin Got It Right the First Time." In *Bakhtin and the Other*, edited by Peter Barta, Paul Allen Miller, Charles Platter, and David Shepherd, 141–64. London: Routledge, 2001.

Mittell, Jason. "Narrative Complexity in Contemporary American Television." *The Velvet Light Trap* 58 (2006): 29–40.

Moores, Shaun. *Media/Theory*. London: Routledge, 2005.

Morgan, Thaïs. "Is There an Intertext in This Text?: Literary and Interdisciplinary Approaches to Intertextuality." *American Journal of Semiotics* 3, no. 1 (1985): 1–40.

Morson, Gary Saul, and Caryl Emerson. *Mikhail Bakhtin: Creation of a Prosaics*. Stanford, CA: Stanford University Press, 1990.

Orr, Mary. *Intertextuality*. London: Polity, 2003.

Ott, Brian L., and Cameron Walter. "Intertextuality: Interpretive Practice and Textual Strategy." *Critical Studies in Media Communication* 17, no. 4 (2000): 429–46.

Otto, Wayne, Sandra White, and Kay Camperell. "Text Comprehension Research to Classroom Application: Developing an Instructional Technique." *Reading Psychology* 1, no. 3 (1980): 184–91.

Parkin, Lance. *AHistory: An Unauthorized History of the* Doctor Who *Universe*. Des Moines, IA: Mad Norwegian Press, 2006.

pellamerethiel, comment on "Four Times Lee Adama Almost Died (4/4)."

Pesce, Mark. "Piracy Is *Good*? New Models for the Distribution of Television Programming." Paper presented at the Australian Film Television and Radio School, Sydney, 6 May 2005. Accessed 23 March 2016. http://hyperreal.org/~mpesce/piracyisgood.pdf.

projectjulie, comment on "Cylons, Shoes and Squirrels."

Rettberg, Jill Walker. *Blogging*. Cambridge, UK: Polity Press, 2008.

rowanjade, "They Took My Towel! And Other Furry Tails. …"

rowanjade, raincitygirl, comments on "They Took My Towel. …"

runawaynun, "Cylons, Shoes and Squirrels."

Schork, R. Joseph, "Acoustic Intratexts in *Aeneid* 7.122 and 4.408." *Classical Philology* 91, no. 1 (1996): 61–3.

Serfaty, Viviane. *The Mirror and the Veil: An Overview of American Online Diaries and Blogs*. Amsterdam: Rodopi, 2004.

Shave, Rachel. "Slash Fandom on the Internet, or Is the Carnival Over?" *Refractory* 6 (2004). Accessed 23 March 2016. http://blogs.arts.unimelb.edu.au/refractory/2004/06/17/slash-fandom-on-the-internet-or-is-the-carnival-over-rachel-shave/.

Shen, Dan, and Dejin Xu. "Intratextuality, Extratextuality, Intertextuality: Unreliability in Autobiography Versus Fiction." *Poetics Today* 28, no. 1 (2007): 43–87.

Shepherd, David, ed., *Bakhtin: Carnival and Other Subjects: Selected Papers from the Fifth International Bakhtin Conference University of Manchester, July 1991*, Atlanta, GA: Rodopi, 1993.

Stallybrass, Peter, and Allon White. *The Politics and Poetics of Transgression*. Ithaca, NY: Cornell University Press, 1986.

Sullivan, Andrew. "Comments." *The Daily Dish*, 03 March 2008. Accessed 23 March 2016. http://andrewsullivan.theatlantic.com/the_daily_dish/2008/03/comments.html.

Thomas, Angela. "Fictional Blogs." In *Use of Blogs*, edited by Axel Bruns and Joanne Jacobs, 199–209. New York: Peter Lang, 2006.

Tulloch, John, and Henry Jenkins. *Science Fiction Audiences: Watching* Doctor Who *and* Star Trek. London: Routledge, 1995.

Turk, Tisha, and Joshua Johnson. "Toward an Ecology of Vidding." In "Fan/Remix Video," edited by Francesca Coppa and Julie Levin Russo, special issue, *Transformative Works and Cultures* 9 (2012). doi: 10.3983/twc.2012.0326.

Walker, Jill. "Distributed Narratives: Telling Stories across Networks." Paper presented at AoIR 5.0, Brighton, 21 September 2004. Accessed 23 March 2016. http://jilltxt.net/txt/Walker-AoIR-3500words.pdf.

Warnick, Barbara. *Rhetoric Online: Persuasion and Politics on the World Wide Web*. New York: Peter Lang, 2007.

Webb, Darren. "Bakhtin at the Seaside: Utopia, Modernity and the Carnivalesque." *Theory, Culture & Society* 22, no. 3 (2005): 121–38.

WHCnelson. Review of "Battlestar Galactica 1980/1985 Finding Earth." *Fanfiction.net*, 20 November 2007. Accessed 23 March 2016. http://www.fanfiction.net/r/3777645/.

wyrdwritere, comment on "Hubris, or the Downfall of Helena Cain, part 4 of 4."

· 3 ·

THE NARRATIVE DATABASE AND THE WEB COMMONS

The serial ... can become a cult both for a naïve audience and for a more sophisticated one.

—Pérez-Reverte, *The Club Dumas* (p. 326)

Heroes Wiki is a reference site for NBC (http://www.nbc.com)'s popular sci-fi drama Heroes. Heroes Wiki currently has 2,348 articles. Come here for the latest information on the show and its characters.

—Heroeswiki Main Page

As Pérez-Reverte describes, serial narratives—that is, narrative plots that continue across more than one installment—can develop a cult following. The antique book collectors at the heart of Pérez-Reverte's *The Club Dumas* have become members of a cult that revere the novels of Alexandre Dumas by seeking out and venerating the serialized chapters of, e.g., *The Count of Monte Cristo*. Indeed, serialized narratives have been common in literature for hundreds of years.[1] One reason these serialized narratives create "cults" is that cult texts can form open-ended narratives that focus on "a singular question or related set of questions."[2] Serial narratives are thus complex and involved: They require the concentration of the reader to piece together narrative events and plot elements inter- and intra-textually separated both by space and by time.

The narrative arc of a serialized television show does not end at the credits of each episode, but rather continues across and through each subsequent episode, building upon itself and generating a complex storyline.[3] As Mittell points out, contemporary prime-time television has embraced the serial form, in shows as generically diverse as *Lost* (science-fiction/mystery), *Desperate Housewives* (soap opera/mystery), *Arrested Development* (sitcom), *The Sopranos* (drama), *Six Feet Under* (drama/black comedy), *24* (action-adventure), and *Curb Your Enthusiasm* (comedy). Add to these a batch of shows that premiered after Mittell's article, and we can see that the trend in contemporary narrative television storytelling seems to be moving toward seriality: *Heroes* (2006), *Dexter* (2006), *Fringe* (2008), *Dollhouse* (2009), *Flashforward* (2009), *Game of Thrones* (2011), and Netflix's *House of Cards* (2013) to name a few. To describe more thoroughly the narrative structure of these complicated shows, Mittell articulates a new category to analyze this particular facet of television storytelling, "narrative complexity," and uses it to measure the serialized plot within an episodic television show: "a true aesthetic innovation unique to its medium."[4] His new work, *Complex Television* (2015), takes this analysis further, arguing for a refined poetics of new complex television narrative. Mittell integrates wikis—specifically, *Lostpedia*—into this poetics, arguing that wikis "blur the boundaries between fan creativity and fan documentation."[5] They are both archives and narratives.

As a new way of examining television narrative, narrative complexity reconceptualizes television studies. However, contemporary scholars have done little work in determining how issues of television's narrative complexity translates to and manifests in fans' narrative creation.[6] Fans of narratively complex television participate in turn in the creation of narratively complex digital documents based on those programs. In an age where "convergence" is the buzzword and television producers create more extra-textual content, it is not enough to examine narrative complexity solely from the standpoint of television. For example, in the second quotation inscribing this chapter, the fan-creators behind the wiki for the cult television show *Heroes* detail the online collection, collation, and categorization of units of narrative information about the program. Fans quote *Heroes* on this wiki as "a completely furnished world," echoing Eco's discussion of cult texts.[7] Utilizing the communal authorship and collective intelligence of the fans of the show, the creators of *Heroeswiki* create a narratological compendium of "narratively complex" content. Many cult shows with "narrative complexity" have equally complicated fan-created wikis, replete with narrative information: The aforementioned

Heroeswiki describes the show *Heroes*; *Lostpedia* examines *Lost*; the *Harry Potter* narrative has *harrypotter.wikia*; the *Doctor Who* narrative is at *tardis.wikia*; and the *Star Wars* narrative has the appropriately named *Wookieepedia*. For digital fandom, these wikis exist as complex archives of narrative information. Wikis force scholars to rethink traditional ways narrative is conceived online, and ask, how does "narrative complexity" work in on online space? How does the web facilitate this narrative construction? And with what theoretical concepts can we describe the creation of these wikis?

In this chapter, I continue my investigation of New Media Studies through an examination of the fan-created wiki as a narrative database. This narrative database is a reflection of a changed media environment, which reconceptualizes narrative from a "chrono-logic" mode to an archival one.[8] Instead of representing "plot" through causality, fans represent it spatially, using the inherent hypertextuality of the web to create connections between narrative elements.[9] To discuss this interactive shift in the understanding of narrative, I examine wikis not solely as portals of information, but also as sites of fans' narrative (re)construction of a cult television program.[10] Wikis exemplify an active and dynamic narrative text; in the academic study of fans, scholars often posit the fan as the epitome of the active audience member, or the productive viewer.[11] Fandom itself means being part of a community, and the writing of fan fiction becomes "a social activity for these fans, functioning simultaneously as a form of personal expression and as a source of collective identity (part of what it means to be a 'fan')."[12] Indeed, Baym's phrase "collaborative interpretation" highlights how fans manifest their collective identity through writing together.[13]

Digital fandom not only encourages, but also relies on a certain level of community and socialization, as we have seen in regards to the Web Commons. Fan fiction is, in a way, both an appreciation and a reappreciation of the object of that fandom. Appreciation involves reading for meaning, but reappreciation is reading the media object again through the lens of the fan community. Narrative databases collect the previously read as well as the never-read, as fans assimilate individual units of narrative knowledge and, as a community, re-enact and reform them in new ways within the database through the communal interactive action I term "narractivity."

Narractivity describes the communal construction of a narrative, or the way that members of a community bring together different elements in different orders to change the narrative of a digital text. Given that narrative emerges from the interaction of the story and the discourse, as I define later in this chapter, narractivity harnesses fans' own interpretations of the discourse or fans' retelling of the story. Narractivity is using group interactive mechanisms to (re)tell a narrative.

Narractivity arrives in an inchoate stage in media theorist Abigail Derecho's concept of archontic texts. For Derecho, a text is archontic if it "allows, or even invites, writers to enter it, select specific items they find useful, make new artifacts using those found objects and deposit the newly made work back into the source text's archive."[14] Derecho brings the term "archontic" to fan studies from Jacques Derrida's analysis of the archive as an ever-expanding inventory for both storage and production. As we have seen, narratives on wikis can be boundless. For example, while *The Club Dumas* is a *text* of approximately 70,000 words, it is also an *archive* of material—characters, situations, books—that can be used in other texts, fan made or non-fan made. Archontic texts "always produce more archive, to enlarge itself," and a narratively complex television show like *Lost* features fan-created elements as part of that archontic text, part of that "tendency toward enlargement and accretion that all archives possess."[15] In order to examine and further refine narractivity, in this chapter I analyze in more detail the wikis for the cult television shows *Heroes* and *Lost*. The results of narractivity manifest in three ways: through the construction of narrative knowledge on the wiki, through the scattering of narrative meaning by spoilers, and through resolution of narrative research. By constructing narrative, fans use narractivity to create knowledge about a text. We can see the necessity for this new narrative form by Ryan, who argues that "digital narrative should emancipate itself from [traditional] literary models."[16] Combing through the database of narrative information on the wiki, the fan community rereads the narrative of the extant media object, and constructs the story in a narrative database. Further, by scattering narrative meaning on a wiki, fans rewrite the extant media object through the speculative fiction of the spoilers of that object. Finally, fans research narrative through a discursive rereading of the story that mirrors academic or scholarly research into the extant media object. In short, the fan community rereads the extant media object using active narractivity to blend that community's interaction around, about, and with a narrative structure.

Wikis and Fandom

Wikis represent a critical shift in our understanding of online interactive texts, authors, and audiences, a change made salient through a reappraisal of our conception of the web. As Paul Levinson shows, the existence of wikis alters even our most traditional assumptions about knowledge, as anyone who encounters Wikipedia while teaching undergraduate research methodologies will attest.[17] Yet, wikis can also engage with a variety of other topics, and are well-suited (and well-used) by fans as well.[18] As fan studies has shown, one important way fans can engage with a cult television program is to rewrite it as fan fiction. As part of this, fan-created fan fiction allows other fans to reread the show in a new way. Rereading is a critical aspect of the fan's online interaction with a media object.

Wikis offer a technology that not only challenges our notion of authorship and creativity, but also encourages new conceptions of readership and interpretation. Specifically, by rewriting *and* rereading the narrative text, fans playfully interact with the created world. Wikis offer a virtual sandbox within which fans can enact a philosophy of playfulness by playing with the narrative within a networked space.

Fan communities interact in many and varied ways with wikis, most notably through rereading other fans' creative works. Wikis provide fans with a technology to match their creative communalization of cult television shows. Mittell's "narrative complexity" is a heuristic device we can use to guide our discussion of these shows; however, a new conception of complex narrative form emerges in the Web Commons, the narrative database. Because the story elements of the extant media object transmediate between episodes of television, the sense of narrative complexity that emerges in the extant media object exists intra-textually between the complex episodes, and "demands an active and attentive process of comprehension" on the part of fans. To meet this demand, fans "extend their participation in these rich storyworlds beyond the one-way flow of traditional television viewing, extending the metaverses of complex narrative creations."[19] In other words, fans employ a series of practices to attend to and develop an understanding and knowledge about the serialized extant media object.

Narrative Database

By forging digital links within the narrative chain of the cult serial, fan wiki writers turn a linear story into a multi-accessed archive of narrative material. Although New Media theorist Lev Manovich decries that the "database and narrative are natural enemies," here I argue that the fan-created wiki actually unites the two.[20] What Manovich calls the database is a collection of unordered information, easily accessible, from which users can pull various entries to assemble a linear narrative. For example, think of an archive of all the shots from a digital film collected in a non-linear editing program like iMovie or Final Cut. Manovich's database is the entire collection of digitized clips. The narrative is the assemblage of the clips into a cohesive order in the timeline. The narrative database, however, differs from this in that the collection of information, created by the community, orders the narrative through communal interaction. It's not the same as taking units from a random archive, but rather of reassembling units in a multiple, new order. However, "order" itself may be an erroneous word in this instance, as it implies a fixture, a set of static coordinates; on a wiki, however, these coordinates are never stable, never fixed, and never completely orderable.

Further, Manovich argues that the database is a technology while a narrative is merely a linear structure. For Manovich, a database is a set of unordered episodes and narrative is casual. In contrast, wikis are databases with organizational principles that offer a non-traditional view of narrative as a lattice of linked narrative events. The process by which fans enact this form of rereading occurs through a communally narrated story; that is, through an interaction between themselves and the wiki, fans extract narrative elements from an extant media object and then rewrite that narrative in a different form by linking wiki-pages via hyperlinks. Any edits themselves remain online, stored in a document accessible by hyperlinks. Clicking any edit will bring up the history of that change, and what the page was before that edit was enacted. Fans redefine the boundaries of "narrative," as they create and recreate the serial cult narrative in a new form.

I define "narrative" as *a community's sum total of knowledge, about, on and/or of a world.* Key to separating my definition from the scores of other definitions lies in the term "knowledge." I differ my definition from Scholes and Kellogg's classic definition of narrative:

> By narrative we mean all those literary works which are distinguished by two characteristics: the presence of a story and a story-teller. ... For writing to be narrative no more and no less than a teller and a tale are required.[21]

Whereas most traditional definitions of narrative indicate the separation of the telling and the tale, in some variation or another, mine highlights a mash-up of the two. I show that narrative within the Web Commons can be conceived differently: not as a story told through discourse, but as a series of communally described and linked events. To "read" a wiki online is to recognize and reconstruct it, to create new readings of the narrative elements every time.

"Narrative elements" are those elements within a narrative that can be readily identified. For example, a character is a narrative element, as are locations, events, conflicts, motifs, themes, props, or any number of things within that narrative world. Narrative elements become visible through "narrative events," which are actions that occur within a narrative that invoke or involve some change of state, and marks a divide between what came before and what comes after. The "story" is the narrative events as they exist ethereally, as non-narrated. In other words, the story is the tale that is told as it is reconstructed chronologically in the mind of the reader. Conversely, the "discourse" signifies the way that the story is told, the narration that describes and places an external order on the elements within the tale.[22] The discourse is shaped by the narrator, who describes the narrative events through the structure of plot.

The narrative wiki is an interactive text, forged through the Web Commons concepts of cooperation and collaboration, and acts as an archontic text in which fans collect narrative information—kernels and satellites, texts and paratexts—about the extant media object. (The term "paratext" means the professionally produced entities within and around a text like the author's name, the title of the work, the introduction, or illustrations that delimit and define the areas of the "text" itself; Genette describes a prescribed boundary, an officially designated border to the narrative.)[23] Much theoretical work on the interactive potential of wikis can be gleaned from the research of Pierre Lévy, for whom the web represents a form of "collective intelligence" and offers fans a community in which to explore information communally.[24] In Lévy's view of collective intelligence, all humans contribute to a collection of humanity's knowledge. On wikis, this is made explicit by variously editing, reworking, changing, or adapting online knowledge spaces. Wikis are collective intelligence writ large, both as examples of mediated sociality on the web and as online repositories of information.

Computer programmer Ward Cunningham invented the wiki as a way of digitally consolidating the knowledge of many computer programmers separated by distance and time.[25] Wiki software forms a metaphorical shell around the space for content, and provides the tools necessary for any user, with any level of technical expertise, to input textual and visual content. The wiki functions, as pointed out by digital scholar Eyal Oren, in three ways: The *authoring* potential of the wiki is a way for fast community interaction, the *retrieval* potential means that anything archived on the wiki is likely recoverable, and the *navigation* potential means that the information is stored in a easily accessible, easily navigable encyclopedia format.[26]

Yet, interactivity and the socialization that comes with it are additional features of wikis, and important aspects of the Web Commons. Even if a single author starts a wiki, that wiki functions through the interactive potential and actualization of the community that reads and writes on it. The wiki records each member's interaction with the wiki, and files the collaboration with the community that contributes to the wiki. Perhaps the best-known wiki, Wikipedia, houses a massive self-regulated online database on which hundreds of thousands of people have contributed encyclopedic entries about millions of topics. Each of these articles, accessible to anyone with an internet connection, can also be *changed* by anyone with an internet connection.

Each Wikipedia entry exists as a document in temporal flux: at any moment the content of a page can be altered, negated, updated, or even removed. As we saw in the previous chapter, this grotesque growth does not limit the ability of the media scholar to analyze the wiki "text," but rather forces a reconceptualization of the manner of that analysis. Benkler describes this communal creation, this "collaborate authorship," as the creation of elaborate structures of mediation online where groups of disparate individuals come together to create representations of their world.[27]

The discussion of how interactive technology changes narrative form is not new. Janet Murray, for example, discusses new digital narrative forms, although she does not describe wikis specifically. She argued that digital forms of television would be remediated onto the computer, and that viewers would be able to log online and watch "a complete digital library of a series," which she calls a "hyperserial."[28] Sites like Hulu or Netflix have realized this hyperserial form, as they allow users to stream television programs. The narrative database combines Murray's concept of hyperseriality with Oren's description of the wiki not to remediate content, but rather to reimagine it in a new space, constructed by a fan community. Narrative form, in other words, "can take a variety of shapes."[29]

Fans' archival on wikis of the cult serial both parses apart the television program by separating out the narrative elements and reconstructs the show's story, through the fan-created discourse of the community of fan writers. Fans tend to take their wiki seriously: At the top of the *Lostpedia* main page, someone has written "Information contained within this article must reference its sources. If not, it will be deleted."[30] This attention to detail is further reinforced by the encyclopedic quality of this fan-created wiki for the television show *Lost*.[31] Each narrative element in this index links to a new wiki page, each of which the fan community has collectively written.

The hypertextual links between each of these items in the index, and within each entry itself, connects narrative elements hypertextually, in the same way the plot connects narrative events causally. For example, on the *Lostpedia* page for the character "Sawyer," fans re-create the narrative of the character named Sawyer by collating information from different episodes of the show, and then rewriting them on the wiki document. Fans can then reread Sawyer's "story" as a new form of narrative.[32] Each element of narrative information in the wiki-page—"James Ford, better known by the alias 'Sawyer,'" ... "known to the DHARMA Initiative as Jim LaFluer" ... "was one of the middle section survivors of Oceanic Flight 815 ..."—comes from different episodes of the show. But when combined and collated into the archive, the show's story has been reread as a new discourse by the community of fans on the wiki.

For fans, the power of the wiki comes not solely from the interactive potential of the text, but also from the narrative potential of the community, made salient and efficient with the Web Commons. Each link in the wiki entry for Sawyer—"DHARMA Initiative," "middle section survivors," "Oceanic Flight 815," etc.—hypertextually connects to another narrative element in the wiki. This intuitive link bridges a connection between these elements, recreating connections made on *Lost* the television show, which may have depicted these connections as plot elements (how is Sawyer connected to the DHARMA Initiative again?) in different ways.

Thus, a narrative database hinges on the interrelated relationship among the audience, the narrative, the story, and the narrator. Traditionally, narratologists define a narrative as the existence of events and the re-telling of those events as ontologically separate components.[33] Often, what these narratologists leave out is the role of the audience in narrative formation. For example, Chatman separates narrative into two timeframes, writing,

> each narrative has two parts: a story (*histoire*), the content or chain of events (actions, happening) ... and a discourse (*discours*), that is, the expression, the means by which the content is communicated.[34]

Chatman describes this as a "chrono-logic," that is, a doubly temporal system, where two chronologies (the time it takes to tell the narrative and the timeframe within the narrative itself) integrate to produce the temporal flow of the narrative.[35] Traditionally, therefore, narratology examines narrative boundaries and limitations.[36] But the boundaries of narractivity illustrate how the *audience* constructs the narrative story from aspects of the narrative discourse. The narrative database emerges from a reconceptualization of the way the story and the discourse function as a digital archive, which, as Derrida shows, is "never closed. It opens out of the future."[37] Narratives on wikis are boundless.

Narrative elements that do not "appear" in the discourse can "appear" in the story: For example, although the character Alvar Hanso appeared only in photographs or dialogue in the television show *Lost*, he has a presence in the diegetic world as the director of the Hanso Foundation. The characters on the show describe his biography, and although his presence has been noted, he does not, as such, exist in the discourse.[38] He does, however, seem to have a presence in the story: as a character in the wiki narrative database of *Lost*.[39] His profile has as much information as do many of the other characters who have actually appeared (in person) on the show.

As Herman describes, there is thus an inherent illogic in this type of story/discourse divide: If Hanso does not appear in the discourse, they would argue, how would viewers know that he is in the story?[40] Something in the discourse cannot by definition be excluded from the story. Indeed, as Richard Walsh states, "fabula [story] is not independent of any sujet [discourse]—it is entirely dependent upon sujet, is nothing other than the permutation and assimilation of sujet features into an ongoing interpretative version."[41] Thus, the story becomes nothing more than the audience's reconstruction of the events in the discourse: The discourse, however, seemingly depicts this already-reconstructed story. Phenomenologically, this is true: Hanso is in the story and the discourse because he is mentioned in the discourse. Yet, the character's visage never appears onscreen, nor does he interact with other characters in the show. Essentially, his "presence" is an audience's reconstruction of pieces of narrative information that have been scattered within the show *Lost* since episode one. However, the audience's reconstruction of the story elements on the wiki make Hanso a discursive character, tied to various others. This act of the reconstruction of narrative elements thus lies at the heart of the narrative database.

New Media Studies can emphasize this divide between story and discourse not by ignoring the paradox between the two, but by absorbing and creating a new narratological concept that exceeds and disturbs the interaction between the story and the discourse. Contemporary analyses of digital narratives need to examine discourse and story as integral components of a mashed up third form. Digital narrative scholar Carlos Duarte de Sena Caires describes how the creation of an "interactive narrative" will be "essential" for the development of computers, technology, and culture.[42] To this extent, wikis disassociate discourse and story and rely on the serialization of narrative complexity and communalization of fans.

The wiki, as the storehouse of a narrative database, becomes one way fans can restructure narrative form to accomplish an active audience-centric reading of the cult serial narrative. In this form of fan writing, fans do not participate in the creation of fiction by rewriting an extant media object, but rather re-create it through rereading the already-written. To envisage narrative as a database is to grant salience to an understanding of narrative not just as a structure, activity or process, but also as a *place*, as an environment upon which meaning can be inlayed. A traditional narrative is temporally connected by plot: the causal association between events. A narrative database, instead, forms from the complex interaction of the audience with the serial narrative, not via an external narrator, but through the connections made by members of the fan community. A narrative database emerges from the sense of cult television narratives as "exotic and ethereal fictional worlds to which the alchemy of textual data and imagination transports the reader, facilitating a pleasurable psychic sense of 'being there' as the action unfolds." The more detail available about the complex cult world, "the denser and more comprehensive it becomes. It assumes the familiar complexity of the material world, sharing its three-dimensional clutter and myriad connective possibilities."[43] This complexity furnishes the fictional world of cult television with the gravitas of authenticity. The rewriting of the cult serial narrative emerges from the rereading of the story by an audience as communal narrator. The rereading of a database is a way of self-narrating, of constructing a discourse for the self.

On the wiki any fan, through hypertextual connections, can link any two pieces of information. This is particularly salient for the narrative of *Lost*, because the television show itself illustrates the myriad connections among characters. For example, during the extensive flashback and flashforward sequences, we often see two or three main characters embroiled in situations that diegetically take place before or after the characters "knew" each other on the island: Further, fans have illustrated these connections hypertextually, through links between wiki-pages. Each character name is a link to another character, representing a connection between the two. Importantly, the audience can create these links and link names even when the extant *Lost* narrative may not make it obvious. For example, *Lostpedia* indicates hypertextual connections among the characters Horace Goodspeed, Benjamin Linus, and Sawyer—all of which the TV viewer can construct from plot connections on the television program, but which are made salient and obvious through the hyperlink on the wiki.

"Plot" here describes the relation between narrative elements. Chatman refines this discussion of plot by delineating two types. "Kernels" are those events that are central to the story: A kernel "advances the plot by raising and satisfying questions." "Satellites," conversely, are minor story events "not crucial … [which] can be deleted without disturbing the logic of the plot."[44] Importantly, Chatman does not delineate exactly who outlines this kernel/satellite distinction. Some events may be considered crucial by all—for instance, Hamlet's murder of Polonius is crucial for advancing the plot—but others are more vague. What about Ophelia's drowning? Is that crucial for the plot, and does that not depend on how the audience constructs what the story of *Hamlet* would be? The narrative database escapes this concern by putting the audience squarely in the center of the narrative construction, and the kernels and satellites thus become differentiated from each other by audience involvement, not authorial intent.

Narrative texts, as Miall shows, can be broken into these constituent parts: Each character's narrative structure on *Lost* is an "episode" of narrative, or the unfolding of the text at the level of the reader's experience.[45] A reader's own experiences with a particular set of narrative events forms an "episode" of that narrative. If any particular connection does not already exist on screen, or appears for the first time in an episode, fans can rewrite that connection in order to reread that discourse. Thus, there are two processes at work. First, fans rewrite the story of the extant media narrative by splitting the kernels and satellites into their own wiki pages. This rewrite forms a narrative database, with elements of the narrative restructured by the fan community. Second, and following from this, fans can then reread the extant media narrative as written through and by the fan community, by hypertextually flipping through the kernels likes a hypertext narrative.[46] The database form of narrative establishes a way for fans to participate in the creation of their own narratives, through communally rewriting and rereading the cult serial narrative. The wiki exists just like a Wikipedia article: to provide details and information about a world. The difference is that the world of the wiki is the cult television world: fictional, but no less "real" to the fan-audience that creates the wiki.

The Wiki as Archive

For Derecho, the advent of digital writing has created a new form of fan literature, archontic literature. Her discussion of archontic texts is itself heavily influenced by Derrida's connection between the archive and Freudian psychoanalysis. Derrida delves into the root of "archive" to show an early sense of the word: a space wherein order is achieved. The Greek word *arkheion* signifies a house, an address, or a residence of the *archons*, the superior magistrates who rule and archive public documents. As Derrida shows, the *arkheion* housed information about the citizens of the *polis*:

> The citizens who thus held and signified political power were considered to possess the right to make or to represent the law. On account of their publicly recognized authority, it is at their home, in that place which is their house (private house, family house, or employee's house), that official documents are filed. The archons are first of all the documents' guardians. They do not only ensure the physical security of what is deposited and of the substrate. They are also accorded the hermeneutic right and competence. They have the power to interpret the archives.

Thus, the archive traditionally has roots not just in collecting information, but in interpreting and reassigning it as well. The *arkheion* were not just keepers, but creators of archived knowledge. The study of archives, of the "Archontic principle," must include the theory "both of the law which begins by inscribing itself there and of the right which authorizes it."[47]

Fans accomplish this dual sense of archival through the inscription of the narrative, in all its meanings, onto the wiki. Indeed, the archive is intimately related to this concept of "inscription": Derrida writes, "can one imagine an archive without foundation, without substrate, without substance …? the inscription … leaves a mark right on the substrate." One inscribes on an archive all the knowledge and information of a community, to both collect and collate a community's place in their culture. Yet, the archive is more than just a *record* (or does more than just *record*):

> No, the technical structure of the *archiving* archive also determines the structure of the *archivable* content even in its very coming into existence and in its relationship to the future. The archivization produces as much as it records the event.[48]

In other words, the archive is not just a thing but a process; is not just a device for reproduced information but for production itself. By archiving something, the archon establishes the importance of the thing being archives: It is tautological preservation.

Derrida describes this capacity for archives to interpret and inscribe knowledge as that of "consignation," or the power of "gathering together signs," of collating knowledge. In many ways, this is similar to what Lévy demonstrates of cyber-knowledge: We do not produce "true knowledge" in a single location, but rather through *interaction* between members of a collective intelligence. This interaction occurs in a network of linked community members: Lévy describes "open hyperdocuments accessible on-line, written and read by a community"—in other words, wikis.[49] In his reading of Lévy, Jenkins articulates the promise inherent in these knowledge communities: "what we cannot know or do on our own, we may be able to do collectively."[50]

The cosignatory capabilities of the archive must always be tempered with its archontic properties. Derrida describes the sense of consignation as the coming together of "a single corpus, in a system or a synchrony in which all the elements articulate the unity of an ideal configuration." He goes on to write:

> By incorporating the knowledge deployed in reference to it, the archive augments itself, engrosses itself, it gains. ... But in the same stroke it loses the absolute and meta-textual authority it might claim to have. One will never be able to objectivize it with no remainder. The archivist produces more archive, and that is why the archive is never closed. It opens out of the future.

The archive constantly grows with the grotesque notion of eternal enlargement, of the type we saw described by Bakhtin in reference to carnival and in terms of blogs and recursive expansion. But there is an inherent paradox at the heart of the archive: by archiving, one must necessarily leave something out. An archive can only record within a culture, within a particular context. Without the place of consignation, "a certain exteriority," there could be no archive: "no archive without outside." At the same time, however, by leaving something out, the archive becomes meaningless: What use is an archive that is not total, is not complete?—"the archive always works, and *a priori*, against itself."[51] For New Media Studies, the archive thus represents the way communities of fans come together on a narrative database to inscribe the extant media object itself. Similarly, fans are rewriting and rereading the narrative parameters of cult shows through this expansive and interactive content creation to collate all information about the shows and to narrativize it.

For the *archon*, archives are places where a community scribes information, not only to unbind that information, but also to unite and impress that community into existence. Indeed, Derrida's connection among the notions of the archive, of inscription, and impressions also highlights a connection to the narrative database. The fans who write *Lostpedia* consolidate, create, and communicate a variety of types of narrative information about these extant media objects. On her blog, Nancy Baym has described wiki like these as "a neglected area of fandom research," and that the popularity of such sites warrants additional scholarship.[52] To demonstrate their popularity, she references an article from *Information Week*, which describes *Lostpedia* as a site that "has grown to nearly 33,000 pages. The site has received 141 million page views. It has 26,000 registered users."[53] Using Derrida's analysis of archives and delineation of three types of archival impressions, we can observe three ways of narractivity occurring in wikis: first, the interactive narrative of the wikis; second, the scattering of narrative knowledge through the creation, deletion, and explanation of the spoilers; and third, the scholarly research of narrative spoilers.

Narractive Impression: Constructing the Narrative

First, through the compilation of narrative information, viewers of the narrative construct their own story bible, their own conception of the backstory and history of the characters and narrative elements for the media text. A show's bible, written by media producers before or during the show's production, highlights the different elements that make up the show's canon: the characters, technologies, back-stories, and other essential components. The term comes from the comic book industry, in which main characters of comics would have detailed back-stories that required background knowledge. Comic books, however, would often have different writers, all of whom would have to work from the same material. Thus, a few key comic designers, in order to ensure continuity and to keep track of details, would write the story bible. The concept of the story bible translated into television usage when televised narratives became expansive and serialized. Writers use the bible to keep track of the expansive back-stories of the characters, and as an extensive reference. In order to construct this version of the bible, fans connect elements of the show's discourse into a cohesive whole by watching the show, or by encountering the elements of the show in various mediated guises.[54] The "collective intelligence" of the fan community becomes the key to this construction as each individual member of the fan community might have different knowledge bases: for example, perhaps one person knows Jack Shepherd's story on *Lost* and another intimately knows John Locke's. Each fan can contribute his/her own unique information to each component of the narrative database. Importantly, both Jack and Locke are themselves also linked, as seen through the hyperlinks on each other's pages, and each links to other characters, situations, events, and narrative elements from the show.[55]

Derrida's first meaning of inscription provides us with a useful heuristic for examining the creation of the narrative on the narrative database. He writes that the inscription of the archive is "scriptural or typographic"; that is, something "which leaves a mark at the surface or in the thickness of a substrate."[56] To impress is literally to inscribe: to etch or to write upon. Applying this measurement of impression to that of the archive, Derrida marks the *foundation* of the archive: the impression of the material, the writing-down of the substance. The archive impresses information into a location, moving it from an ethereal location of a "knowledge space" to that of a specific site. Inscription is thus the most basic function of the archive, and one of the constituent components of the narrative database.

For the wiki communities on *Lostpedia*, fans impress their own discourses to the narrative story, in order to construct information about that narrative. They construct this narrative through the addition of paratextual information about the show (titles, credits), secondary textual information (reviews, interviews), extra-textual information (DVD extras and commentaries), intertextual information (novels and tie-ins), and intra-textual information contained in the show and wiki to the narrative rewritten on the wiki, as in the case of the aforementioned Alvar Hanso.[57]

A second method of narrative construction on a wiki occurs as the community-formed wiki offers fans a non-narrated story in which they can construct their own discourse. By providing narrative elements in a series of hypertextual links, wiki illustrate how all the parts of a narrative exist *without* an external narrator providing a cohesive order. *Lost*, as we have seen, is the perfect example of an extant media object in which all elements are kernels, and all kernels are linked. For example, the stories of Jack and Locke intertwine with those of many other characters, each of whom, as we also saw in the last chapter, intersects with other characters, situations, and motifs. On the main *Lostpedia* page, a series of graphically and textually active "portals" hyperlink to narrative elements and underscore their relevance to the *Lost* narrative.[58] Listed at the top of the main page, the different portals include "Survivors Camp, Supporting Characters, The Others, Mysterious Happenings, Locations, Themes" and others. Clicking on *The Others* portal leads to a page that details the characters that populate the camp of "the Others," the particular themes and motifs that are associated by the community with "the Others," and information about the history of "the Others." Each portal leads to different narrative elements ostensibly removed from the process of narration. They seem pure story, and indicate a shift away narrative discourse.

Because there appears to be no discourse on a wiki, no order of events present, it would appear as through a narrator did not exist. Although this would make a wiki seem more like a story than a discourse, there does exist, in fact, an obvious discourse at the heart of a wiki: The narrator is the fan community itself. By presenting the narrative in a wiki, the community of fan writers presents a narrative that makes obvious the narration, an obviousness that comes from the mutability of the text itself. Through the interaction with each other by building a wiki narrative database, fans make obvious the narrative and make opaque their narration. The expansion of the narrative information by the interaction between members of the wiki community highlights narractivity. The wiki community thus writes a discourse for the wiki reader, but at the same time, also rewrites all the textual elements with which the reader can reread the narrative in his/her own discursive fashion. By following any hyperlinks, readers can use the narrative elements enumerated on the wiki in any order and can create their own discourse through the story. Although Cobley establishes that "hypertext in written computer narratives makes explicit the degree of choice involved in the ways in which readers construct narratives," wikis do more.[59] They do not just "make explicit" the reader's choices, they leave open the possibility for any reading at all. As we have seen, fan readers can hypertextually link to any narrative kernels rewritten on the wiki, and if these links do not exist, they can create them.

Narractive Scattering: Trace-ing Narrative Futures

Second, fans also scatter the narrative by rewriting future elements of the extant media object's story on the narrative database, and then rereading that story as a fan-created "imaginative discourse." Using spoilers as a form of speculative fiction, fans provide details and assertions about the narrative itself. Then, by reading that narrative again, fans re-interpret the way the narrative information moves across the flow of the wiki.

Key to this discussion of narractivity on the narrative database is the paradoxical relationship between the archival form of the database and the spoilers on the archive.[60] Archives both record and represent the past; they inscribe, as Derrida would say, "on historiography."[61] Spoilers attempt to reveal key pieces of information for a media object's narrative before the producers of that narrative release that information to the public. The archive is documentation of the previously happened, a record of the contextual has-been. Conversely, a spoiler is a future event: a piece of narrative that hasn't-yet, or might-not.

To record a spoiler on an archive like a narrative database is to historicize the future; to chronicle the forthcoming. It is to make the future the past, and to assert the truth of the may be. Each spoiler represents a form of speculative fan fiction, promoting and proposing elements of the narrative that have not happened, but still might. It's a theory based in actuality.

Spoilers for television narratives are pieces of narrative information that producers reveal or audience members discover before the show airs. As Rebecca Williams states, they "allow fans to sustain a reading formation based on narrative speculation."[62] They can be as innocuous as the title of an upcoming episode or as narratively important as who will live and who will die. Spoilers exist as communal hypotheses of future events, built through interactivity and centered on narrative exposition. Spoilers are also a way for fans to actively construct meaning in the extant media object, and to refute the dominant interpretation of the media text. Spoilers are therefore a form of what Sconce calls "narrative conjecture" for television shows, elements that audiences imagine could happen in a cult universe.[63] Rather than reading the text "as 'intended' by the producers or interpreted by mainstream critics, fan spoilers can offer unique, alternate, and sometimes quite elaborate new readings of the text."[64]

There are a number of different ways scholars have theorized spoilers in studies of fandom. The most common conception can be seen in literary critic Laura Carroll's observation: Spoilers offer clues as to "abrupt and sensational narrative developments."[65] In this view, spoilers are a way for viewers of a show to find out information about the plot before producers reveal it to the general population. Another theorization shows that spoilers become a way of fans to gain cultural capital in issues of power dynamics in their fan communities. Research has shown that fans use them as a way of establishing a hierarchy of knowledge within the subculture of the fan community: "subcultural capital," according to Williams, "can foster fan social capital." Thus, some fans write spoilers and while

> the majority of fans openly express the egalitarianism and equality of spoiled and unspoiled fan factions, there are differences in the forms of capital that each group possesses. The spoiled have greater subcultural and, in particular, fan social capital and therefore occupy the more dominant position in the fandom, dominating the fandom with their fan knowledge, textual interpretations and discursive power.[66]

In short, fans may not just spoil to reveal unreleased narrative moments, but also to dominate the discourse surrounding the extant media object, to demonstrate their knowledge of that text's story, or to gain a powerful tool for social influence in the fan community hierarchy. Foster mirrors this assertion, describing those who spoil *Survivor* on the show's website as "fans [who have] attempted to elevate their status amongst other fans by convincing them of the accuracy of the reading of their text."[67] The spoiler acts as capital, with which fans assert their supremacy over other fans. The pleasures that come from spoiling in this sense come from the interaction with other fans, just as much as from the text itself.

Another conception of the spoiler is that of extra-textual pleasure. For Jenkins, spoiling is "a giant cat and mouse game that is played between the producers and the audience."[68] This game brings people together in a community of spoiled fans. Gray and Mittell argue that the extra-textual pleasure of the spoiler goes beyond what Jenkins articulates, by "conceptualizing narrative and textuality as entailing much more than plot exposition." The pleasures that fans derive from spoilers come from experiencing the plot as known, with "the emotional pulls of the text" emanating from "that which they know is coming, enjoying the moments leading up to events they have never seen, but feel known."[69] Spoiling is a way of experiencing a narrative differently—not as what will happen, but rather as how something will happen, or why.

A further analysis of spoilers presents them not just as a way to build community or to elevate the individual status of fans, but also as a means to create narrative possibilities. First, the spoiler extra-textually builds on the extant media object, as a form of speculative fan fiction. According to Derecho, "in fan fiction, there is an acknowledgement that every text contains infinite potentialities, any of which could be actualized by any writer interested in doing the job."[70] Speculative fan fiction sees these possibilities as future narrative events. Perceptually, to recognize a spoiler as a spoiler is to recognize the narrative possibility that that spoiler offers to the community, which often comes in the guise of communal discussion. For Williams, the spoiler is intimately tied to both the narrative of the extant media object and to the community of fans surrounding the show. Gray and Mittell conclude their argument by implicitly referencing the virtuality of the cult text:

> Most importantly, a well-told tale lives and thrives after its telling, and in the gaps within its telling. Any given reader's path through this story may bring the reader into contact with only a small portion of what it has to offer, and hence textual studies have often been too quick to assume a unitary, unified text, nailing one textual path down as "normal" or even "the text itself," and marginalizing other paths as extra-textual or abnormal. Instead, we might more properly conceive of the text as an active space with a varied terrain and numerous potential pathways.[71]

Thus, the spoiler becomes a way of unifying this text, of fans choosing different pathways to tread across the extant media object's terrain.

A final conception of spoilers is as a form of researched narrative information, which mirrors the scholarly work of academics.[72] To research and compile spoilers, fans must engage in an intense academic program, snooping on websites, scouring journals and other sites of arcane knowledge about the show, photographing set locations, even examining Google Earth webpages to see evidence of filming in remote locations.[73] Keith Johnston shows that there is already an academic sensibility in fandom: "Unlike any other time ... content analysis [is] now something that [can] be done by the most casual fan with access to a computer." He goes on to show how fans critically analyze film trailers to reveal potential spoilers. Indeed, fan analysis can even be more complicated and intense than academic scholarship: as "film companies ... add in more images and increase the pitch of editing to a point where the casual view might miss a piece of information."[74] Fans read and research on narrative elements, just as a scholar might spend a great deal of time textually analyzing a particular literary work. For example, on the spoiler page for *Lostpedia*, a contributor has written:

> This page contains a summary of unconfirmed season 4 spoilers from **unofficial** sources. The source is mentioned for each spoiler. If spoilers repeat or expand upon earlier information, the more reputable or more complete sources are preferred. For season 4 spoilers confirmed by official sources, see Season 4. [sic][75]

What we find, therefore, is that the community of online fans promotes a narrative text that is multi-faceted and complex. It is not just the text seen on the television, but also the text as experienced through conversation, though extra-textual comments, and through spoilers. Spoilers become part of a narrative constructed through interactivity: a narractive experience, created both by spoiling to create fan fiction, and by spoiling to create fan scholarship. In both these instances, the form of the narrative database influences the content of the spoiled narrative information.

Ultimately, spoilers on wikis make the future events present, as the narrative database records the interpretation of the events as if they had happened—or, to be more precise, as if they were happening. However, spoilers do not automatically and necessarily delimit the way fans have of looking at a text. Although of the archive, Derrida says that one can "no longer think otherwise," in truth spoilers are representations of fans literally thinking otherwise: of thinking outside the paradigm of the traditional narrative.[76] Spoilers represent a break with narrative unity, as the temporal displacement of future narrative information is spatially united with present narrative elements.

Derrida's second definition of "impression" provides us with a useful heuristic for examining this narrative scattering. This impression is something felt in anticipation, for concerning the archive, "we have no concept, only an impression, a series of impressions associated with a word."[77] This impression is a form of a promise, a deferment of desire, a trace of a thing left forever behind. Fan authors leave spoilers on the page even after producers have revealed the spoiled event on the show. A spoiler is this trace, but with one important difference: It is a trace of a future event. What is a spoiler once the spoiled event has occurred? Once the spoiler has been "revealed" as either true or false (that is, as either accurate as to what happened in the show, or inaccurate), then what is left of the "spoiler" is a *trace* of that hypothesis of future event.

The wiki for the cult show *Heroes* provides a relevant example of this trace. The first page of spoilers on *Heroeswiki* indicates a number of possibilities for the finale that had not yet aired.[78] Some of these spoilers include general statements, as in "Sylar will kill Molly or Maya or Bob or Elle." Others are much more specific: "Niki will get trapped in the fire, but D. L [sic] will somehow be alive (possibly by Claire's blood) and save Niki (and possibly die himself)." In total, there are 24 "fan theories" presented on the wiki page, two days before the premiere of the episode. Once the episode aired, we might assume that the spoiler page would be erased, or would at least include additional information about which fan theories were "correct" and which were not. What we find, however, is that the spoiler page continues untouched. In other words, these spoilers did not disappear from the page once the episode has aired. Fans did not adjust them, nor did the community of fans change or delete them. Although it is possible that fans merely forgot to change the wiki, or ignored this page once the episode aired, the continued existence of the spoilers indicates that these spoilers also still exist as narrative discourse possibilities. (Going back to check in 2016, the spoiler page is still present, although now it is labeled as past. Interestingly, the heading at the top of the page reads: "This article contains archived spoilers about Powerless. See the main article to see how they panned out." These spoilers' continued archived presence seems to indicate that they *deliberately* still exist as narrative possibilities.) Inaction on a wiki is a choice. Fans examine the finale of *Heroes*, summarized on its own page, as one possible narrative in a galaxy of other possibilities. Each spoiler, each fan theory, becomes a possible outcome set in "a universe composed of a plurality of distinct … worlds."[79] These distinct worlds, what Ryan references as the "Possible World Theory," exist as discrete narrative entities in the minds of the audience. For example, in Possible World Theory, two narrative worlds based on *Heroes* can exist simultaneously in the mind of the viewer: in one, D. L. rescues Niki, and in the other, he does not. Perhaps the discourse presented onscreen in the show *Heroes* depicts one of these, but the fact that an audience member can imagine this other world, and impart the knowledge of that narrative world to others, indicates that its presence is one in a vast array of possibilities.

For fans on wikis, this spoiling goes beyond an attempt to "compete" with the producers of a show to uncover narrative knowledge.[80] Spoiling can now be a further step on the path to creating original fiction. The rhetorical awareness of the spoiler as narrative possibility exists in the fan community. In some small way, in a minute and seemingly innocuous manner, one way fans could use spoilers is to produce paths to other possible narrative outcomes. By *not* erasing the spoilers once the "revelation" of the episode airs, the community appears to voice a collective appreciation for the narrative possibilities that are not, but could be, aired.

Narractive Research: Memory of the Narrative Moment

A third method of narractivity lies in the way fans use spoilers as a form of narrative research. Fans scour websites and news sources to find information with which they can construct a model of what is to come in the show. Narrative research demonstrates the emergence of fan-scholars, fans that use scholarly research methods to "express their love for a text."[81] Fan-scholars use content, textual, and rhetorical analysis, among other methodologies, to examine fragments of paratexts, secondary texts, and transmediated intertexts to assemble a spoiled reading of the extant media object's narrative story. In this way, fans attempt to construct a meaningful articulation of the "facts" of the narrative. By "facts," I refer to spoiled events that eventually occur in the extant media object. As O'Neill shows, however, narrative "factuality" can have a different meaning. For O'Neill, a narrative story always appears factual, while a narrative discourse always appears fictitious, because the narrator's "multiple discursive possibilities" preclude a "discursive authority."[82] In other words, any discourse, shaped through a narrator's focalization, appears as though it is a reconstruction of events, and readers cannot trust it to be "true" to the story. Fan-scholars, however, can investigate the database narrative as if it were "factual," because it displays the story through the community's discourse. It is as "real" as fans want it to be.

Derrida provides us with another useful reading of the term "impression" that proves fruitful for investigating the fan-scholar. In his analysis, with which he uncovers and analyzes the archontic properties of Sigmund Freud and psychoanalysis, he describes this third reading of the term as "the impression *left* by Sigmund Freud" that is "in memory of the signs" of Freud.[83] In other words, it is a memory of a subject, which when described and inscribed leaves a mark in the mind. Spoilers as areas of research activity become these impressions, as they record memories of narrative, marks of research. For fans, the act of spoiling is not just finding out what is yet unknown, but exploring the possibilities of future memories, or impressions of future events.

Thus, spoiling as research impresses upon the fan community that fan research is a legitimate and worthy activity. In undertaking this research, fans must verify and confirm all the narrative kernels they find, just as if they were researching an academic tome. As Ryan states, "we read fiction as nonfiction when we extract ourselves from its world and, switching reference worlds, assess its viability as a document of real-world events."[84] We can witness this on *Lostpedia* as fans research spoilers about the show. These kernels of narrative information become verified through fan-scholar research: They aren't fan fiction, they're fan fact. These fan-scholars may also face the realization that they might be encountering narrative spoiler kernels that have been falsely planted by the producers of the show, also called "foilers." As Carleton Cuse, one of the executive producers and writers of *Lost*, described in *Variety*: "We always write fake sides for the stuff we send to talent agencies and (casting) breakdowns. ... We are casting characters under false pretenses."[85] Fans thus sift through kernels that could be purposefully misleading to find the kernels of story that might be "true"; or, rather, might accurately reflect the fans' view of the show.

In contrast to the *Heroeswiki* spoilers, which enumerate different fan theories without reference to accompanying citation data, the *Lostpedia* spoilers have extensive, referenced organization. The *Lostpedia* spoiler organizational page catalogs the spoilers into different categories: The first are spoilers about the meta-narrative of *Lost*, the second are spoilers for each individual episode of *Lost*, the third are spoilers about main characters, the fourth are spoilers about each recurring guest characters, and the fifth are general spoilers about themes and motifs that run throughout the narrative. Additionally, each category is subdivided into units: The episode spoilers are divided by each separate episode, while the main character spoilers are divided by the name of the character. The actual spoilers themselves are organized within each category, subcategory and unit chronologically by when fans discovered them. Next to the date of the discovery, fans have written the source of the information.

So, for instance, instead of randomly searching for spoilers about the recurring character of "Frank Lapidus," one need only examine the menu at the top of the spoiler page and find that any spoilers about him will be found in section 4, unit 3. The information is meticulously researched and shows that "Frank Lapidus" is to be played by "Jeff Fahey" and was learned on August 28, 2007 from an article in *Entertainment Weekly*, found online.[86]

Such organization highlights a great deal of research and collection of data. One source of information, for example, comes from the DVD collection of Season 3 of *Lost*. The fan community acted quickly in adding the extra-textual features of the DVD to the wiki after its December 11, 2007 release. The extra-text on the DVD allows the viewer to learn more about the narrative through running commentaries by both those behind and those in front of the camera, as well as other special features like deleted scenes, featurettes, and documentaries about the making of the film. This extra-text represents knowledge *about* the narrative, not *in* it.

We can observe the emergence of the knowledge garnered by fans from the *Season 3* DVD through the spoiler site on *Lostpedia*. By December 18, a mere week after the release of the DVD set, fans had added six new pieces of information to the spoiler site. We learn, for instance, that the character of Annie, seen only once before in the episode "The Man behind the Curtain" in the third season, "is an extremely important part of the island's backstory, and it is a planned chapter to come," and that "what happened to Annie is another chapter of Ben's story that will be seen at some point."[87] As more people watch the *Lost Season 3* DVD, more information is added to the spoiler sites. On December 21, a contributor cites actor M. C. Gainey, who says that "Whoever is coming to this island next, they're gonna wish they had the Others back. We were scientists and humanitarians compared to what kind of animals are coming next." Fans label each of the new spoilers with a citation of where it was found: The December 21 contribution, for instance, is labeled "Season 3 DVD." In short, fans uncover and display units of information via an academic method of citation and verification. They spoil to make the show seem more researched, and provide a location for this research in the form of a wiki.

Narractivity and the Narrative Database

By interactively contributing to a growing enunciation of the meaning of a narrative, members of the wiki community integrate their communal/collective knowledge into a narrative framework. The narrative builds the community, just as the community builds the narrative. In the same way as blog authors intra-textually construct blog fan fiction, so too do communities in and on a wiki integrate narrative and interactivity in a narractive process to create wiki fan fiction. This integration of narrative knowledge and interactivity heralds a shift in the way we think about online textuality and authorship; but even more so, it becomes a reflection of a changed media environment, one where individual texts—or individual viewers—become subsumed under a larger corpus.

These fans construct individual narrative episodes in their own narractive style. Each wiki page devoted to a particular episode, a synopsis or summary will retell the narrative discourse using, through hyperlinks, the story that has already been constructed. Far from negating the story-telling, however, this discursive practice instead illustrates the narrative possibilities that exist on a wiki. Because of the interactive elements of a wiki, the discourse is always in flux. One fan's construction of the narrative discourse might be different from another fan's construction of the narrative discourse. The narractive elements of the wiki provide the means for this interactive narrative database creation. This narrative creation also highlights the spoilers on the wiki as separate elements of fan fiction. They are not just guesses as to what might happen, but rather indications of what would happen in a different possible narrative. For every possible narrative world that exists, there exists every other narrative possibility as well. Spoilers represent narrative possibilities, indications of a plot that did not happen, but could.

Notes

1. Fans also tend to publish fan fiction in serial format: Many works of fan fiction arrive chapter by chapter.
2. Hills, *Fan Cultures*, 134.
3. See Thompson, *Storytelling in Film and Television*, 98–105 for a description of this process; and Johnson, *Everything Bad Is Good for You*, 62–71.
4. Mittell, "Narrative Complexity," 30.
5. Mittell, *Complex TV*, 278.
6. However, see Stein, *Millennial Fandom*.

7. Eco, "*Casablanca*," 3.
8. Chatman, *Coming to Terms*, 9.
9. *Heroeswiki* Main Page.
10. Jenkins, *Convergence Culture*; Baym, "The Lost Librarians of National Defense"; Hellekson, "SF Fan Wikis"; and Mittell "Sites of Participation."
11. See, Jenkins, *Textual Poachers*, 280–3; Tulloch and Jenkins, *Science Fiction Audiences*, 10–15; Hills, *Fan Cultures*, 178–81; Sandvoss, *Fans*, 164; Gray, Sandvoss, and Harrington, "Introduction," 1–16.
12. Jenkins, *Textual Poachers*, 154.
13. Baym, *Tune In, Log On*, 83.
14. Derecho, "Archontic Literature," 65.
15. Derrida, *Archive Fever*, 26, 64.
16. Ryan, "Beyond Myth and Metaphor," 581.
17. Levinson, *New New Media*, 84.
18. Gane and Beer, *New Media*.
19. Mittell, "Narrative Complexity," 32.
20. Manovich, *The Language of New Media*, 225.
21. Scholes and Kellogg, *The Nature of Narrative*, 4.
22. Both my "story" and "discourse" come mainly from Chatman, *Story and Discourse*, 19.
23. Genette, *Paratexts*, 1; see also Gray, *Show Sold Separately*.
24. Lévy, *Collective Intelligence*, 61–5.
25. Leuf and Cunningham, *The Wiki Way*, 3–4.
26. Oren, "SemperWiki, Sec 5.
27. Benkler, *Wealth of Networks*, 73.
28. Murray, *Hamlet on the Holodeck*, 256.
29. Ryan, *Avatar*, xviii.
30. *Lostpedia* Season 4 Spoilers.
31. *Lostpedia* Main Page.
32. James (Sawyer) Ford.
33. Genette, *Narrative Discourse*, 27–9, in contrast, uses three, including narration, in his list.
34. Chatman, *Story and Discourse*, 19.
35. Chatman, *Coming to Terms*, 9.
36. Bal, *Narratology*, 3.
37. Derrida, *Archive Fever*, 68.
38. The ARG *The Lost Experience* features Hanso, although only in background video and pictures.
39. Alvar Hanso.
40. Herman, *Story Logic*, 211.
41. Walsh, "Fabula and Fictionality," 604.
42. de Sena Caires, "Towards the Interactive Filmic Narrative," 801.
43. Gwenllian-Jones, "Virtual Reality and Cult Television," 83, 93.
44. Chatman, *Story and Discourse*, 53–4.
45. Miall, "Episode Structures in Literary Narratives."
46. Ryan, *Narrative*; see Landow, *Hypertext* 3.0 for more on hypertext narratives.

47. Derrida, *Archive Fever*, 1–4.
48. Derrida, *Archive Fever*, 26–7, 16–17.
49. Lévy, *Cyberculture*, 65.
50. Jenkins, *Convergence Culture*, 27.
51. Derrida, *Archive Fever*, 3, 68, 11–12.
52. Baym, "The Lost Librarians of National Defense," ¶3.
53. Wagner, "Lost Fans Find Internet Thrills Via Wikis, Games, Second Life," ¶9.
54. We can see the ways, specifically with *Lost*, that narrative kernels and satellites appear transmediated in Mittell, "The Loss of Value (or the Value of *Lost*)"; Mittell, "The Value of *Lost*, Part Two"; and Mittell, "Narrative Complexity."
55. Jack Shephard; John Locke.
56. Derrida, *Archive Fever*, 26.
57. See, Gray, *Show Sold Separately*.
58. *Lostpedia* Main Page.
59. Cobley, *Narrative*, 205.
60. *Heroeswiki* Spoilers.
61. Derrida, *Archive Fever*, 5.
62. Williams, "'It's about Power,'" ¶6.
63. Sconce, "What If?," 108.
64. Soukup, "Hitching a Ride on a Star," 327.
65. Carroll, "Cruel Spoiler, That Embosom'd Foe," ¶3; as quoted in Gray and Mittell, "Speculation on Spoilers, ¶20.
66. Williams, "'It's about Power,'" ¶3, ¶42.
67. Foster, "'Jump in the Pool,'" 286.
68. Jenkins, *Convergence Culture*, 25.
69. Gray and Mittell, "Speculations on Spoilers," ¶22, ¶47.
70. Derecho, "Archontic Literature," 76.
71. Gray and Mittell, "Speculations on Spoilers," ¶47.
72. See Hills, *Fan Cultures*.
73. Jenkins, *Convergence Culture*, 25.
74. Johnston, "'The Coolest Way to Watch Movie Trailers in the World,'" 148.
75. *Lost* Season 4 Spoilers.
76. Derrida, *Archive Fever*, 10; see, also Gunkel, *Thinking Otherwise*.
77. Derrida, *Archive Fever*, 29.
78. *Heroeswiki* Spoiler: Powerless.
79. Ryan, *Narrative*, 99.
80. Jenkins, *Convergence Culture*, 25–31.
81. Hills, *Fan Cultures*, 19.
82. O'Neill, *Fictions of Discourse*, 159.
83. Derrida, *Archive Fever*, 30.
84. Ryan, *Narrative*, 105.
85. Quoted in Littleton, "'Lost,'" ¶27.
86. Jenson, "'Lost' Adds Fifth New Cast Member"; *Lostpedia* Season 4 Spoilers.
87. *Lostpedia* Season 4 Spoilers.

Bibliography

Alvar Hanso. <http://lostpedia.wikia.com/wiki/Alvar_Hanso>

Bal, Mieke. *Narratology: Introduction to the Theory of Narrative*, 2nd ed., translated by Christine Van Boheemen. Toronto: University of Toronto Press, 1997.

Baym, Nancy. "The Lost Librarians of National Defense." *Online Fandom: News and Perspectives on Fan Communication and Online Life*, 30 April 2008. Accessed 22 March 2016. http://www.onlinefandom.com/archives/the-lost-librarians-of-national-defense.

Baym, Nancy. *Tune In, Log On: Soaps, Fandom, and Online Community*. London: Sage, 2000.

Benkler, Yochai. *The Wealth of Networks: How Social Production Transforms Markets and Freedom*. New Haven, CT: Yale University Press, 2006.

Carroll, Laura. "Cruel Spoiler, That Embosom'd Foe." *The Verve: A Literary Organ*, 09 October 2005. Accessed 21 March 2016. http://www.thevalve.org/go/valve/article/cruel_spoiler_that_embosomd_foe.

Chatman, Seymour. *Coming to Terms*. Ithaca, NY: Cornell University Press, 1990.

Chatman, Seymour. *Story and Discourse: Narrative Structure in Fiction and Film*. Ithaca, NY: Cornell University Press, 1978.

Cobley, Paul. *Narrative*. London: Routledge, 2001.

Create Your Own Hero. <http://heroeswiki.com/Create_Your_Hero>

de Sena Caires, Carlos Duarte. "Towards the Interactive Filmic Narrative: 'Transparency': An Experimental Approach." *Computers and Graphics* 31 (2007): 800–8.

Derecho, Abigail. "Archontic Literature: A Definition, a History, and Several Theories of Fan-Fiction." In *Fan Fiction and Fan Communities in the Age of the Internet*, edited by Karen Hellekson and Kristina Busse, 61–78. Jefferson, NC: McFarland, 2006.

Derrida, Jacques. *Archive Fever: A Freudian Impression*, translated by Eric Prenowitz. Chicago: University of Chicago Press, 1996.

Eco, Umberto. "*Casablanca:* Cult Movies and Intertextual Collage." *SubStance* 47 (1984): 3–12.

Foster, Derek. "'Jump in the Pool': The Competitive Culture of *Survivor* Fan Networks." In *Understanding Reality Television*, edited by Su Holmes and Deborah Jermyn, 270–89. London: Routledge, 2004.

Gane, Nicholas, and David Beer. *New Media: The Key Concepts*. Oxford, UK: Berg, 2008.

Genette, Gérard. *Narrative Discourse: An Essay in Method*, translated by Jane Lewin. Ithaca, NY: Cornell University Press, 1980.

Genette, Gérard. *Paratexts: Thresholds of Interpretation*, translated by Jane Lewin. Cambridge, UK: Cambridge University Press, 1997.

Gray, Jonathan. *Show Sold Separately: Promos, Spoilers, and Other Media Paratexts*. New York: New York University Press, 2010.

Gray, Jonathan, and Jason Mittell. "Speculation on Spoilers: *Lost* Fandom, Narrative Consumption and Rethinking Textuality." *Particip@tions* 4, no. 1 (2007). http://www.participations.org/Volume%204/Issue%201/4_01_graymittell.htm.

Gray, Jonathan, Cornel Sandvoss, and C. Lee Harrington. "Introduction: Why Study Fans?" In *Fandom: Identities and Communities in a Mediated World*, edited by Jonathan Gray, Cornel Sandvoss, and C. Lee Harrington, 1–16. New York: New York University Press, 2007.

Gunkel, David. *Thinking Otherwise*. West Lafayette, IN: Purdue University Press, 2007.

Gwenllian-Jones, Sara. "Virtual Reality and Cult Television." In *Cult Television*, edited by Sara Gwenllian-Jones and Roberta E. Pearson, 83–98. Minneapolis: University of Minnesota Press, 2004.

Hellekson, Karen. "SF Fan Wikis: Source, Reference, World." *Res gestae—Documentary and Digital Evidence of the Trace*, 20 July 2008. Accessed 21 March 2016. http://khellekson.wordpress.com/2008/07/20/sf-fan-wikis/.

Herman, David. *Story Logic*. Lincoln: University of Nebraska Press, 2002.

HeroesWiki. Episode: Powerless <http://Hereoeswiki.com/Episode:Powerless>

HeroesWiki. Main Page <http://Hereoswiki.com/Index>

HeroesWiki. Spoiler: Powerless <http://Hereoeswiki.com/Spoiler:Powerless>

Hills, Matt. *Fan Cultures*. London: Routledge, 2002.

Jack Shephard. <http://www.Lostpedia.com/wiki/Jack>

James (Sawyer) Ford. <http://www.Lostpedia.com/wiki/Sawyer>

Jenkins, Henry. *Convergence Culture: Where Old and New Media Collide*. New York: New York University Press, 2006.

Jenkins, Henry. *Textual Poachers: Television Fans and Participatory Culture*. New York: Routledge, 1992.

Jensen, Jeff. "'Lost' Adds Fifth New Cast Member." *Entertainment Weekly*, 29 August 2007. Accessed 23 March 2016. http://www.ew.com/ew/article/0,,20053479,00.html.

John Locke. <http://www.Lostpedia.com/wiki/John_Locke>

Johnson, Keith M. "'The Coolest Way to Watch Movie Trailers in the World': Trailers in the Digital Age." *Convergence* 14, no. 2 (2008): 145–60.

Johnson, Steven. *Everything Bad Is Good for You*. New York: Riverside, 2005.

Landow, George. *Hypertext 3.0: Critical Theory and New Media in an Era of Globalization*, rev. ed. Baltimore, MD: The Johns Hopkins University Press, 2006.

Leuf, Bo, and Ward Cunningham. *The Wiki Way: Quick Collaboration on the Web*. Boston: Addison-Wesley, 2001.

Levinson, Paul. *New New Media*. Boston: Allyn & Bacon, 2009.

Lévy, Pierre. *Collective Intelligence: Mankind's Emerging World in Cyberspace*, translated by Robert Bononno. New York: Perseus, 1997.

Lévy, Pierre. *Cyberculture*, translated by Robert Bononno. Minneapolis: University of Minnesota Press, 2001.

Littleton, Cynthia. "'Lost': The Weight of the Wait." *Variety.com*, 12 October 2007. Accessed 23 March 2016. http://www.variety.com/article/VR1117974014.html?categoryid=2641&cs=1.

Lost: Mysterious Happenings. <http://www.Lostpedia.com/wiki/Portal:Mysterious>

Lostpedia Main Page. <http://Heroeswiki.com/Main_Page>

Lostpedia Main Page. <http://www.Lostpedia.com/wiki/Main_Page>

Lostpedia. Season 4/Spoilers <http://www.Lostpedia.com/wiki/Season_4/spoilers>

Manovich, Lev. *The Language of New Media*. Cambridge, MA: The MIT Press, 2001.

Miall, David S. "Episode Structures in Literary Narratives." *Journal of Literary Semantics* 33 (2004): 111–29.

Mittell, Jason. *Complex TV: The Poetics of Contemporary Television Storytelling*. New York: New York University Press, 2015.

Mittell, Jason. "The Loss of Value (or the Value of Lost)." *Flow* 2, no. 5 (2005). Accessed 23 March 2016. http://flowtv.org/?p=165.

Mittell, Jason. "Narrative Complexity in Contemporary American Television." *The Velvet Light Trap* 58 (2006): 29–40.

Mittell, Jason. "Sites of Participation: Wiki Fandom and the Case of Lostpedia." *Transformative Works and Cultures* 3 (2009). http://dx.doi.org/10.3983/twc.2009.0118.

Mittell, Jason. "The Value of Lost, Part Two." *FlowTV* 2, no. 10 (2005). Accessed 23 March 2016. http://flowtv.org/?p=435.

Murray, Janet. *Hamlet on the Holodeck: The Future of Narrative in Cyberspace*. Cambridge, MA: The MIT Press, 1997.

O'Neill, Patrick. *Fictions of Discourse: Reading Narrative Theory*. Toronto: University of Toronto Press, 1994.

Oren, Eyal. "SemperWiki: A Semantic Personal Wiki." Semantic Desktop Workshop 2005 @ ISWC. 2005. Accessed 23 March 2016. http://ceur-ws.org/Vol-175/27_oren_semper- wiki_final.pdf.

Ryan, Marie-Laure. *Avatars of Story*. Minneapolis: University of Minnesota Press, 2006.

Ryan, Marie-Laure. "Beyond Myth and Metaphor: Narrative in Digital Media." *Poetics Today* 23, no. 4 (2002): 581–609.

Ryan, Marie-Laure. *Narrative as Virtual Reality*. Baltimore, MD: The Johns Hopkins University Press, 2001.

Sandvoss, Cornel. *Fans: The Mirror of Consumption*. Malden, MA: Polity Press, 2005.

Scholes, Robert, and Robert Kellogg. *The Nature of Narrative*. London: Oxford University Press, 1968.

Sconce, Jeffrey. "What If?: Charting Television's New Textual Boundaries." In *Television after TV*, edited by Lynn Spigel and Jan Olsson, 93–112. Durham, NC: Duke University Press, 2004.

Soukup, Charles. "Hitching a Ride on a Star: Celebrity, Fandom, and Identification on the World Wide Web." *Southern Communication Journal* 71, no. 4 (2006): 319–37.

Stein, Louisa Ellen. *Millennial Fandom: Television Audiences in the Transmedia Age*. Iowa City: University of Iowa Press, 2015.

Thompson, Kristin. *Storytelling in Film and Television*. Cambridge, MA: Harvard University Press, 2003.

Tulloch, John, and Henry Jenkins. *Science Fiction Audiences: Watching* Doctor Who *and* Star Trek. London: Routledge, 1995.

Wagner, Mitch. "Lost Fans Find Internet Thrills Via Wikis, Games, Second Life." *InformationWeek*, 26 April 2008. Accessed 23 March 2016. http://www.informationweek.com/news/personal_tech/virtualworlds/showArticle.jhtml?articleID=207401542.

Walsh, Richard. "Fabula and Fictionality in Narrative Theory." *Style* 35, no. 4 (2001): 592–606.

Williams, Rebecca. "'It's about Power': Spoilers and Fan Hierarchy." *Slayage: The International Journal of Buffy Studies* 3, no. 3–4 (2004). Accessed 23 March 2016. http://slayageonline.com/Numbers/slayage11_12.htm.

· 4 ·

INTERREALITY AND THE DIGI-GRATIS

At this point in the story, the line between fantasy and reality appeared rather tenuous. The flesh-and blood Corso ... was increasingly tempted to see himself as a real character in an imaginary world. But that wasn't good. From there it was only a small step to believing he was an imaginary character who thinks he's real in an imaginary world. Only a small step to going nuts ...

—Pérez-Reverte, *The Club Dumas* (p. 260)

Jim's Interests: General: Working out, hanging out at the local bars, expanding my mind, eating Tuna Sandwhiches ... or so I'm told and poker ... Television: ... this show that's on Thuresday nights at 8:30pm ... I can't place the name of it but it has this crazy interview style thing ... [all sic]

—Jim, on MySpace

The Club Dumas, while on the surface a novel about the horrors of occult fanaticism, also delves into discussions about the nature of identity, self, and fictionality. Identity, an intangible and complex topic, emerges as a theme in *The Club Dumas*, as the book collectors base part of their identity on the characters they read about in Dumas' oeuvre. Indeed, fans of any media text—be it book, opera, film, or television show, base part of their identity on that text: As Sandvoss explains, we constantly use media to define us, for a text "is intrinsically interwoven with our sense of self, with who we are, would like to be, and think we are."[1] Fans can use media texts as a call-sign, identifying themselves as a fan through the clothes they wear, the paratextual products they purchase, or the discussions they have with other fans.[2]

One important aspect of this definition of the self can take place online, in virtual worlds like Facebook and MySpace. As online environments grow more and more populous—as of December 2015, Facebook has 1.59 billion monthly active users[3]—it becomes crucial to understand the relationship between our virtual lives and our physical lives. For viewers of media objects, the Web Commons offers a variety of inter-, intra-, and extra-textual ways of constructing an external identity. For example, many media producers have created fictional websites for television show characters that appear "real" (that is, they appear as though they were made by the character in the diegetic universe of the show). Fans can thus seemingly interact more directly with the extant media object by clicking on fictional webpages that purport to depict the personal webpages of the characters.[4] As Jenkins and others have shown, extant media objects no longer encompass just the movie, TV show, or video game: Instead, through the spread of narratives across different media, the audience's understanding of that narrative comes through a reconstruction of the fluid nature of this mediation.[5] In fact, by distributing narrative across media boundaries we find that fans encounter more intra-textual ways to construct their "senses of self" than in the past.

In particular, social network sites (SNS) present virtual spaces in which fans can construct profiles of television characters and then interact with those characters as if they were autonomous people. In this chapter, I will be specifically examining MySpace as the social network site where such character creations occur. There are, however, other social network sites that feature interactive fan-created content: Facebook, for example, allows users to become "fans" of particular media objects, and then notice of this fandom is inserted into the profile of the fan—imbuing the fan's Facebook page with salient identifying markers. MySpace has the capacity for more fictionalization than does Facebook.[6] Of course, as I mentioned in the introduction to this second edition, this focus seems the most dated. MySpace is more of a punchline than a space for fannish roleplay. But as times change, it's important to remember and reflect on previous fan experiences. Unfortunately, unlike the wikis and LiveJournals that I wrote about in earlier chapters, many of the MySpace profiles I discuss in this chapter no longer exist. I continue to reference them here for posterity's sake, and hope that the lessons learned from them will apply across other types of digital fandom.

For example, the second quotation that inscribes this chapter comes from the MySpace profile of the character Jim from the television show *The Office*.[7] Although some of the narrative information in the quotation comes directly from *The Office* (for example, there is a running joke about one character's obsession with Jim eating a tuna sandwich), other information comes from the fan's rewriting of the character (for example, Jim's musical taste). Other narrative information is itself non-diegetic, as when the fan writes that Jim's favorite television show airs "Thuresday [sic] nights at 8:30pm"—the night and time that *The Office* was on. The construction of these fan/character personas on MySpace represents a playful way for fans to interact with a character from the television show: Instead of identifying with the character as an external narrative object, the fan can literally roleplay the character as the fan sees him/her. This new form of interaction with a digital document takes place in a dynamic *interreality* of physical and virtual identities. This "mash-up" identity represents a created object, itself poised between subversion and support of a commercial economy. A mash-up here is not mere hybrid or convergence, in which the separate components lose their individual distinctiveness as they are subsumed in/by the converged culture; rather each maintains its distinctiveness while, at the same time, the combination creates something valuable and new. Value, in a mash-up culture becomes a concept thus steeped in both finance and in relationships.[8] In terms of fandom and New Media Studies, the mash-up of the virtual and the physical identity of fans and users of New Media exemplify an interreal examination of online self.

I'm defining *interreality* as a concept poised between the polarities of "fictional" and "factual." Interreality is a state of betweenness; an ever-shifting conception where fictionality and reality mingle. Character roleplay on MySpace, as I define, reflects this connection, as characters must be seen as fictional and factual at the same time. In many ways, interreality is the space between the "muse" and the "mun," or the roleplayed character and the authorial self.

The fans I discuss on MySpace present an interreal display of two types of virtual identity: fan persona and character persona. These virtual identities change the ontological status of the fictional characters into what Espen Aarseth describes as "simulations": personas that exist "somewhere in between reality and fiction."[9] This alteration occurs because fans on MySpace see the social network site as a space apart from that of the de Certeauan strategic space of the producer. In a video podcast, Jenkins argues that online spaces are seen, specifically by youth culture, as separate from other spaces: "they often present themselves as a utopian space that is totally removed from the world of school and the world of home. They're described as a place where people can have their own identities … ."[10] The dialectical nature of "consumption" and "production," previously conceptualized as the "strategies and tactics" of reading by de Certeau, changes with the advent on online social network sites. MySpace interacts with, complements, and even sometimes overrides "real" space, and forces the dichotomy between de Certeau's twined conceptions of strategies and tactics to break down. Fans create interreal personas online to swap narratives, to create communities, and to show their devotion to an extant media object. This rewriting leads other fans on MySpace to reread this creative content as examples of virtual identity in a digital environment. As Poster argues, "electronic media are supporting a … profound transformation of cultural identity."[11] By using the terms "strategies" and "tactics," de Certeau makes his metaphor obvious: Producers and consumers are seemingly at war with one another over meaning. Producers encode meaning through strategies of power while consumers tactically decode their own meaning through "movements that change the organization of a space."[12] MySpace personas, however, do not just change the relationship between strategies and tactics—they change the field of battle entirely.

By creating MySpace profiles of characters, fans build a space where the virtual identity of the fan and the virtual identity of the character mash up into an ontologically new entity through what I'm terming "identity roleplay." By the phrase "identity roleplay," I refer to the sense that fans both act as if they were the character and "play" with the characteristics that define that character. To enact this identity roleplay, fans insert aspects of their own personality into the character to rewrite the narrative around this fan/character amalgam. In doing so, they construct a space for a community's rereading of extant narratives and the reproduction of the fan community represented viscerally through the social nature of fans' creative content.

This form of identity roleplay augments parasocial research into the relationship between fans and extant media characters, in which audiences are seen to form social bonds with both fictional and nonfictional television personalities by identifying with the characters as if they were real.[13] In other words, fans' interaction with characters is seen as an interpersonal relationship. This form of parasocial interaction is complicated on MySpace, however, as fans enact this identity roleplay not by seeing characters as if they were real, but by inhabiting them; by becoming and roleplaying them.

Through an analysis of a number of MySpace profiles for the television shows *Veronica Mars*, *Gilmore Girls*, and *Torchwood*, I illustrate three ways identity roleplay functions in online profiles: (1) narrative affects the crafting of an identity; (2) profiles not only build on the narrative of the show, but branch from it as well; and (3) fan fiction on MySpace enacts an intra-textual form of fan/character roleplay. With this intertwining of narrative and identity, MySpace creates an interreal space where profiles of offline "real people" and profiles of online "simulated characters" can mingle, interact, and create digital fan fiction, and, in the process, problematizes the de Certeauan dichotomy of strategic and tactical readings.

Identity Definitions

Although it is beyond the scope of this book to explore all the intricacies of identity theory (especially in terms of the post-structuralist vein of this book's methodology), it is useful here to pause to lay out some important ways that I am using the terms "identity" and "virtual identity," and how they relate to the larger concepts of interreality and digital fandom, and especially how both of these function in New Media Studies.[14] One tension that repeatedly comes to a head is that between the physical world and the virtual, or the way in which virtual activities can play out in real-world situations.

By the term "identity," I refer to Stuart Hall's description of the concept as a doubled referent, a "point of *suture*," between "the discourse and practices which attempt to 'interpellate,' speak to us or hail us into place as the social subjects of particular discourse," and "the processes which produce subjectivities, which construct us as subjects which can be 'spoken.'" Hall sees identity as positioned in between a prescribed identity built from external forces and a self-realized, internal identity, a "sense of self." Identities are always in flux, as "identity" itself is a complex and multifaceted concept that uses "the resources of history, language and culture in the process of *becoming* rather than being."[15] One never is, but always is in process.

The term "virtual identity," on the other hand, comes from Dutch legal scholar Jacob van Kokswijk, who has defined the virtual identity as a "representation of an identity in a virtual environment, [that] can exist independently from human control and can (inter)act autonomously in an electronic system." Van Kokswijk's definition of "virtual identity" is useful because it mirrors, in many ways, the construction of a MySpace persona: When other people post comments on a MySpace persona, that persona becomes a continually updatable identity, which exists independently of the person who created it. Importantly, van Kokswijk's definition implies a *mechanical* alternation of the virtual identity, although I use the term more generally to refer to any change of a persona not enacted by the creator of that persona. Like identity itself, a virtual identity is not necessarily a stable construction, for "when a user creates an identity, it can be a conscious construction, it can evolve subconsciously over a period of time, or it could simply be a reflection of the user in real life."[16]

Finally, the term "interreal" comes from electronic/interactive artist Roy Ascott, who describes our digital culture as a "post-biological age" where each computer or digital terminal becomes part of a "complex and often widely distributed system, in which both human and artificial cognition and perception play their part." All human knowledge is connected and accessed through a series of interconnected, interactive discussions. Indeed,

> we have abandoned the model of a continuous reality whose essence is revealed to us in greater or lesser profundity according to some mental skills or tricks we have picked up in the schoolroom or the laboratory. … Insight then is derived from access to these layers, seeing all hypotheses as transient, looking only for usefulness in the metaphors they variously present, rather than abiding truths.

This space of stacked layers of knowledge, he claims, is an "inter reality set between the virtual and the 'real.'"[17] Ascott's "inter reality" is a theoretical concept suited for examining the connections between different concepts in the sciences, arts, and technologies, if not for making concrete conclusions.

My use of the term "interreality," however, differs in important ways from Ascott's. While I embrace Ascott's use of the term "inter real" as a concept between the virtual and the "real," I extend the term to describe not only a particular knowledge space, but also the creation of a particular set of fictional constructs. As van Kokswijk shows, a virtual identity exists also in a form of the interreal that mixes "the virtual and physical realities [of people online] into a hybrid total experience."[18] Thus, I put van Kokswijk's virtual "face" on Ascott's conceptual inter reality to show the "interreal" as a conceptual arena where fictional and factual identities can commingle and create a new "space," and its application to the Digi-gratis economy. By mashing up the identities of the television character and the fan, the MySpace profile functions both as fan gifting (giving a form of fan-created fiction to the fan community reproduced on MySpace) and of fan marketing (as the profiles themselves become forms of advertising for the media product).

Interreal Identities and the Digi-Gratis Economy

Fans' identity roleplay on MySpace highlights the emergence of the Digi-gratis economy, especially in relation to the transmediation and serialization of contemporary media products. As contemporary cult texts defer narrative closure, they create narrative desire through unfulfillment.[19] Thus, to construct a cult text, a media producer need only focus on a few key elements to give the text a sense of lack. A cult text focuses on a serial plot, an extended narrative structure, and on far-reaching ideas and imagination.[20] Through the development of these characteristics, media producers hope to raise questions about the narrative: questions that fans will have to tune in to answer. As we have seen, Hills extends this analysis, arguing that the cult text "focuses its endlessly deferred narrative around a singular question or related set of questions."[21] Answering these questions is the fan community, a culture of fans that exists around the cult text to fill in these deferred narrative elements.[22] This narrative deferral has antecedents in gift-giving, commodity fetishism, and the contemporary media environment.

Early in *The Gift*, Mauss asks the question that plagues him throughout his investigation: "What power resides in the object given that causes its recipient to pay it back?" This power he eventually describes using the Maori word *hau*: the power of the object, the inherent individual essence, or the spirit of the object.[23] Mauss found that the *hau* was, for many participants in a gift economy, an innate quality that necessitated its exchange. We can see this *hau* for the participants in a gift economy as the supernatural elements of the gift itself. Following Durkheim's description of the sacred as society's unexplainable events, Mauss investigated the way this socio-religious power contributed to the social significance of gifts.[24] Those who participate in a gift economy find that "things given have a soul that compels them to return to their original owner."[25] This feeling resonates for some fans, as the "power" of the media object can seem to come from that text—and in response, fans can insert their selves into that object through digital technology.

Like Mauss, Marx too finds power in things, although his value comes from the social significance of the object itself. This he calls "commodity fetishism," or the process by which a product, "as soon as it emerges as a commodity, it changes into a thing which transcends sensuousness"—in other words, the value-added of the commodity has inherent worth.[26] The "commodity fetishism," however, perverts the notion of an exchange. While gifts have a quality—a *hau*—that makes them worth owning, this quality—the fetish—also causes them to be more than mere constructed objects: They become socio-cultural products.[27] Participants in the market economy embrace this fetishized version of the commodity, taking it into the "misty realm of religion," where commodities become worshiped, sought after, and revered, as if sacred.[28] Baudrillard extends Marx's argument, showing that this religious fetishism is deployed in society in order to turn commodities into status that center on the hierarchical, power-based society created around the object. When objects are exchanged, the object loses its "objectness" and becomes part of the relations of exchange, or "the transferential pact that it seals between two persons."[29]

In modern cultures, according to Baudrillard, this notion of community-through-exchange has disappeared and been replaced by the sense of a "semiotic exchange." *Hau* has become fetishized. Semiotic exchange sees objects not as integral in *representing* social arrangements, but rather as *replacing* social dynamics. To consume a product is not to enter into a social arrangement with another, but to engage in a form of outclassing one another. Objects have more significance, more cultural worth, and more inherent value than do the people that use them. Apple's iPhone serves as a good contemporary example of this symbolic meaning: Although other phones may be more powerful or better designed than this Apple product, ownership of this product is important because of the symbolic status it bestows upon the owner.

In his famous critique of Mauss, however, Lévi-Strauss argues that the *hau* "is not the ultimate explanation for exchange; it is the conscious form whereby men … apprehended an unconscious necessity whose explanation lies elsewhere."[30] In this, the *hau* represents not a mystical force, but rather, like Marx's assessment of fetishism, a feeling inscribed by the community to that object. Whereas Mauss sees the *hau* as emanating from some inherent, mystical value in the object itself, Lévi-Strauss argues that it comes from the value others place upon that object.

Consequently, the concepts of *hau* and fetishism both deal with the Lacanian concept of displaced desire. The fetishized object is never enough: The desire to own contributes to the feeling of lack that can never truly be fulfilled. Yet, the desire for a *hau* appears to emerge, as Mauss would indicate, from the object itself. Trading objects back and forth promotes social relationships, but leads to a lack during the expected exchange. In the gift economy, it is the lack caused by the giving of the gift (a gift kept is a gift ungiven); in the market economy, it is the lack caused by the desire to own, to become part of the commodity culture prescribed by the media producers themselves. But as we've seen, cult narratives emphasize this lack: They promote it as a way of garnering and keeping viewers. Ironically, the market economy thus values the community formed via the fetishized product while the gift economy values the object itself as a commodity. Our economy, in this way, has always been hybridized.

To define a text as a cult, as an "endlessly deferred narrative," is to assert an active negation: A show is what it constantly defers, what it constantly does not show. A cult show's meaning exists not in any one place, but rather in the ethereal location in-between answer and question, in-between desire and the fulfillment of desire.[31] From the psychoanalytic perspective, this eternal displacement of "lack" represents nothing more and nothing less than reality itself. For Žižek, reading Lacan, "the realization of desire does not consist in its being 'fulfilled,' 'fully satisfied,' it coincides rather with the reproduction of desire as such, with its circular movement."[32] Desire always occurs through a postponement of the realization of that desire: we always must desire, for what we desire is desire itself. For the cult narrative, the fan realizes a desire to see a full rendering of that cult world, "the creation of a vast and detailed narrative space, only a fraction of which is ever directly seen or encountered within the text."[33] This expansiveness of desire was crucial to Mauss's argument about the gift as well: The cycle of gift-giving perpetuates because we desire the desire for giving, for receiving, and for reciprocating. The cult text is the literary equivalent of that desire, brought to the fan by the media producer as a *product*, but experienced by the media fan in the same way she might experience a gift.

Cult fans thus encounter their lack and their desire in two ways: first, in the deference of the narrative; second, in the deference of having watched the show and feeling the need to "give" back. In the market economy, we might emphasize the financial contribution: buying DVDs, shirts, or digital downloads. In the gift economy, we might emphasize fan fiction, gossip, or conversations. But in the Digi-gratis economy, other factors become apparent that mash up the very notions of monetary exchange and fan fiction.

This mash-up can be seen on MySpace character personas. For example, Gwen Cooper, a character from the BBC television show *Torchwood*, has a number of profiles created for her on MySpace. One of these profiles in particular demonstrates clear-cut interreality: this profile is a not-for-profit expression of a devotion to a media text; a love for the show *Torchwood*.[34] The creator doesn't just make the profile in order to talk to her friends, but rather there is a deliberate use of the profile to demonstrate an identity of the creator as a "fan" of the show *Torchwood*. The profile becomes a way of "putting on"

Torchwood—and Gwen Cooper especially—so that the fan identifies not just as *fan* but as *fan-of-Gwen* (and thus identifies *as* Gwen). However, the profile also works as free advertising for *Torchwood* itself, as anyone who stumbles across it, or any of the creator's friends who are also on MySpace, may discover the show through this profile. Thus, because the profile was made by a fan, it allows other friends of the fan to experience *Torchwood* not just as a television show, but as a narrative dispersed across different mediations.

The creator of this profile posts information about the character ("I'm part of a top secret 'Special Ops' team") in the About Me section, describing attributes of the fictional character. The profile does purport to be a persona of Gwen Cooper from the show—as if she had written it herself. This writing becomes the character identity. Further down the profile, however, the fan then breaks the "fourth wall" and reveals to her readers that she is "not, and do[es] not claim to be, either Eve Miles [the actress] or Gwen Cooper [the character]."[35] This disclaimer is obviously a product not of the character's profile, but of the fan herself: With it, she directly addresses the reader of the MySpace profile. This is the fan identity, and with this disclaimer, the MySpace fan/character persona becomes neither just a profile of the character, nor of just the fan, but of a mash-up of the two. Profiles as representations of this identity exist neither wholly online, nor entirely in the "real world," but rather in the interreal spaces between the online and the offline world. Through the creation of these fan/character amalgams, fans redefine the boundaries of identity online as they both create and contribute to the development of extant media characters.

Social Network Sites and Fandom

MySpace has a long history as an online social network site (SNS) on which users create profiles, as well as view the profiles of others. Although in 2016 it is used more for musical artists to promote their work, during its heyday in the mid-2000s, it was one of the most popular social sites in the world. There are three elements of MySpace that affect fan studies and the myriad ways fans used the social network site for fan/character amalgamation: persona creation, persona mutability, and virtual community formation. Importantly, each of these concepts feeds into the larger structure of MySpace: One cannot exist without any of the others. Each of these conceptions thus functions as a component of the interreal space created for, and by, MySpace, and an element in the design and existence of the Digi-gratis.

First, MySpace offers its users an interreal space in which they create virtual identities of themselves ("personas") that pictorially and textually represent their senses of self. A persona is a digital representation of the person ontologically dependent on the computer, the creator, and the community for its existence. Being aware of one's own identity is an important part of MySpace persona creation, as the creator must understand the public visibility of the profile.

While the creator of a profile may include information he/she feels is relevant to the digital world, there are other aspects of the persona that the person creating it may not have control over. For example, the playful comments that other personas post on the profile become aspects of that virtual identity that influence how others will see it. In the aforementioned profile of Jim from *The Office*, two of the comments demonstrate this playfulness and this virtual identity affect.[36] As readers, we can interpret the profile of "Jim" using these comments: The first, from L., reads "hey do you know john krasinskis real myspace? [sic]."[37] Through this comment, we understand that Jim is a character (we may already know that he is portrayed by John Krasinski) and that therefore, this profile must not be ontologically based on a "real" person. L. wants to know his "real" MySpace page (as though this one were "fake"?), not the "real" MySpace page for the *fictional* character of Jim. Real, here, is a problematic term for L.: Does she mean the "real" page or the "real" person? Which is "real," anyway, when, as we see, all personas are virtual? However, this real/virtual description is again confused by the playfulness of the next comment, from Knoxy!, who writes "Hey Jim when is season 5 on DVD?:)." This comment also creates a juxtaposition between what we (as savvy viewers of *The Office*'s documentary-style filmmaking) know of Jim and what we (as MySpace denizens) know of the profile of Jim.[38] We know that the profile is of a fictional character, and the comment seems to take a tongue-in-cheek tone with that knowledge.

The profile creator of Jim did not have influence over these comments: Although Jim can include information in his profile that hints at his identity as a character on *The Office*, the comments can confuse or even directly contradict that assertion. The playful mix of the comments also references the twin concerns of the Digi-gratis economy. L.'s request for information about John Krasinski's MySpace page is a non-monetary assertion, a search for a digital gift in the form of information. In this way, she strives to explore *The Office* not as a product, but as a multi-modal process that is transmediated across television and MySpace. Knoxy!'s comment, however, directly ties into a market economy, as she wants to know where to purchase Season 5 of *The Office* on DVD. This MySpace persona in general, in fact, represents a mash-up of the gift and market economies, as it seems to be created to generate a sense of community around Jim's character, but that character is a media product marketed, and sold, to the public.

If we examine MySpace for its persona creation abilities, then we must also examine it for the way these personas can be changed online, for the persona mutability. In the past, any public display of identity would have been intimately tied to the physical body, and therefore difficult to change. The body provides a visual sense of who we are, as throughout humankind's history, we have been identified by physical, observable bodily factors: our race, our sex, our facial features, et al. Online, however, there is "no body" upon which to base identity. No longer linked to the body, an online identity can be constructed in a variety of ways. Sherry Turkle has shown us the multiplicity of identity construction in her description of online textual gaming, and describes how we use different identities for different effects: Online, I can have a powerful avatar, even if I am a passive person offline.[39] Alluquere Rosanne Stone echoes this, arguing further that online, we present *multiple* identities and personalities.[40] Psychologist Wolfgang Kraus illustrates this multiplicity when he writes, "belonging [to an identity] becomes—generally speaking—a question of choice, which must be answered by the individual."[41] Our online identity "is not grounded in history and politics, but is a pastiche, a mosaic, made up of ephemeral fragments," all of which coalesce around a sense of the mutability emerging from our offline sense of self.[42]

One way this identity pastiche forms on MySpace is through the visible representation of a social community. A MySpace persona is constructed not just with textual details about the person who made it, but also as a node on that persona's social network. This public display of connections is a crucial component of SNSs, as the Friends list "contains links to each Friend's profile, enabling viewers to traverse the network graph."[43] Users of MySpace can "friend" other users of MySpace, and this friendship then becomes visibly represented by a hyperlink between each persona. The more "friends" one has on MySpace, the larger the social network, and the more connected the persona becomes in his or her network. As more personas connect, a larger social network forms. This visible community illustrates what Bruns terms "equipotentiality," or the sense that all participants can make unique contributions to a group project. If we think of MySpace as an infinite group project—a community intra-textually built by the users of MySpace, then this equipotentiality operates "through shared discursive, communicative spaces."[44] Each connection a persona establishes not only joins it with another persona, but also helps to establish a shared space between the two personas. Jim may be the profile owner, but his visible connection to both L. and Knoxy! makes his community—those who share and contribute information—part of this identity. MySpace visibly represents community *as* identity.

For users of MySpace, creating a profile of a character is more than fan fiction, and more than textual poaching: It is a space where identities mingle. This examination of the interreal identity illustrates a reconceptualization of the theoretical binary between the fan-as-consumer and the fan-as-producer. Through the structure of the Digi-gratis economy, these fans articulate a notion of digital gift-giving: The persona roleplay enacted between two interreal personas indicates a gift of *time* and *attention* to another fan. Only in an economy that had characteristics both of the commodity market and of the gift economy could MySpace fan/character roleplay happen. Fans use these profiles as "gifts" to give both to other fans and to the media producers themselves: These profiles are a way to articulate notions of appreciation, thankfulness, and gratitude for the characters. At the same time, however, they function also as free advertising for the media producer, and serve to promote the product as much as they do the fan.

MySpace Roleplay and Parasocial Theory

Following Jenkins's lead, many traditional fan studies have used implicit examples of textual poaching as the basis for analyzing fan practices. In these types of analyses, the fan identifies with characters through an active reading of the text, and operates within and from the space of the producer.[45] To interact with characters from this extant media object in traditional fan studies, therefore, is to see them as inhabiting a separate space from that of the fan. The character becomes a vehicle, a means of (re)presenting their own selves in fan fiction. In other words, the fan and the character are distinct.

One way to differentiate between fan and character on MySpace is to rely on parasocial interaction theory, Horton and Wohl's influential statement that stresses that the social bond formed between viewers and characters is "analogous to and in many ways resembles social interaction in ordinary primary groups."[46] In other words, we might say that viewers often form bonds with the characters they witness onscreen, and these bonds mirror the social bonds formed between people that meet in the "real world." This mediated character could be a non-fictional person (such as a news anchor or a weatherman), or the character could be a fictional entity (such as a character in a television series). Either way, the mediated personality becomes a figure with whom a fan identifies. This use of parasocial interaction theory has been used to examine a fan's level of involvement with online, digital characters with virtual identities, as the interaction between the person sitting at her computer screen and the persona at which she looks mirrors the relationship between the viewers of television and the emotional attachment those viewers feel toward the characters.[47]

There are two main limitations to parasocial interaction theory when using it to examine fans, however. First, Horton and Wohl's use of the theory, which works because the mediated character "uses the mode of direct address, [and] talks as if he [sic] were conversing personally and privately" with the audience, focuses more on the individual watching and interpreting the program than on the media text.[48] Parasocial "interaction" is therefore a misnomer, for the characters within the text can affect the viewer, but the viewer has little recourse in affecting the characters. One of the reasons the internet so dramatically influences parasocial theory is because of its inherent interactive characteristics, as Thorson and Rodgers have demonstrated.[49] They refer to what Rafaeli calls the "ortho-social," or an extra-textual participation with a media text wherein fans can write a letter to the editor of a newspaper, call a talk show, or use, in some way, a "traditional unidirectional mass media in a new, reactive, or interactive manner."[50] Although fans have interacted with media texts in this ortho-social manner for decades, the inherent interactive capabilities of the internet have increased dramatically both the means and the ability for fans to directly influence the media texts, which we can perhaps best see in the interactions of fans organizing to save or resurrect television shows like *Roswell*, *Firefly*, and *Chuck*.

Second, additional research on parasocial theory has found that it needs to "distinguish between media figures who are direct representations of real people (such as newscasters), and [those who are] fictional creations."[51] Parasocial theory exists to identify a relationship between the viewer and any persona on television. In the digital, however, such individualized relationships are problematized, for as we have seen, *any* persona on MySpace, be it character or fan, fictional or authentic, is in itself a complete construction. The interreality of the internet, especially on MySpace, means that, by definition, one cannot inherently *know* whether a virtual identity is of a "real" person or of a character: External knowledge (about the media object, about the source canon, or about the show itself) is necessary for viewers to grasp the ontological distinction. This knowledge is created by the fan community on MySpace, not through parasocial interaction, but through identity roleplay. As fans roleplay by creating profiles of fictional characters and then insert their own identities into them, they forge a new relationship between virtual identity and narrative. Importantly, this fan/character amalgam is unique for each fan, and contributes not to the extant media object itself, but rather to each fan's individual rereading of that object.

Although traditional media studies assume this divide between the fan and the character, the amalgamation of fan and character in online spaces challenges that assumption. MySpace denizens force a revised reading of de Certeau's poaching as they make MySpace a space where the strategies of the producers, the "elaborate theoretical places (systems and totalizing discourses)," are rewritten as libratory creative content. Thus, instead of tactically reading the source text to "appropriate popular texts and reread them in a fashion that serves different interests," these fans transgress the boundary between de Certeau's strategies and tactics of reading. Specifically, de Certeau conceptualizes the production of a text spatially, using the metaphor of the city: Created with certain routes laid out, the city allows pedestrians to walk their own path, to define their own "rhetoric of walking."[52] In this way, producers can intend a path, but pedestrians can tactically traverse another. Pedestrians cannot create a new path, but can reappropriate a city path for their own use. Unlike de Certeau's text-as-city, however, the creation of character profiles does not just allow pedestrians to walk, it encourages them to rebuild the roads and repave the sidewalks as they go. It is in the working together, in the community of a social network, that the narrative dimensions of a text can become fleshed out and expansive.

Fans on MySpace, interacting within the interreal, answer Hills, who asserts, "the difficulty with de Certeau's model is that it seems too rigid to deal helpfully with any blurring of consumer and consumer-as-producer identities."[53] The fan community reproduces their interpretation of the extant media narrative by reasserting their influence in a digital space. This influence creates a unique digital confusion: Since it is impossible to know who the creator is behind any one profile, every profile itself becomes a distinctive identity. Without knowledge of the extant media object's canon, or direct textual markers on the persona itself, readers of MySpace could not distinguish a profile of a fictional character from a profile of a person existent offline. In other words, there is nothing inherent to MySpace that indicates fictionality or factuality in persona presentation.

This shift in the sense of fan identity calls into question traditional fan studies' use of textual poaching. By "textual poaching," de Certeau argues that

> The reader takes neither the position of the author nor an author's position. He invents in the text something different from what they [the authors] "intended." He detaches them from their (lost or accessory) origin. He combines their fragments and creates something un-known in *the space organized by their capacity for allowing an indefinite plurality of meanings*.

As indicated, this "poaching" of meaning necessarily occurs in the space of the producer any time one reads. As de Certeau makes clear, any particular reading follows a different trajectory than any other: "readers are travelers; they move across lands belonging to someone else, like nomads poaching their way across fields they did not write."[54] For Jenkins, reading de Certeau, textual poaching offers "an alternate conception of fans as readers who appropriate popular texts and reread them in a fashion that serves different interests."[55] Fandom is a form of continual poaching on others' works. Further, this poaching, this movement, can only take place in a space of power prescribed by the producer, who strategically makes and inserts intended meanings into a text, but from which readers can tactically make their own interpretations.

De Certeau argues that readers poach meaning from producers and make "reading thus … an 'art' which is anything but passive." In short, the act of reading does not mean passively consuming another's text, but rather "making it one's own, appropriating or reappropriating it."[56] Jenkins extends this argument by showing how fans appropriate the works of producers into their own creativity, poaching on the works of others.[57] According to Harrington and Bielby's meta-analysis of fan studies, Jenkins's conception of this tactical reading makes up the basis of most traditional studies of fan practices.[58] For example, Jenkins demonstrates that fans create amateur music videos from footage "poached" from television shows, in order to interact with the characters parasocially: "fan viewers are often totally disinterested in the identity of the original singer(s) but are prepared to see the musical performances as an expression of the thoughts, feelings, desires, and fantasies of the fictional character(s)."[59] By "poaching" the meaning of the song, the fan music video creator reappropriates and reinscribes that meaning to the character by poaching content from two different sources. In effect, without directly referencing Horton and Wohl, Jenkins has described how parasocial theory functions within a fan's reception of a media text: Like in parasocial interaction, fans create a connection to a character or a media text.

In this view of fandom, fans and characters are immediately divided from each other. Fans identify with the character. The fan externalizes her own feelings, her own sense of self, and inscribes these feelings into an external source, the extant media character. We can see this fan separation from an idealized version of a character in a quote from one of Jenkins's students, interviewed for *Science Fiction Audiences*. John, the student, writes:

> "I always wanted to be like Spock. I suppose I was like him—an outsider. He was an alien in the ship and I was an outsider in my town. I grew up there but I didn't belong. He was in science and things like that. He would go to investigate things. He would solve all the problems. He was someone I looked up to at the time. ... I think that being in internal conflict with yourself about being outwardly emotional is something that is common to many people in science and technology and especially something I have experienced."

For Jenkins's student, the character of Mr. Spock, the emotionless science officer on the *Enterprise* in *Star Trek*, represents an ideal model, someone he always wanted to be like. Even in this connection, however, the fan establishes a distance between himself and Spock: It is not that he wants to *be* Spock, but rather he wants to be *like* Spock. This is not mere semantics, for part of any investigation of fandom hinges on the intimate connection the fan feels with the extant media object. For John, as for many fans, the closest they could hope to get to a text was, as Tulloch and Jenkins state, to "have particularly strong feelings ... towards characters which allow them to explore conflicts they face in their own lives."[60]

What some fans practice on MySpace, however, is not the identification *with* the fictional character, but rather identification *as* that character. As fans use the Web Commons for exploring extra-dimensions of characters, a reconceptualization of fan studies must take into account this form of performativity. Fans have always identified closely with a narrative text, but online the "relationship between fans and objects of fandom goes beyond forming a symbolic basis for fan communities" and the "perception of the external object as a part of the self is in turn based on the recognition, consciously or unconsciously, of aspects of the self in the external object."[61] The generative possibilities—the ways that fans can generate their own content, their own media objects—offers not only a way to appropriate cultural material into a relevant fan practice, but also a way that cultural material can be tied to the fan's own "sense of self."

Identity Roleplay

By "putting on" the persona of the extant media character, fans participate in a form of identity roleplay that goes beyond the "impersonation" described by Hills as the act of "becoming" a fan; or the "performativity" described by Sandvoss as self-reflection.[62] When fans create MySpace character personas, they intimately, and necessarily, become entangled in an identity conflict. Offline, we can tell the difference between a TV character and a physical person because of the screen; one is mediated, the other is not. When a character becomes the object of affection of a cult television fan, he/she becomes mediated through a technology that clearly delineates a separation from the spectator and the character. On MySpace, however, the personas of users exist in the same space as those of the personas of the characters. All these characters are equally mediated. When two personas interact, we find that fans not only connect their virtual identity with aspects of the extant media object, but also that they incorporate their selves into that extant media object's narrative text.

Within this connection, however, fans not only place themselves within the media product, but also roleplay as the character seemingly in order to invest themselves with aspects of that media character. Because social network sites are online spaces that

> allow individuals to (1) construct a public or semi-public profile within a bounded system, (2) articulate a list of other users with whom they share a connection, and (3) view and traverse their list of connections and those made by others within the system,

they facilitate this fan (re)construction of extant narratives in a way unlike that which we have seen in blog fan fiction or wikis.[63] Interreality, and the virtual identity that exists within it, are publicly created through fan interaction in the Web Commons. MySpace allows fans to explore this public identity formation in a conceptual space of their own creation.

Fans incorporate themselves into an extant media object on MySpace in three ways: through a fragmentation of virtual identities, through a reconstruction of narrative texts, and through character acting with reconstructed characters, which is integral to fandom on MySpace. These three methods center on the issue of the transmediated narrative text: The more dispersed a text is in time and space, the more a fan can find places to insert herself into the text.[64] As Gwenllian-Jones shows, "the appeal of … vast, transmedia fictions lies precisely in their invitations to immersion and interactivity; they are constructed, marketed, and used by fans not as 'texts' to be 'read' but as cosmologies to be entered, experienced and imaginatively interacted with."[65] In the Web Commons, fans understand that these cult texts can be reproduced on MySpace as vast, interreal worlds. By tying together identity formation, narrative construction, and character acting, all these personas relay specific identity information not just about fans or characters, but also about MySpace personas; not only about virtual identities, but also about identities that mash-up traditional notions of character, fan, and identification.

Fragmented Identity

The first method scholars can use to examine the virtual identity of the fan in an interreal space is to view the fan's integration of their persona with the extant media character's persona. A fan can also create myriad personas, each one corresponding to a different set of identity markers. Each persona, therefore, becomes an "embellished … personality" within a realm of other embellished persona-alities.[66] By extricating different aspects of their own senses of self, and then portraying those aspects in different personas online, fans learn not only how to mark their virtual identity as a unique form from other virtual identities online, but also how to represent their own conceptual sense of self as it applies in the "real world." Through this representation, the fan identifies the aspects of a virtual identity that can be related to that of an extant media character.

This form of virtual identity writing becomes a way to digitally self-identify. The space of this post says a lot about what that fan wants known about his/her persona. In terms of virtual identity construction, comments can be a way for the commenter to forge a virtual identity, though the interaction with the profile on which one comments. For example, in the comments section of a character persona from *Gilmore Girls*, the poster Gilmoregirls3 writes "DUDE i love gg its my fav show ever!!!!!! I love ur space its so kute!!!! [all sic]."[67] Gilmoregirls3 here establishes her identity as fan of *Gilmore Girls*, and helps to establish characteristics of this identity on the "[My]space" site that she visits—"its so kute" expresses her opinion about the site, confirming her distinctive traits. As Judith Donath has shown, the choice of where to post is a vital one for online communication, as personas gain credibility through the location of their posts.[68]

The fan author of a MySpace character persona then actively rewrites his/her virtual identity by posting comments that respond to, and elicit a response from, the other members of the community: This presents a difference from other personas. As literary scholar Kenneth Burke has written, identity is negotiated by examining what one is not. To form an identification with another, one needs to entice the other to dialogue, and within this dialogue, find both places of commonality and places of difference.[69] Through differences we can learn both what we are and, importantly, what we are not. We can see how the posting of a particular statement can be contrasted to the other statements made on and by the persona. A rereading of that persona's comment might reveal a difference between the two interlocutors, or might reveal a similarity—but either way, it would reveal an aspect of the virtual identity for both personas. Importantly, as pointed out by Mark Poster, this formation of identity hinges on the interactive response of others, on the back-and-forth responses of the community.[70]

This communal interactivity between fan participants functions to tie narrative to virtual identity. We can see this on the official *Veronica Mars* MySpace site, created for the CW, in which a number of salient identity issues become apparent.[71] This official CW site depicts a form of parasocial attachment to these characters, but the actual use of the site by fans indicates rather that the fans are creating virtual identities in the interreal through identity fragmentation. In terms of parasocial interaction, the site appears to be created by the CW with the intent of giving MySpace users and fans of *Veronica Mars* the ability to communicate with the character. The more authentic they can make them seem—that is, the more they can make the persona "Veronica Mars" appear to represent an offline person—presumably the more attachment audience members will feel. Veronica's profile displays her virtual characteristics including her sex, age and location. However, the large "Catch *Veronica Mars* Tuesday 9PM CW" in the "About Me" section seems to negate this "reality." Here is another aspect of the Digi-gratis economy: The digital mash-up of the character identity as both fictional—catch her on TV—and "real"—"female, 19 years old, Neptune High"—becomes an interreal aspect of the profile.

The first nine "friends" in Veronica's social network come from the show *Veronica Mars*. Each link demonstrates how the CW makes personas out of these characters. The social network of the characters creates a further identity for "Veronica Mars," because it situates her within a complex series of relationships that mirror the relationships the personas of non-character, non-mediated users and fans on MySpace. As boyd and Ellison show, connecting with friends on MySpace is an "aspect of self-presentation" that "forms through the articulation of friendship links, which serve as identity markers for the profile owner."[72]

Further down the page, however, fans appear to subvert this parasocial persona in an interreal manner: A more scattered list of non-character friends becomes apparent.[73] A mixture of personas of characters and personas of fans, these friends confuse the issue of "Veronica Mars's" personality. Of note, one "friend" is the CW itself, while another friend is the show that immediately preceded *Veronica Mars* on the CW network, *Gilmore Girls*. This normalizes the television world into an interreal juxtaposition of identities: Suddenly, everything that exists on the CW seems to exist in the *same virtual world*. Importantly, this virtual world seems to mix easily with our own world. To become friends with fan personas, or any persona of a non-fictional, non-mediated person, is to make the Veronica Mars CW persona a nexus of interreal identities, each a fragmented form of an individual interreal persona connected by a virtual community.

A second indication of this identity roleplay comes through the interaction of the posted comments with the profile itself, which helps build the identity credibility of the character of Veronica Mars.[74] Each comment is displayed in reverse order, like blog comments. For example, a comment on the *Veronica Mars* MySpace page from Aneeda burrita? reads, "so about the new pic … LoVe iT!!!" Although the comment may seem banal, it is anything but: It begs the question, why does Aneeda burrita? comment to a character she realizes is fictional? Whom does she think it going to respond? While the communication that occurred here is obvious (that one persona changed his/her persona picture and a second persona responded), the complicated meaning behind the posting of the comment—the "why," not the "what" of the comment—offers insight into fan identity roleplay.

To change a persona picture is not necessarily just a cosmetic change, but also a way of "throwing on" a new identity, as one might wear a different style of clothing to school. Aneeda burrita? here identifies with "Veronica Mars" and tries to build her own virtual identity through the interreal combination of the space of "Veronica Mars" and the content of her own posted comments. In this way, Aneeda burrita? navigates her identity through another's profile, perhaps hoping that when others see her comment, they will click over to view her profile. In this way, the persona of "Veronica Mars" seemingly acts as a surrogate persona for the identity construction of other personas, through the relationship between the sites. By creating a new space, an interreality, in which to construct her own identity, Aneeda burrita? has forged the relationship between character persona and the fan personas. Additionally, she has contributed a digital "gift"—of attention, of comment, of time—to the profile, further establishing the commercial popularity of *Veronica Mars* for other viewers of the profile, and bolstering a Digi-gratis economy.

Thus, by conversing on these sites, not only does Aneeda burrita? create an identity centered on *Veronica Mars*, but she also rewrites her presence within the world of the shows. Aneeda burrita? could be Veronica's friend—just as the nine "character" personas are on "Veronica's" persona. Given that SNSs enable individuals to connect with one another, it is not surprising that they have become deeply embedded in user's lives, as boyd has demonstrated.[75] The conversation between two personas is more than a form of parasocial desire: it is identification and roleplay in an interreal space. By placing herself in this particular space, Aneeda burrita? creates a unique interaction between the producer's space of *Veronica Mars* and her own virtual identity, her own interreality.

Reconstruction through Branched Narratives

A second way scholars can use an interreal analysis is by examining the reconstruction of extant media objects on MySpace. Fans identify more readily with characters through interacting with extant media objects, as Ng shows: Creating fan fiction or fan videos creates an "experience that cannot be derived from watching the show."[76] MySpace offers fans a space in which they can reproduce characters and situations from the extant media object, and then use those characters to reproduce fan culture. Through imaginative dialogue, the rewriting of characteristics seen on the television show, or reliving the narrative events that have been most dramatic, the fan encounters the extant media object's narrative as a mutable text. This mutability is what Jenkins calls the "silly-putty" of the story: Fans end up "stretching [the extant media object's] boundaries to incorporate their concerns, remolding its characters to better suit their desires."[77] Each fan-created character persona becomes a way of both reliving and altering the extant media object's narrative text, opening up new possibilities by recreating those that have already occurred onscreen.

This method of identity roleplay on MySpace occurs through the interaction between narrative elements on the page. One's persona fits within the social framework produced by the fan community on MySpace. As van

Kokswijk shows, virtual identity is not just defined as the place, but also the "*plot* you are in."[78] Plot, here, can mean two things: either a plot of land (i.e., the space you construct online) or a "plot" of a narrative (i.e., the narrative you create for yourself online). Thus, an extant media character's persona provides both a space and a narrative framework around which fan personas can extend narrative possibilities.

We can see this narrative dimension in the way the character persona of gwencooper, another profile of the character Gwen Cooper from the television show *Torchwood*, describes herself in the "About Me" section on her MySpace persona.[79] Gwencooper first uses this space to define her persona, (re)writing "My story starts off with me as a police officer in Cardiff but then i get involved in the Torchwood team [all sic]." In this statement, the writer of gwencooper has described the extant media character as media producers and the actress have made her on the show—this is, therefore, a diegetic description, or one that repeats narrative/character information we already know from the canon of the extant media object. This is an aspect of the character's persona.

The "About Me" goes on, however, to rewrite *Torchwood*: "I'm a very down-to-earth girl, kind and generous, but extremely ambitious, feisty, intelligent and witty, if i do say so myself. But im really the girl next door [all sic]." These characteristics do not all diegetically describe the character of Gwen Cooper from *Torchwood*. Instead, these characteristics point to an amalgam identity, one created in an interreality by *adding* aspects to the diegesis of the narrative. The insertion of alternative narrative elements into a fan's rewritten version of an extant media object alters the interpretation of that narrative, and creates what Jenkins calls "I Wonder Ifs": narratives that "explore 'possibilities' that are hinted at but not developed within the [texts]."[80] By adopting a different narrative from what is considered canonical by the viewers, fan persona creators can tie their own virtual identities more directly with a set of values that may or may not differ from other fans. They become identified through their differences. In the gwencooper example, the fan aspects of the persona are inserted into the character aspects of the persona elements, creating a virtual identity that is not necessarily supported by the show (or, least, are opinions about the character that may not be shared by all viewers). This subjectivity creates a form of branching narrative, in which a rereading of the additions by fan writers creates an alternative view of the diegetic narrative text, another possible outcome from the one observed on the television.

By refocusing the narrative, gwencooper rewrites the virtual identity of the persona through a separate narrative thread, which exists in an interreal narrative space. This new identity, built from pieces of *Torchwood*, from pieces of the textual elements gwencooper has posted elsewhere online, and from the elements of her own life that she may have included in this persona, bypasses the mechanics of the media producer. Gwencooper removes herself from the *Torchwood* narrative not by writing fan fiction directly, but by rewriting the space of the text itself. Her *Torchwood* is one of her own creation, to be reread as such by other fans.

Roleplay through Fan-Created Dialogue

The desire for an individual identity is shaped and structured by a tension between communal expectations and individual desires. This tension is made most salient on MySpace, as the tension between the fan and the character in one interreal persona creates a difficult tangle of identity formations. As Bolter suggests, in digital writing one needs to "construct her identity," yet any persona creation becomes "a cooperative effort."[81] In order to have an effective post, a poster must necessarily construct an identity that is different from others, but similar enough to warrant being in/on the same page.

Scholars can thus use a third method of interreal analysis: by observing the identity roleplay of the fan's fragmented, virtual identities, the scholar can observe these reconstructed characters as representation of the fan herself. Soukup identifies a similar process at work in fan-created celebrity websites:

> via photograph galleries, the designer of the fansite has, to an extent, the means to control the physical representation of the celebrity via the inclusion or exclusion of specific photographs. These fans are not merely interpreting the celebrity via existing media texts, they are also creating a new text.[82]

As fans integrate into their fan-created MySpace character personas, they insert identity characteristics to deepen the character, and flesh out additional aspects of the extant media object. Through connections made with other personas on MySpace—some character, some fan, some neither—the fan-creator incorporates the virtual community as an additional aspect of the virtual identity of the character. This incorporation indicates the formation of an interreal space, for the fan to post not only characteristics of the character, but characteristics of the fan as well. Through the posting of comments back and forth between two different personas, written in the characters' voice, fans act with their character persona creations. For instance, if John, from Tulloch and

Jenkins's example, had created a Mr. Spock character persona on MySpace, not only would he have been able to be "like" Spock, but he also would have been able to post comments *as* Spock, interact with characters who are friends *with* Spock, and write blogs in the voice *of* Spock.[83] Roleplay involves a definition of the self as something "other," or what Torvill Mortensen calls "another you."[84] We are always performing, as Goffman describes, but the roleplay on MySpace takes a decidedly different turn: for when we deliberately perform another already-extant identity on MySpace, we highlight and underscore the different identity characteristics that we feel are important, and put on a presentation of a different self.[85] MySpace virtual personas exist neither in the space of the other, nor in the space of the reader, but rather in a different, interreal space. These personas foster fan/character roleplay as MySpace provides surrogate experiences for fans to experience the narrative text.

This third form of virtual identity construction on MySpace can also be seen in a fan-constructed site for persona "Veronica Marz." Fans not only refer to this character as one who exists in a non-mediated, physical reality, but also live vicariously through this character on MySpace. Instead of simply *thinking* that characters can exist external from their own mediation, fan personas roleplay and *create* an interreal space where they interact to write a narrative, utilizing a form of narractivity. The "Veronica Marz" profile uses additional fan/character personas to rewrite a more individualized identity.[86] The layout of the profile represents a more personal approach to the website. In the case of "Veronica Marz," the site is designed as if it represented the personality of Veronica Mars. Using the idea of a distributed narrative of the *Veronica Mars* text allows the creator to imagine that the narrative of the show continues online, transmediated with MySpace.

Further, the roleplay of the fan/character amalgam creates a new form of interactive fan fiction: dialogue between "Veronica Marz" and other character personas.[87] The construction of fan fiction on MySpace differs from the way it is constructed on the blog. On MySpace, the fans write fan fiction through interreal dialogue posted as comments between personas. In this way, the fiction appears as an interaction, written through the narractive dialogue between two interreal personas. For example, the persona of "Dick" interacts on the profile of "Veronica Marz" to rewrite the narrative of *Veronica Mars* as a new story. "Dick" narrates, in the third person:

> Dick rolled his eyes. God, she was pathetic. And Logan was pathetic, too. He couldn't believe that something so 'wonderful' (as Logan had previously described it) as love could turn two people into completely different people than they'd been before.

Further, the two people that "Dick" refers to had broken up in the show before this post had been made; therefore, the events referred to are unique to the MySpace fiction.

In order to function as fan fiction, however, the fan community must necessarily separate this dialogue from the show that exists diegetically as text of *Veronica Mars*. In writing this new text, "Veronica Marz" creates a digital environment outside the reach of the television show, the CW network, or the other fan-readers. Although the characters are shared between the network and the fan, they are rewritten with a new virtual identity that neither asserts the strategic input of the media producer, nor demonstrates the tactical response of the fan to others in the fan community. Instead, the fan/character amalgam echoes the Digi-gratis in an interreal realm. Although aspects of the market economy, in the guise of the media producer, may "own" the character, the fan also controls the fan/character persona and provides the profile as a type of gift to the other fans of *Veronica Mars*.

Through their dialogue, Dick and Veronica Marz are not textually poaching material from *Veronica Mars*, but making their own material, created both from the remnants of an extant media object, and from their own rereading, their own sense of community. Their interaction on MySpace allows an ever-evolving space for the creation of meaning from extant narratives. Whereas traditional fan studies would use de Certeau's conception of textual poaching and tactical reading to indicate a form of character interaction and identification, MySpace, the communalization of the Web Commons, and the mashed up economic structure of the Digi-gratis, denotes a need for new scholarship into digital fandom, as a characteristic analysis of New Media Studies.

Identity Roleplay and Fandom

By eliding the role of the consumer from a tactical one to one that mashes up strategies and tactics in a new space, MySpace dramatically represents the Digi-gratis economy. By melding virtual identity production and narrative completion, MySpace personas merge two worlds, fan identities that exist offline and character identities that exist in a mediated relationship to the fan, into the interreal. The producer's "power" over the characters, intuitively assumed in a market economy, is negated by the fan's own determinism over the media object. Yet, the fan's work—given freely and in order to build a community—participates within the realm of the gift.

The fan's participation in the creation process, facilitated by the creation of the interreal, results in a new form of media text, that itself could become the object of fandom from others. Fans use MySpace as a constructive arena of media creation: They change the object of their fandom into something *new*. By combining elements of characters and fan personas into one interreal text, fans not only circumvent the products they have grown to love, but also create new products that others can love. In this way, they articulate a fandom in which they are willing to alter their favorite texts in new spaces of fandom. In these new spaces—fan spaces—the virtual identity becomes part of the fan's interpretation of the extant media object. Because MySpace depicts not just the ability to create new profiles, but also the capacity for visually representing whole communities, it represents a new way of conceptualizing fandom. Through the active rewriting and the interactive rereading of extant media objects, fans reproduce their fandom in new contexts. The communities of fans that form and roleplay together on MySpace are not just meeting to discuss the object of their fandom, but are also creating a whole new way to interact.

Notes

1. Sandvoss, *Fans*, 96.
2. See Fiske, "Cultural Economy of Fandom."
3. Facebook Newsroom—http://newsroom.fb.com/company-info/
4. Caldwell, "Convergence Television," 51.
5. Jenkins, *Convergence Culture*, 97; see also, Walker, "Distributed Narratives," ¶1.
6. Levinson, *New New Media*, 122, I use the term "social network site," instead of the more common "Social Networking Site," in terms referenced by boyd and Ellison, "Social Network Sites," ¶5–6.
7. Jim.
8. Andrejevic, "Watching Television without Pity," 32.
9. Aarseth, "Nonlinearity and Literary Theory," 78–9.
10. Jenkins, "Lecture 06."
11. Poster, *The Information Subject*, 72.
12. de Certeau, *The Practice of Everyday Life*, 38.
13. Horton and Wohl, "Mass Communication and Parasocial Interaction," 229.
14. The reader is invited to explore this topic in more depth in: Turkle, *Life on the Screen*; Rutherford, *Identity*; du Gay, Evans, and Redman, ed., *Identity*; Gauntlett, *Media, Gender and Identity*.
15. Hall, "Introduction," 5–6, 4, my emphasis.
16. van Kokswijk, *Digital Ego*, 13, 63.
17. Ascott, "Homo Telematicus," ¶2.
18. van Kokswijk, *Digital Ego*, 27.
19. Hills, *Fan Cultures*, 134.
20. Gwenllian-Jones, "Virtual Reality and Cult Television," 83.
21. Hills, *Fan Cultures*, 134.
22. See, Le Guern, "Toward a Constructivist Approach to Media Cults," 19.

23. Mauss, *The Gift*, 3, 11–12.
24. Durkheim, *The Elementary Forms of Religious Life*.
25. Godelier, *The Enigma of the Gift*, 15.
26. Marx, *Capital*, 163.
27. Godelier, *The Enigma of the Gift*, 20.
28. Marx, *Capital*, 165; see also Pérez-Reverte.
29. Baudrillard, *For a Critique of the Political Economy of the Sign*, 64.
30. Lévi-Strauss, "Selections from Introduction to the Work of Marcel Mauss," 55–6.
31. Derrida, *Positions*.
32. Žižek, *Looking Awry*, 7.
33. Hills, *Fan Cultures*, 137.
34. Gwen Cooper.
35. Gwen Cooper. Also, Gwen Cooper wouldn't advertise that she worked for a "secret agency" on MySpace. Not exactly secret anymore.
36. Jim.
37. For the sake of anonymity, I use an abbreviation here instead of the name on the MySpace page.
38. See Booth and Ekdale, "Translating the Hyperreal."
39. Turkle, *Life on the Screen*, 188.
40. Stone, *The War of Desire and Technology*, 18.
41. Kraus, "The Narrative Negotiation of Identity and Belonging," 108.
42. Agger, *The Virtual Self*, 110.
43. boyd and Ellison, "Social Network Sites," ¶10.
44. Bruns, *Blogs, Wikipedia, Second Life, and Beyond*, 47.
45. Fiske, "Cultural Economy of Fandom," 30–3.
46. Horton and Wohl, "Mass Communication and Parasocial Interaction," 299.
47. See Thorson and Rodgers, "Relationships between Blogs and eWOM and Interactivity, Perceived Interactivity, and Parasocial Interaction," 42–3 for an excellent literature review of parasocial theory.
48. Horton and Wohl, "Mass Communication and Parasocial Interaction," 215.
49. Thorson and Rodgers, "Relationships between Blogs and eWOM and Interactivity, Perceived Interactivity, and Parasocial Interaction," 46.
50. Rafaeli, "Interactivity, 124.
51. Giles, "Parasocial Interaction," 286.
52. de Certeau, *The Practice of Everyday Life*, 38, 100.
53. Hills, *Fan Cultures*, 39; see also Hills, "Strategies, Tactics and the Question of *Un Lieu Propre*."
54. de Certeau, *The Practice of Everyday Life*, 169, my emphasis, 174.
55. Jenkins, *Textual Poachers*, 23.
56. de Certeau, *The Practice of Everyday Life*, xxii, 166.
57. Jenkins, *Textual Poachers*, 24–7.
58. Harrington and Bielby, "Global Fandom/Global Fan Studies," 188.
59. Jenkins, *Textual Poachers*, 235.
60. Tulloch and Jenkins, *Science Fiction Audiences*, 233.
61. Sandvoss, *Fans*, 96–7.

62. Hills, *Fan Cultures*, 171; Sandvoss, *Fans*, 97. For Hills, impersonation can also entail the "constant management of fluid boundaries between self (cult impersonator) and other (icon)" (167). Hills uses the example of Elvis impersonators, who have to negotiate their identity as a fan of Elvis and their identity as Elvis. By impersonating Elvis, the fan inhabits the role of the musician, subsuming his/her own identity in order to complete the transformation.
63. boyd and Ellison, "Social Network Sites," ¶4.
64. Dena, "Patterns in Cross-Media Interaction Design," ¶1; see also, Walker, "Distributed Narratives"; Dena, "Player-Created Tiers."
65. Gwenllian-Jones, "The Sex Lives of Cult Television Characters," 84.
66. van Kokswijk, *Digital Ego*, 56.
67. Gilmoregirls3.
68. Donath, "Identity and Deception in the Virtual Community," 34; also Donath and boyd, "Public Displays of Connection," 77.
69. Burke, *A Rhetoric of Motives*, 21–2; also Crusius, "A Case for Kenneth Burke's Dialectic and Rhetoric."
70. Poster, *The Information Subject*, 78.
71. *Veronica Mars* CW.
72. boyd and Ellison, "Social Network Sites," ¶41.
73. *Veronica Mars* CW.
74. *Veronica Mars* CW.
75. boyd, "Why Youth (Heart) Social Network Sites," ¶14–15.
76. Ng, "Reading the Romance of Fan Cultural Production," 110.
77. Jenkins, *Textual Poachers*, 156.
78. van Kokswijk, *Digital Ego*, 74, emphasis mine.
79. Gwencooper.
80. Jenkins, *Convergence Culture*, 181.
81. Bolter, *Writing Space*, 116, writes before the popularity of social network sites, but his description of the dialogic nature of online communication applies to MySpace and other SNSs as well.
82. Soukup, "Hitching a Ride on a Star," 329.
83. Tulloch and Jenkins, *Science Fiction Audiences*, 233. There are Spock personas on MySpace: see closefriendofjamestkirk as an example.
84. Mortensen, "Me, the Other," 297–8.
85. Goffman, *Presentation of Self*.
86. Veronica Marz.
87. Veronica Marz.

Bibliography

Aarseth, Espen. "Nonlinearity and Literary Theory." In *Hyper/Text/Theory*, edited by George Landow, 51–86. Baltimore, MD: The Johns Hopkins University Press, 1994.

Agger, Ben. *The Virtual Self*. London: Blackwell, 2007.

Andrejevic, Mark. "Watching Television without Pity: The Productivity of Online Fans." *Television and New Media* 9, no. 1 (2008): 24–46.

Ascott, Roy. "Homo Telematicus in the Garden of A-Life—Editorial." *TightRope 1/95* (1994). Accessed 21 March 2016. http://www.geelhaar.de/TR/artikel/text/asc/index.html.

Baudrillard, Jean. *For a Critique of the Political Economy of the Sign*, translated by Charles Levin. St. Louis, MO: Telos Press, 1981.

Bolter, Jay David. *Writing Space: Computers, Hypertext, and the Remediation of Print*, 2nd ed. Mahwah, NJ: Lawrence Erlbaum Associates, 2001.

Booth, Paul, and Brian Ekdale. "Translating the Hyperreal (Or How *The Office* Came to America, Made Us Laugh, and Tricked Us into Accepting Hegemonic Bureaucracy)." In *American Remakes of British Television: Transformations and Mistranslations*, edited by Carlen Lavigne and Heather Marcovitch, 193–210. Lanham, MD: Lexington Books, 2011.

boyd, danah. "Why Youth (Heart) Social Network Sites: The Role of Networked Publics in Teenage Social Life." In *MacArthur Foundation Series on Digital Learning—Youth, Identity, and Digital Media Volume*, edited by David Buckingham. Cambridge, MA: The MIT Press, 2007. Accessed 22 March 2016. http://www.danah.org/papers/WhyYouth-Heart.pdf.

boyd, danah, and Nicole Ellison. "Social Network Sites: Definition, History, and Scholarship." *Journal of Computer-Mediated Communication* 13, no. 1 (2007). Accessed 22 March 2016. http://onlinelibrary.wiley.com/doi/10.1111/j.1083-6101.2007.00393.x/epdf.

Bruns, Axel. *Blogs, Wikipedia, Second Life, and Beyond*. New York: Peter Lang, 2008.

Burke, Kenneth. *A Rhetoric of Motives*. Berkeley: University of California Press, 1969.

Caldwell, John. "Convergence Television: Aggregating Form and Repurposing Content in the Culture of Conglomeration." In *Television after TV*, edited by Lynn Spigel and Jan Olsson, 41–74. Durham, NC: Duke University Press, 2004.

captain_j_harkness: Myspace.com/captain_j_harkness. (2008). <http://profile.myspace.com/index.cfm?fuseaction=user.viewprofile&friendid=119494970>

Closefriendofjamestkirk: MySpace.com/closefriendofjameskirk. (2008). <http://profile.myspace.com/index.cfm?fuseaction=user.viewprofile&friendID=310509123>

coffeeluvr4life: Myspace.com/coffeeluvrlife. (2007). <http://profile.myspace.com/index.cfm?fuseaction=user.viewprofile&friendid=110833970>

Crusius, Timothy, W. "A Case for Kenneth Burke's Dialectic and Rhetoric." *Philosophy and Rhetoric* 19, no. 1 (1986): 23–37.

de Certeau, Michel. *The Practice of Everyday Life*, translated by Steven Randall. Berkeley: University of California Press, 1984.

Dena, Christy. "Emerging Participatory Culture Practices: Player-Created Tiers in Alternate Reality Games." *Convergence: The International Journal of Research into New Media Technologies* 14, no. 1 (2008): 41–57.

Dena, Christy. "Patterns in Cross-Media Interaction Design: It's Much More than a URL … (Part 1)." *Crossmediaentertainment*, 2007. Accessed 21 March 2016. http://www.christydena.com/wp-content/uploads/2007/11/dena_morethan_part1.pdf.

Derrida, Jacques. *Positions*, translated by Alan Bass. Chicago: University of Chicago Press, 1981.

Doctor Who: Myspace.com/leerainford. (2009). <http://www.myspace.com/leerainford>

Donath, Judith. "Identity and Deception in the Virtual Community." In *Communities in Cyberspace*, edited by Marc A. Smith and Peter Kollock, 29–59. London: Routledge, 1999.

Donath, Judith, and danah boyd. "Public Displays of Connection." *BT Technology Journal* 22, no. 4 (2004): 71–82.

du Gay, Paul, Jessica Evans, and Peter Redman, eds. *Identity: A Reader*. London: Sage, 2000.

Durkheim, Emile. *The Elementary Forms of Religious Life*, translated by Karen Fields. New York: Free Press, 1995.

Facebook Newsroom—http://newsroom.fb.com/company-info/

Fiske, John. "The Cultural Economy of Fandom." In *The Adoring Audience: Fan Culture and Popular Media*, edited by Lisa A. Lewis, 30–49. London: Routledge, 1992.

Gauntlett, David. *Media, Gender and Identity: An Introduction*. London: Routledge, 2002.

Giles, David C. "Parasocial Interaction: A Review of the Literature and a Model for Future Research." *Media Psychology* 4 (2002): 279–305.

gilmoregirls3: Myspace.com/gilmoregirls3. (2007). <http://profile.myspace.com/index.cfm?fuseaction=user.viewprofile&friendid=193528050>

Godelier, Maurice. *The Enigma of the Gift*, translated by Nora Scott. Chicago: University of Chicago Press, 1999.

Goffman, Erving. *The Presentation of Self in Everyday Life*. New York: Doubleday, 1959.

GwenCooper: Myspace.com/Gwen. (2009). <http://profile.myspace.com/index.cfm?fuseaction=user.viewProfile&friendID=122033336>

Gwencooper: Myspace.com/gwencooper. (2008). <http://profile.myspace.com/index.cfm?fuseaction=user.viewprofile&friendid=90956481>

Gwenllian-Jones, Sara. "The Sex Lives of Cult Television Characters." *Screen* 43, no. 1 (2002): 79–90.

Gwenllian-Jones, Sara. "Virtual Reality and Cult Television." In *Cult Television*, edited by Sara Gwenllian-Jones and Roberta E. Pearson, 83–98. Minneapolis: University of Minnesota Press, 2004.

Hall, Stuart. "Introduction: Who Needs 'Identity'?" In *Questions of Cultural Identity*, edited by Stuart Hall and Paul du Gay, 1–17. London: Sage Publications, 1996.

Harrington, C. Lee, and Denise D. Bielby. "Global Fandom/Global Fan Studies." In *Fandom: Identities and Communities in a Mediated World*, edited by Jonathan Gray, Cornel Sandvoss, and C. Lee Harrington, 179–97. New York: New York University Press, 2007.

Hills, Matt. *Fan Cultures*. London: Routledge, 2002.

Hills, Matt. "Strategies, Tactics and the Question of *Un Lieu Propre*: What/Where Is 'Media Theory'?" *Social Semiotics* 14, no. 2 (2004): 133–49.

Horton, David, and Richard Wohl. "Mass Communication and Parasocial Interaction." *Psychiatry* 19 (1956): 215–29.

Jack Harkness™: Myspace.com/122039243 (2009). <http://profile.myspace.com/index.cfm?fuseaction=user.viewProfile&friendID=122039243>

Jenkins, Henry. *Convergence Culture: Where Old and New Media Collide*. New York: New York University Press, 2006.

Jenkins, Henry. "Lecture 06: Media Literacy as a Strategy for Combatting Moral Panic [sic]." *MIT OCW: CMS.930/21F.034 Media, Education, and the Marketplace, Fall 2001*. Video Podcast on iTunes University. Accessed 21 January 2008.

Jenkins, Henry. *Textual Poachers: Television Fans and Participatory Culture*. New York: Routledge, 1992.

Jim: Myspace/45053624 (2009). <http://profile.myspace.com/index.cfm?fuseaction=user.viewProfile&friendID=45053624>

Kraus, Wolfgang. "The Narrative Negotiation of Identity and Belonging." *Narrative Inquiry* 16, no. 1 (2006): 103–11.

Le Guern, Philippe. "Toward a Constructivist Approach to Media Cults," translated by Richard Crangle. In *Cult Television*, edited by Sara Gwenllian-Jones and Roberta E. Pearson, 3–26. Minneapolis: University of Minnesota Press, 2004.

Lévi-Strauss, Claude. "Selections from *Introduction to the Work of Marcel Mauss*," translated by Felicity Baker. In *The Logic of the Gift: Toward an Ethic of Generosity*, edited by Alan D. Schrift, 45–69. New York: Routledge, 1997.

Levinson, Paul. *New New Media*. Boston: Allyn & Bacon, 2009.

Marx, Karl. *Capital: A Critique of Political Economy*, translated by Ben Fowkes. London: Penguin Classics, 1990.

Mauss, Marcel. *The Gift: The Form and Reason for Exchange in Archaic Societies*, translated by W. D. Halls. New York: Norton, 1990.

Mortensen, Torvill Elvira. "Me, the Other." In *Second Person: Role-Playing and Story in Games and Playable Media*, edited by Pat Harrigan and Noah Wardrip-Fruin, 297–306. Cambridge, MA: The MIT Press, 2007.

Ng, Eve. "Reading the Romance of Fan Cultural Production: Music Videos of a Television Lesbian Couple." *Popular Communication* 6, no. 2 (2008): 103–21.

Pérez-Reverte, Arturo. *The Club Dumas*, translated by Sonia Soto. New York: Vintage, 1997.

Poster, Mark. *The Information Subject*. London: Taylor and Francis, 2001.

Rafaeli, Sheizof. "Interactivity: From New Media to Communication." In *Sage Annual Review of Communication Research: Advancing Communication Science: Merging Mass and Interpersonal Processes, 16*, edited by R. P. Hawkins, J. M. Wiemann, and S. Pingree, 110–34. Beverly Hills, CA: Sage, 1988.

Rutherford, Jonathan. *Identity: Community, Culture, Difference*. London: Lawrence & Wishart, 1998.

Sandvoss, Cornel. *Fans: The Mirror of Consumption*. Malden, MA: Polity Press, 2005.

Soukup, Charles. "Hitching a Ride on a Star: Celebrity, Fandom, and Identification on the World Wide Web." *Southern Communication Journal* 71, no. 4 (2006): 319–37.

Stone, Alluquere Rosanne. *The War of Desire and Technology at the Close of the Mechanical Age*. Cambridge, MA: The MIT Press, 1996.

Tulloch, John, and Henry Jenkins. *Science Fiction Audiences: Watching* Doctor Who *and* Star Trek. London: Routledge, 1995.

Turkle, Sherry. *Life on the Screen: Identity in the Age of the Internet*. New York: Simon and Schuster, 1995.

van Kokswijk, Jacob. *Digital Ego: Social and Legal Aspects of Virtual Identity*, rev. ed. Delft, The Netherlands: Eburon Academic Publishers, 2008.

Veronica Mars CW: Myspace.com/veronicamarscw. (2006). <http://profile.myspace.com/index.cfm?fuseaction=user.viewprofile&friendID=92251231>

Veronica Marz: Myspace.com/mizzmarzx. (2006). <http://profile.myspace.com/index.cfm?fuseaction=user.viewprofile&friendid=130922322>

Veronica_Mars: Myspace.com/veronica_mars. (2006). <http://profile.myspace.com/index.cfm?fuseaction=user.viewprofile&friendid=14789920>

Walker, Jill. "Distributed Narratives: Telling Stories across Networks." Paper presented at AoIR 5.0, Brighton, 21 September 2004. Accessed 23 March 2016. http://jilltxt.net/txt/Walker-AoIR-3500words.pdf.

Zachandstaceforever:Myspace.com/zachandstaceforever. (2007). http://profile.myspace.com/index.cfm?fuseaction=user.viewprofile&friendid=72513851

Žižek, Slavoj. *Looking Awry: An Introduction to Jacques Lacan through Popular Culture*. Cambridge, MA: The MIT Press: 1991.

· 5 ·

DIGITAL FANDOM, ALTERNATE REALITY GAMES, AND DEMEDIATION

In essence, games are the only universally serious activity. They leave no room for skepticism, wouldn't you agree?

—Pérez-Reverte, *The Club Dumas* (p. 314)

These are the [books] I love best. They shine above the rest for their beauty, for the love they have inspired. These are the ones I walk hand in hand with to the brink of the abyss. ... Life may strip me of all I have. But it won't turn me into a miserable wretch.

—Pérez-Reverte, *The Club Dumas* (p. 152)

As Pérez-Reverte's protagonist Corso progresses through *The Club Dumas*, the events he encounters seem constructed to confound him, as if he were a player unwittingly led to play an artificially constructed game. Corso believes that what he could consider "reality" and what he sees as the "ludicity," or game-like characteristics of the plot machinations of the collectors, have mashed up and have become inseparable from one another. Corso's confusion of his reality for a game represents an ontological break, a tension between the ludic and the non-ludic that we can see represented in the first quotation that inscribes this chapter: Games are serious. For the book aficionados, this game is serious business, not ludic escapism.

Games are serious for New Media Studies too, and in this chapter I examine Alternate Reality Games (ARGs), perhaps the ultimate expression of New Media. ARG players are not necessarily fans, and fans are not necessarily ARG players, but digital fandom and ARGs are related by the transgression of boundaries that both the fan and the ARG enact. ARGs, as we will see, transgress a number of boundaries, the most salient of which is that between the mediated and the non-mediated. Fans, similarly, transgress the line between production and consumption, and importantly, alter our interpretation of that boundary. The transgressive nature of both fans and ARGs highlights a connective thread between them. In this conclusion, I want to examine how the ubiquitous mediation of ARGs functions within an already-saturated media society, through a paradoxical demediation of media technology. Further, I want to show that the philosophy of playfulness encountered within fandom has larger ramifications across the media environment.

The mechanics of the ARG represent a shift in our understanding of key media studies concepts, as participants in ARGs often work long hours doing repetitive and often tedious tasks, yet consider this to be play within these games. Play becomes normal. By looking at these shifts through the lens of fan studies, we not only see an explicit version of New Media Studies but also an implicit idea of how the mechanics of an ARG structure and affirm new concepts while destabilizing traditional ones.

An ARG is a game-like narrative played both off- and online, which uses multiple modes of mediation to immerse the player in the game's play and narrative. Players can participate in these narratives by roleplaying and acting with non-players in the "real world." The form of the ARG is "defined by intense player involvement with a story that takes place in real time and evolves according to player's responses. Subsequently, it is shaped by characters that are actively controlled by the game's designers."[1] These designers are called "Puppetmasters," as they metaphorically represent people "pulling the strings" of the players' actions. It is near impossible to collate the complete "text" of an ARG, because so much of it depends on user involvement, unlike a film or a website, which can often be studied in and of itself as a singular text. ARGs are similar to transmediated texts, in that multiple media outlets cohere intra-textuality to create one narrative. When playing an ARG, players receive clues disguised as these outlets—phone calls, e-mails, text messages, or other modes of mediated communication.[2] Developed by the Puppetmasters, these communiqués mediate the game.[3] Not only do the communications mediate, but the mediation also communicates, for the ARG can only be seen through this mediation.

Because of this constant mediation, ARGs offer us a mirror with which we can view the digital inebriation of our contemporary mediated culture.

However, just as digital fandom offers a new way of analyzing cult media fans, so too does it provide us with a new way of analyzing larger cultural changes related to fandom and the media. How do we analyze what happens when the playing stops and the game becomes real, when the game itself is happening in "real" life? The ARG found its way into popular media before it became realized in ludic form: For example, in the 1980 Disney film *Midnight Madness*, a technologically savvy college student invites a number of students to a clandestine meeting where he informs them that they have each been picked to participate in a "real-world" game. Each team scrambles across Los Angeles at night to solve puzzles and complete tasks. As they traverse the city, the participants of the game interact with real-world environments, in order to fulfill this ludic goal. Importantly, the film *Midnight Madness* portends a conceptual blend of the ludic with the non-ludic, as characters encounter "real-world" people not part of the game that become embroiled within the "game-world." Further, David Fincher's 1997 aptly titled *The Game* depicts the protagonist thrown into a series of events that seem to be random, but are in fact different tasks he must perform in order to proceed to the "next level" of the reality game. For the heroes of *Midnight Madness* and of *The Game*, much like for Corso in *The Club Dumas*, part of the challenge of the narrative is not just to navigate the events, but also to determine the ontological significance of those events: Are they part of their non-ludic world or part of the ludic world? Is life just, as Pérez-Reverte states, a giant intertextual game? Or does the game represent particularly salient moments of everyday life?

Media/Technology

The ARG is a manifestation of the "technologization" of our society, a massive representation of a mediated population, and thus can act as a microcosm of the intersection between work and play in contemporary fandom.[4] It is difficult to determine the popularity or the exact numbers of players who play ARGs, for games can take place over months and players are often dispersed around the globe. New media scholar Christy Dena has made an admirable attempt at compiling statistics about the most widely played ARGs, and her statistics come from a variety of sources, both academic and popular.[5] Still, true numbers are hard to come by, as the way ARGs are played limits the ability to tell who might be involved at any minor level as full-time players or as mere lurkers. As Dena also suggests, play can take many forms, and players can come and go from the game without consequence. She describes different levels of commitment to the game, and defines two types of players: the casual players, who simply follow the game's unfolding on webpages and wikis, and the hard-core players, who are involved with actually solving the puzzles.[6] Thus, while Dave Szulborski estimates that 10,000 people worked collectively to solve the first ARG, *The Beast*, Elan Lee, one creator of *The Beast*, estimates the total number of users of that game as "somewhere in the neighborhood of 2.5 million," with 100,000 people playing in the first three hours.[7] For hundreds of thousands of people, playing an ARG is an important, and time-consuming, passion.

The ARG is a mediated reality that mashes up, but does not blur, the line between the non-ludic and the ludic. While there is no clear distinction between the two, it is not that a player of an ARG exists either in or not in the non-ludic or the ludic, but rather is perpetually in-between them. ARGs mash up the concepts of a ludic world, where the rules of play are discrete and unique, with the world as non-ludic, where gameplay does not affect a player's mental construct of the world.[8] Players work at play in an ARG. For players, this mash-up occurs as they take the game seriously. Jane McGonigal believes that part of the reason people want to play an ARG is to lose their sense of reality, to surrender control to the Puppetmasters who determine the in-game missions.[9] The ludic regulations of ARGs, the rules and policies set down by the Puppetmasters who control what the players can and cannot know, become virtual codes and mores enacted by the players in the game. These virtual codes mirror the "magic circle" hypothesized by Johan Huizinga, the "temporary worlds within the ordinary world, dedicated to the performance of an act apart," or the representation of the playspaces where game rules supersede rules by which society is organized.[10] As Marcus Montola persuasively illustrates, an ARG "is a game that has one or more salient features that expand the contractual magic circle of play socially, spatially or temporally."[11] The magic circle of the ARG, in other words, expands its virtual codes to include the entire "real" world. For players of ARGs, the phenomenological experience of the game can subsume their experiences outside the game.

Specifically, Alternate Reality Games utilize ubiquitous web technology to help players participate in a game that is both constructed through and effaced by mediation. On the one hand, these games are constructed though what Jay David Bolter and Richard Grusin would describe as hypermediated media content: ARG "texts" can only be experienced through mediation.[12] Because elements of the ARG are spread out across different spatial environments and temporal locations, they literally exist in many locations at different times. They are ubiquitous, multitudinous, and global. On the other hand, despite the overtness of their mediation, ARG elements are also effaced by the very mediation they construct, and attempt to immerse the player in an immediate environment by appearing to be non-mediated. As McGonigal has shown, the Puppetmasters attempt to make the players forget that they are playing a game: the ARG exists between "what [players] know to be feigned with what they feel to be real."[13] The mediation used in ARGs—the e-mail, instant messages, phone calls, websites—are ubiquitous technologies, and create what Bolter and Grusin would describe as an immediate experience. They disappear, fade into the background of the game, and exist in order that players, who are used to such media technology, can easily forget the mediated nature of these in-game missives. In effect, the media mechanisms of the game allow players to forget that it is mediated, by mashing up the *ludic* into a *non-ludic* environment.

As Schneider and Kortuem describe, the apparent tension between technology and mediation has been conceptualized as a tension within the ARG (they refer to it as a persuasive game):

> technology is not the focus of the game but rather the technology supports the game. Although technology is ubiquitous in a Pervasive Game, its role is a supporting one and thus the technology is kept as unobtrusive as possible.[11]

This dichotomy between the ubiquity and the unobtrusiveness of media and technology forces a reconceptualization of our understanding of contemporary mediation and technology. Specifically, the ARG problematizes Bolter and Grusin's examination of remediation as an outgrowth of traditional media studies. They claim that every media technology depends on previous media technology for its content. I argue that key components of New Media they elide, including interactivity, dialogue, and community, are equally important.[15] Bolter and Grusin neglect to investigate how the user's experiences of mediation—so crucial for an ARG—can change with the use of different online media technologies. Specifically, an ARG highlights a reality for players as the mediation of the game disappears in what I'm calling a *demediation* of digital culture.

I am defining demediation as a state where the ubiquity of mediation hides that selfsame mediation. In other words, demediation is what happens when the obviousness of mediation becomes so overt it hides mediation. This concept lies in contrast to Baudrillard's simulacra, for demediation does not, in fact, itself hide the real.[16] Instead, despite the appearance that they blur the boundaries between the ludic and the non-ludic, ARGs actually reinforce that boundary; and in doing so, they also reinforce the constituent elements of the ludic/non-ludic binary, and provide a reassurance and strengthening of what contemporary media studies might call the unmediated reality of physical existence.

Remediation/Demediation

ARGs exist in a "demediated" reality, a Derridean "undecidable" position that mashes up the hypermediate and immediate conceptions of mediation.[17] For Derrida, an undecidable concept cannot exist at either end of a binary or a dialectic. For example, in *Specters of Marx*, Derrida describes the concept of the ghost as existent in an undecidable state between living and dead, two normally mutually exclusive polarities.[18] The undecidability of demediation lies in opposition to Bolter and Grusin's decidable definition of remediation, which in contrast underscores the dialectic of hypermediacy and immediacy. Mediation, as Bolter and Grusin write, is contradictory and impossibly alternative: "our culture wants both to multiply its media and to erase all traces of mediation: ideally, it wants to erase its media in the very act of multiplying them."[19]

At the heart of their argument lies the paradoxical notion that media both represent and replace human communication. We assume that media exist in this constant tension, and we must always negotiate between our looking at media (hypermediacy) and looking through media (immediacy). The ARG exists in a position inside this tension, between hypermediacy and immediacy. This position, the demediate, represents a change in our conception of reality as both mediated by and detached from technology.[20] As such, the demediacy of ARG becomes a potent metaphor for New Media Studies as well, as the complete integration of all media technologies into one media environment heralds an inescapable connection between media and technology in contemporary society.

Immediacy is the logic of erasure, as Murray describes.[21] We expect our media to disappear; we want to become enveloped in the moment presented by the media. "Immediacy is supposed to make [the] computer interface 'natural' rather than arbitrary" and will "erase itself, so that the user would no longer be aware of confronting a medium."[22] In this, the nature of immediacy

provides an interface through which the viewer is supposed to have an unhindered view of the "real" world. For example, when a film viewer momentarily forgets she is in the theater and becomes "lost" in the screen, enveloped and absorbed by the mediation to the point that the mediation replaces reality for that viewer, she is immersed in an immediate environment.

Conversely, hypermediacy is the logic of overt mediation. According to Bolter and Grusin, "hypermediacy offers a heterogeneous space, in which representation is conceived of not as a window on the world, but rather as 'windowed' itself—with windows that open on to other representations or other media."[23] An appreciation of the cinematography of a film, or the acknowledgment of the composition of chiaroscuro in a photograph, are moments of hypermediacy, where the representation of the media itself becomes more visible than the "reality" that media attempts to portray. Hypermediacy is the state of the obviousness, of extreme mediation. When we admire the special effects in a film—effectively losing our immersion and appreciative of the craft—we become enamored with hypermediacy.

Bolter and Grusin argue that the resolution of this tension between immediacy and hypermediacy is remediation, a "double logic" that both erases and makes visible the act of mediation.[24] Remediation is the reflection of one type of media content in another medium, the re-presentation of a medium. The key to Bolter and Grusin's argument lies in their citation of Marshall McLuhan's assertion that "the content of any medium is always another medium."[25] Rather than the obvious denotative meaning of his statement, that a medium literally displays another and repurposes it, Bolter and Grusin show that McLuhan was thinking of "a more complex kind of borrowing." Specifically, they argue that digital technologies are today's quintessential remediated devices, which constantly reflect the media of the television, film, photography, and painting. This reflection has two consequences. First, "the digital medium wants to erase itself, so that the viewer stands in the same relationship to the content as she would if she were confronting the original medium." Digital technology, therefore, tries to "disappear," to become immediate. Conversely, "the very act of remediation … ensures that the older medium cannot be entirely effaced; the new medium remains dependent on the older one."[26] In this way, digital technology represents both the immersive experience of mediation and the obvious of the act of that mediation as the overt "repurposing" of one medium in another: "[E]very medium includes its predecessors within itself even as it appears new and improved."[27]

Their process of contemporary (re)mediation, however, evolves from more than just the dichotomy between hypermediacy and immediacy. It actually centers on three distinct tensions, each of which emerges from the combination of three apparatuses: a medium, media content, and technology. As media become more complex, complications occur in the way these three apparatuses overlap as an audience's participation with both the tools and the presentation of media content complicates mediation. For example, a tension becomes apparent in the intersection of the medium and the media content, as we can recognize media and media content either as objects themselves or as representations of real objects in the world. A photograph, for example, is both a tangible artifact and a representation of a real object. In this way, media always present an absence: The object in a photograph is not that object, but rather a representation of the absence of that object.

A second tension occurs when a form of technology physically intersects with a medium used to present it. We can see this tension most saliently in print, as the medium must be both present and absent in order for the technology to create the media content. If a medium, in this case the paper (or papyrus, or other flat surface), is marked by a technology—pen and ink, pencil and graphite, quill and ink, etc.—then etchings on the surface of the medium leave content in the form of letters, symbols, pictograms, or other forms of symbolic communication. In order to interpret these symbols, a reader must construct meaning from the tension between the presence of the ink and the absence of ink—or, the visibility of the technology and the disappearance of the technology. It is equally important where one does not write, as it is where one does. The markings on a paper are noticeable because they are not uniform, and because they illustrate an absence of blank space. In the digital too we can see this presence/absence most literally in the code used to program. One of the constituent aspects of digital writing lies in its binary base of 1/0 and the constituent metaphors this binary represents: yes/ no, on/off, presence/absence.

A third tension, between the media content and the technology, occurs when technology develops as both a tool for producing media content and also a means for displaying media content. For example, the computer is both the technology and also the medium through which one views the media content.[28] As Jenkins writes, the computer has ushered in a culture of media convergence where users can both consume and produce media on the same device and "participation is understood as part of the normal ways that media operate."[29] Such tension is noted by W. J. T. Mitchell, who summarizes Barthes's description of photography by stating, "Everyone knows a picture of their mother is not alive, but they will still be reluctant to deface or destroy it."[30] Even though photography itself is a technology, it's also content marked with additional symbolic meaning from object it represents.

The tensions among the medium, the content, and the technology illustrate one particularly salient absence in Bolter and Grusin's *Remediation*: the effect of the social/cultural on mediation. Although they discuss in brief how the social and cultural logic of mediation, what they call the "authenticity of experience," can affect our interpretation of media, they spend little of their book articulating these authenticities. They propose to "explore digital technologies … as hybrids of technical, material, social and economic factors." Yet, their basic thesis remains deterministic: As they demonstrate, media are always based on other media as "each act of mediation depends on other acts of mediation. Media need each other in order to function as media at all."[31] This determinism limits the scope of their analysis by eliminating the user from the process of mediation, the unique element of each medium that forms a user's experiences. As Jensen states, "the theory of remediation tends to choose sides, inviting an analytical gaze at the surface of the interface, *bracketing technologies, users and social contexts*."[32] For example, a unique property of the televisual medium comes not from any individual element onscreen, but from the "flow" of programming, the progression of television programming, commercials and station identifications that creates a new type of text. Flow is not just "hypermediated television, but also the result of the viewer's *experience* of the medium."[33]

In contrast, and pertinent to contemporary digital fandom, ARGs exist solely as user experiences and not entirely as remediations. To take part in an ARG is to experience an intangible, untenable, ethereal media event. As we have seen, there is no "text" of an ARG; Barthes's definition of "text" is apropos here. Barthes defines the "text" as existent between works, creating an intertextual web of meaning.[34] Because elements of the ARG are spread out across different spatial environments and temporal locations, they literally exist in many locations at different times. Only through the immediate as well as the hypermediate experiences of the users/players of ARGs can the ARG "text" itself be known, encountered, and played.

The Demediation of the ARG

As a perceptible example of this inherent fallacy of remediation's double logic, Alternate Reality Games challenge and overturn the binary division between hypermediacy and immediacy. The media used in ARGs represent breaks in the logic of remediation, as the poles of immediacy and hypermediacy have been flipped and the binary between the two has been elided. In fact, reality-blurring games, as Sherry Turkle points out (although her analysis pre-dates the ARG), "stand betwixt and between the unreal and the real; they are a game and something more."[35] She calls this a form of "blurred genre," which underscores the way ARGs overturn enact two contradictory tenets: They do not so much remediate as they do demediate, effacing the media as they conversely rely on them.

As McGonigal describes, ARGs are "digitally distributed" across different media technologies, and are only visible through the obviousness of their mediation. They are hypermediated because they do happen entirely because of mediation. By existing solely through mediated technology, ARGs make that mediation the basis of their ontological state of being. At the same time, ARGs attempt to hide through mediation, to use "natural settings as the immersive framework."[36] To experience the mediation of an ARG is to become aware of its contextual use in real life. If clues come in a mediated form, they mirror real-life communication. ARGs are immediate because they thus seem to happen outside of mediation. Players of ARGs use everyday technology to interact with the game: This technology becomes part of the game—part of the immersion. By attempting to perform both sides of this dichotomy, ARGs participate in a particular sense of hyper-immersion, as the complete ubiquity of media technology envelops the players just as it becomes invisible.

The ARG thus uses media to hide its mediation; as such, the ARG seems immersive, but effaces the mediation it necessarily presents. ARGs are not just dependent on media, they are media. The only way we could experience mediation as mediation is if there is something outside of that mediation. As philosopher Slavoj Žižek shows, "what is crucial for us here is … the very borderline separating the outside from the inside."[37] The total effacement of one end of a dialectical polarity serves only to sever the knowledge of that polarity itself from humankind. For example, Žižek later notes that if humankind were to be completely mechanized, we would lose the ability to be able to tell machine from flesh: The complete effacement of one pole (the "flesh" of humankind) standardizes identity and "the human being will gradually lose its grounding in the concrete life-world—that is to say, the basic set of coordinates which determine its (self-) experience."[38] To become completely engulfed in a mediated world, to extrapolate, is to lose the ability to tell mediation from reality.

ARGs take place at the level of mediation: There is no ARG outside of that which is mediated. Thus, media encompass the ARG. If there is no ARG outside of media, then there can be no perception of the ARG without media, and media anneal our understanding of what the ARG actually is, or can be. Further, once puzzles in ARGs are solved, they can be mediated again on wikis, which house all the information collected about the game.[39] Even so-called "real" world events, non-ludic stagings that take place in the "real" world, become mediatized as they are re-broadcast to the rest of the players on these encyclopedic wikis. This hyper-immersion of ARG content hides the "event" of the ARG, which becomes a compilation of mediations collected elsewhere.[40]

ARGs, because of this blend of immediacy and hypermediacy, inhabit an interreal space that has characteristics that can be described as both immersive and as obvious. In an ARG, the awareness of the media serves only to reinforce the immediacy of the experience. In real life, one is constantly surrounded by media. E-mail routinely interrupts a workspace. A phone call will disrupt a face-to-face conversation. Although there is still mediation, it has become so common that to not have mediation would be bizarre: the ARG highlights the hypermediacy as an immediate experience. It is important to note this elision, for as Žižek demonstrates, "the simple direct reference to external reality … as well as the opposite attitude of 'there is no external reality,'" are both totalizing and limiting determinations. What is necessary for Žižek is a space in which external reality and the mediation of that reality can coexist "without lack and obstacles."[41]

The ARG and demediation indicate a shift to the reestablishment of a non-mediated reality. By playing games, and externalizing the mechanisms by which we recognize the game as game, the ARG, in fact, reveals the underlying non-ludicity of the situation. Players of ARGs experience the completely mediated, and in doing so, reestablish the phenomenological experience of the "real world" outside the ARG. By uncovering the ludic, we can recognize the non-ludic, and demonstrate the reestablishment of reality itself. Because the ARG reinforces the boundary between the constituent binaries, they must be both ludic and non-ludic at the same time. Thus, by reinforcing the boundary, the ARG also reinforces each side of that boundary: As Žižek might say, because we have a window, we know there is both an inside and an outside. Players of ARGs cause the "game world" to seem to disappear into the non-ludic "window" of the "real world," and the experiences of the players playing the game become nothing more—or less—than participation in a form of real life. Because the ARG reinforces the boundary, it also reinforces a divide between the ludic and the non-ludic, and the existence of both the media and the absolutely unmediated. In the next chapter, I will explore this demediation in greater detail by discussing the live-action game *MagiQuest*.

ARGs, therefore, represent an exemplar of a New Media Studies paradigm: instead of focusing on "media" as separated from the reality of daily life, New Media Studies assumes media as integral. The further we advance the less important the boundary between the two becomes. By demediating reality through the ARG, the Puppetmasters and other ARG creators construct an experience for players in which they have a greater ability to control their use of media. When a mediated message is used in an ARG, the confusion over its "ludism" (is this e-mail part of the game or not?) both detracts and enhances the player's power over the media. The use of media in an ARG is so ubiquitous, it both mirrors and exaggerates the use of media in non-ludic life. Just as we do not think about the ubiquity of the "real world," so too do ARGs cause us to forget their own mediation. Although McGonigal argues that one reason users play ARGs is for the passivity of the game to overtake their decision-making in the non-ludic world, the opposite may also be true: When a mediated message is used in an ARG, the confusion over its "realism" enhances a player's power over the media.[42] Players choose whether to accept the e-mail, even whether or not to play the game at all. In effect, ARGs assuage the passivity of postmodern simulation.

The unique power of the ARG to exist solely through the player's experiences of the media separates it from Bolter and Grusin's concept of remediation. By displaying its media, the ARG becomes more opaque: the more the media we use to play become obvious, the more we cannot know if we are playing. This hypermediate immediacy is countered by an immediate hypermediacy. The more obvious the mediation in an ARG, the more immediate the experience. Paradoxically, to make the media more obvious is to make the game less so.

Demediation and Digital Fandom

I conclude this chapter with reference to the final quotation from Pérez-Reverte inscribing this chapter as a way of relating ARGs and digital fandom. In it, he reflects for the book collectors in *The Club Dumas* the same type of passion that fans hold for the extant media object of their affection: "These are the ones I walk hand in hand with to the brink of the abyss …" Despite all the travails, all the horrors and pain he experiences throughout the novel, Pérez-Reverte's hero concludes his journey by reasserting his deep affection for the antique books he collects and trades: He takes them to his grave for eternal contemplation.

In much the same way, we have experienced the way players of Alternate Reality Games embrace the mediation, remediation, and demediation of contemporary media culture. Similarly, by exploring new concepts in media studies, digital fandom does not examine a new type of fan, but rather defines a new way of looking at fans. As Christy Dena writes, some media content created by an audience's reaction to an extant media object has actually become

> the main product of consumption for mass audiences. Although observable in many forms of cultural production, this phenomenon is interrogated specifically through the genre of alternate reality games.[43]

ARGs present cult worlds *as* the real world: not just cult worlds fans can watch and integrate into their lives, but also cult worlds fans can actually inhabit and populate. ARGs exist between and betwixt the ludic and the non-ludic: Indeed, as pointed out by Gosney, this divide between the ludic and the non-ludic is already eroding.[44]

ARGs are creeping into our everyday lives in ways both familiar and unique, as ARGs are gaining widespread popularity. For instance, a National Public Radio newscast described how some museums, in order to draw in visitors, are constructing Alternate Reality Games around works of art, culture, or history in the museum.[45] The curators construct a narrative around many of the objects and artworks within a collection, and when visitors come to the museum, they can try and "solve" this real-world puzzle in the space of the exhibit. Entire genres of ARGs exist that are neither based on extant media objects nor are viral marketing schemes. Instead, these new games, called "games for social change," or "serious games," articulate experiments in socially active and civically minded participants. For instance, the game *A World without Oil* detailed a fictitious 32 weeks of a global oil crisis.[46] Participants were asked to spend each of 32 weeks acting as if the world was running out of petroleum. During the first week, the game asked participants to pretend gasoline had risen above $4.00 a gallon.[47] By week 16, the in-game gas prices had risen to over $7.00 a gallon and participants were told that riots were taking place. By demonstrating a possible future, these games become a way of exploring contemporary issues with a more ludic sensibility. A more recent serious game, *Superstruct*, used the idea of *A World without Oil* but forecasted a more general future of the world as a whole.[48] The game asked participants to imagine what would happen in the year 2019 if many of the nascent social ills of today had been magnified and expounded upon. The game asks, to what degree can we change our present by investigating our future? By using Web Commons technologies like blogs, wikis, social network sites, and videos, the Puppetmasters of *Superstruct* made the game—and our reality itself—hyper-immersive.[49]

As ARGs make their way into the mainstream, more participants will start to play. Whether they be serious games, narrative add-ons like *Why So Serious*, viral advertisements like *ilovebees*, or interactive narrative elements like *The Lost Experience*, these Alternate Reality Games will provide players with active and powerful communities: This form of communal creation, as we have seen, is paramount to fans in general and digital fandom more specifically. We can see this change illustrated persuasively in Jane McGonigal's description of the Cloudmakers. This group of online fans, who met online to try and solve the Alternate Reality Game *The Beast*, were puzzling over the game on September 11, 2001, when terrorists piloted planes into the World Trade Center buildings. McGonigal describes how the Cloudmakers' conversations at first mirrored many others at that time: They exchanged words of hope and inspiration. However, the online interaction between the people on the listserv quickly evolved into something else: Wanting to parlay their expertise at solving online Alternate Reality Games, the Cloudmakers decided to "solve the puzzle of who the terrorists are."[50] As one participant wrote, "Let's become a resource. Utilize your computer & analytical talents."[51] The online group formed an online collective using the communal capabilities of the Web Commons to solve a real-world issue. The Web Commons fostered the interactions between the group members, and became a way for the group to parlay their specific talents in a real-world situation.[52] What ARGs offer, therefore, is not just a way of examining reality, or a way of playing a game, but also a means through which we can see how communities form and populate the Web Commons.

ARGs represent a demediate arena of culture, where mediation becomes a way of life. Fans embrace this notion: Through entering the hyperdiegesis of the cult worlds to which they attach their fandom, fans accept this dualism. It is with a philosophy of playfulness that participants in New Media enact the mediated communities they join. The examples in this book—ludic comments on blog posts; playful narrative progressions on wikis; roleplay on social network sites—are just those: examples that represent a larger trend. Many more are out there, and in the following new chapters, I explore Tumblr, Twitter, crowdfunding, and fan-created spaces.[53]

Notes

1. "Alternate Reality Game." Wikipedia is one of the better sources for definitional work on ARGs, because the players of ARGs utilize online Web Commons tools like Wikipedia and Google, to interactively work in a Digi-gratis economy to solve the puzzles in the games for Alternate Reality Games. See Jenkins, *Convergence Culture*, 127.
2. McGonigal, "I Love Bees."
3. Szulborski, *This Is Not a Game*, 266.
4. Ong, *Orality and Literacy*, 80–2. His description of "technologization" describes the way technology shapes consciousness and culture.
5. Dena, "ARG Stats."
6. Dena, "Emerging Participatory Culture Practices," 43.
7. Szulborski, *This Is Not a Game*, 98; Meadows, *Pause and Effect*, 136. Over 600,000 people participated in the "real world" excursions promoted by the ARG *ilovebees*, but Szulborski doubts this number, and attributes a minor rise in popularity due to gamers' excitement over the release of *Halo 2*, which *ilovebees* was created to promote, see McGonigal, "I Love Bees," ¶18; Szulborski, *This Is Not a Game*, 166.
8. Eskelinen, "The Gaming Situation," ¶2
9. McGonigal, "The Puppetmaster Problem," 262.
10. Huizinga, *Homo Ludens*, 10.
11. Montola, "Exploring the Edge of the Magic Circle," sec. 3.
12. Bolter and Grusin, *Remediation*, 53–4.
13. McGonigal, "A Real Little Game," ¶44.
14. Schneider and Kortuem, "How to Host a Pervasive Game," 2.
15. Bolter and Grusin, *Remediation*, 28, 45–48; Murray, *Hamlet on the Holodeck*, 71, shows how interactivity is a key aspect of New Media; Jenkins, *Convergence Culture*, 18–19 describes the dialogic quality of New Media; Smith and Kollock, ed., *Communities in Cyberspace*, look at online community.
16. Baudrillard, *Simulacra and Simulation*; Baudrillard, "Aesthetic Illusion and Virtual Reality," 199.
17. Derrida, "Afterward," 116.
18. Derrida, *Specters of Marx*, 5.
19. Bolter and Grusin, *Remediation*, 5.
20. Bolter and Grusin, *Remediation*, 28; also, Gunkel, *Thinking Otherwise*, 27–9.
21. Murray, *Hamlet on the Holodeck*, 98–9.
22. Bolter and Grusin, *Remediation*, 23–4.
23. Bolter and Grusin, *Remediation*, 34.
24. Bolter and Grusin, *Remediation*, 5, 45.
25. McLuhan, *The Medium Is the Message*, 23–4.
26. Bolter and Grusin, *Remediation*, 45–50.
27. Allen, "Telling Our Stories in New Ways," 192.
28. Manovich, *The Language of New Media*, 64–5.
29. Jenkins, *Convergence Culture*, 246.
30. Mitchell, *What Do Pictures Want?* 31.

31. Bolter and Grusin, *Remediation*, 71, 77, 55.
32. Jensen, "Mixed Media," 11, my emphasis.
33. Williams, *Television*, 93; Thompson, *Storytelling in Film and Television*, 8, disagrees with the importance placed on "flow."
34. Barthes, "From Work to Text," 156–8.
35. Turkle, *Life on the Screen*, 188.
36. McGonigal, "'This Is Not a Game,'" 2.
37. Žižek, *Looking Awry*, 15.
38. Žižek, "Cyberspace," 135.
39. Dena, "Player-Created Tiers."
40. What Boorstin, *The Image*, 9–44, might call "pseudo-events."
41. Žižek, "Cyberspace," 132–3.
42. McGonigal, "The Puppetmaster Problem," 262.
43. Dena, "Player-Created Tiers," 42.
44. Gosney, *Beyond Reality*, xii.
45. Blair, "Interactive Games Make Museums a Place to Play."
46. http://worldwithoutoil.org/.
47. Ironically, the next year it did.
48. http://superstructgame.org.
49. McGonigal, *Reality Is Broken*.
50. McGonigal, "'This Is Not a Game,'" 1.
51. Quoted in McGonigal, "'This Is Not a Game,'" 1.
52. The Cloudmakers stopped after two days, however, for fear of appearing as if they were not taking the tragedy seriously.
53. Sadly, my hypothesis that Google Wave mash-up video/text/audio amalgams would become big was disproven quite quickly after publication of the first edition.

Bibliography

Allen, Nancy. "Telling Our Stories in New Ways." Review of *Remediation*, by Jay David Bolter and Richard Grusin, and *The Language of New Media*, by Lev Manovich. *Computers and Composition* 18 (2001): 187–94.

"Alternate Reality Game." *Wikipedia*. Accessed 21 March 2016. http://en.wikipedia.org/wiki/Alternate_reality_game.

Barthes, Roland. "From Work to Text." In *Image-Music-Text*, translated by Stephen Heath, 155–64. New York: Hill and Wang, 1977.

Baudrillard, Jean. "Aesthetic Illusion and Virtual Reality." In *Reading Images*, edited by Julia Thomas, 198–206. Houndsmills, UK: Palgrave, 2000.

Baudrillard, Jean. *Simulacra and Simulation*, translated by Sheila Faria Glaser. Ann Arbor: University of Michigan Press, 1994.

Blair, Elizabeth. "Interactive Games Make Museums a Place to Play." *NPR*, 12 January 2009. Accessed 21 March 2016. http://www.npr.org/templates/story/story.php?storyId=99244253.

Bolter, Jay David, and Richard Grusin. *Remediation*. Cambridge, MA: The MIT Press, 1999.

Boorstin, Daniel. *The Image*. New York: Random House, 1961.

Dena, Christy. "ARG Stats." *Universe Creation 101*, 08 December 2008. Accessed 21 March 2016. http://www.christydena.com/online-essays/arg-stats/.

Dena, Christy. "Emerging Participatory Culture Practices: Player-Created Tiers in Alternate Reality Games." *Convergence: The International Journal of Research into New Media Technologies* 14, no. 1 (2008): 41–57.

Dena, Christy. "Patterns in Cross-Media Interaction Design: It's Much More than a URL … (Part 1)." *Crossmediaentertainment*, 2007. Accessed 21 March 2016. http://www.christydena.com/wp-content/uploads/2007/11/dena_morethan_part1.pdf.

Derrida, Jacques. "Afterward: Toward an Ethic of Discussion." In *Limited Inc.*, translated by Samuel Webber, 111–54. Evanston, IL: Northwestern University Press, 1988.

Derrida, Jacques. *Specters of Marx: The State of the Debt, the Work of Mourning, and the New International*, translated by Peggy Kamuf. London: Routledge, 1994.

Eskelinen, Markku. "The Gaming Situation." *Game Studies 0101* 1 no. 1 (2001). Accessed 23 March 2016. http://gamestudies.org/0101/eskelinen/.

Gosney, John W. *Beyond Reality: A Guide to Alternate Reality Gaming*. Boston: Thompson, 2005.

Gunkel, David. *Thinking Otherwise*. West Lafayette, IN: Purdue University Press, 2007.

Huizinga, Johan. *Homo Ludens: A Study of Play Element in Culture*. Boston: Beacon Press, 1955.

Hutchens, B. "Great Wolf Lodge Resort Much More Than Just a Water Park." *News Tribune*, 2008. Accessed 26 February 2011. Link no longer available.

James, Carrie, with Katie Davis, Andrea Flores, John M. Francis, Lindsay Pettingill, Margaret Rundle, and Howard Gardner. *Young People, Ethics, and the New Digital Media: A Synthesis from the GoodPlay Project*. Cambridge, MA: The MIT Press, 2009.

Jenkins, Henry. *Convergence Culture: Where Old and New Media Collide*. New York: New York University Press, 2006.

Jensen, Klaus Bruhn. "Mixed Media: From Digital Aesthetics towards General Communication Theory." *Northern Lights* 5 (2007): 7–24.

Manovich, Lev. *The Language of New Media*. Cambridge, MA: The MIT Press, 2001

McGonigal, Jane. "I Love Bees: A Buzz Story." Paper presented at AD:TECH, San Francisco, 25–27 April 2005. Accessed 23 March 2016. http://avantgame.com/McGonigal_42%20Entertainment_ADTECH_April%202005.pdf.

McGonigal, Jane. "The Puppetmaster Problem: Design for Real-World, Mission-Based Gaming." In *Second Person: Role-Playing and Story in Games and Playable Media*, edited by Pat Harrigan and Noah Wardrip-Fruin, 251–63. Cambridge, MA: The MIT Press, 2007.

McGonigal, Jane. "A Real Little Game: The Performance of Belief in Pervasive Play." Level Up. Proceedings of DiGRA conference, Utrecht, Netherlands, November 2003.

McGonigal, Jane. *Reality Is Broken: Why Games Make Us Better and How They Can Change the World*. New York: Penguin, 2011.

McGonigal, Jane. "'This Is Not a Game': Immersive Aesthetics and Collective Play." Digital Arts & Culture Conference Proceedings, 2003. Accessed 23 March 2016. http://edgaps.org/gaps/wp-content/uploads/mcgonigal_ting_03.pdf.

McLuhan, Marshall. *The Medium Is the Message*. New York: Random House, 1994.

Meadows, Mark Stephen. *Pause and Effect*. Indianapolis, IN: New Riders, 2003.

Mitchell, W. J. T. *What Do Pictures Want?: The Lives and Loves of Images*. Chicago: University of Chicago Press, 2006.

Montola, Marcus. "Exploring the Edge of the Magic Circle: Defining Pervasive Games." Proceedings of DAC, Copenhagen, Denmark, 2005. Accessed 23 March 2016. http://www.iki.fi/montola/exploringtheedge.pdf.

Murray, Janet. *Hamlet on the Holodeck: The Future of Narrative in Cyberspace*. Cambridge, MA: The MIT Press, 1997.

Ong, Walter. *Orality and Literacy*. London: Routledge, 2002.

Schneider, Jay, and Gerd Kortuem. "How to Host a Pervasive Game: Supporting Face-to-Face Interactions in Live-Action Roleplaying." UbiComp Workshop on Designing Ubiquitous Computing Games, 2000. Accessed 23 March 2016. http://citeseer.ist.psu.edu/viewdoc/summary?doi=10.1.1.29.3258.

Smith, Marc A., and Peter Kollock, eds. *Communities in Cyberspace*. London: Routledge, 1999.

Szulborski, Dave. *This Is Not a Game: A Guide to Alternate Reality Gaming*. New York: New Fiction Publishing, 2005.

Thompson, Kristin. *Storytelling in Film and Television*. Cambridge, MA: Harvard University Press, 2003.

Turkle, Sherry. *Life on the Screen: Identity in the Age of the Internet*. New York: Simon and Schuster, 1995.

Williams, Raymond. *Television: Technology and Cultural Form*. London: Routledge, 1990.

Žižek, Slavoj. "Cyberspace, or the Unbearable Closure of Being." In *The Plague of Fantasies*. London: Verso, 1997.

Žižek, Slavoj. *Looking Awry: An Introduction to Jacques Lacan through Popular Culture*. Cambridge, MA: The MIT Press: 1991.

PART 2

DIGITAL FANDOM 2.0

· 6 ·

DEMEDIATION AND HYBRID FANDOM

He realized that he'd thrown the dice. That he'd moved to the first square in a dangerous game ... but he felt like playing.

—Pérez-Reverte, *The Club Dumas* (pp. 62–3)

All throughout history Magi have wandered the globe, spreading their teachings of powerful magic. The magic they wield comes from the enchanted Runes they have collected over the ages. In order to learn this magic, a Magi must complete a Rune Quest for the specific power that they seek. Only those that complete all Rune Quests can be given the illustrious title of Master Magi.

—Magiquest, *Runes of the Questmaster*

In the conclusion to the first edition of *Digital Fandom*, I introduced a concept I called "demediation," drawn largely from my experiences with gaming, alternate reality, and pervasive games.[1] Demediation drew from postmodern theories as well as the book *Remediation* by Jay David Bolter and Richard Grusin. I defined demediation in relation specifically to alternate reality games, those games played in a combination of physical and digital worlds, where the game world would often overtake and encompass the non-game world. Demediation is certainly not limited just to alternate reality games, but in the first edition I felt that ARGs were the most obvious and exemplary examples. Demediation was what happened when all component media of ARGS (web pages, wikis, tangible books, puzzles, etc.) immerse the player in an environment where mediation is forgotten.

In a digital environment, and viewed through what I've called New Media Studies, however, the ARG can be analyzed as a whole, each of its multiple mediations contributing to a demediated experience. The ARG represents a type of all-encompassing game, of the type Corso meets in *The Club Dumas*. Demediation is a concept of ubiquitous mediation: where digital technology becomes so rampant that its *absence* becomes noticeable. Mediation becomes the ubiquitous state, the underlying experience of media shaping our experienced reality. (The converse occurs with demediation as well: for example, when watching a film or television show from before the widespread adoption of cell phones or tablets and noting how different the narrative would be with the use of a phone; e.g., *Home Alone* would be radically different if Kevin could just call his parents.) ARGs function precisely because mediation is so commonplace in our lives that it would be suspect to have an all-encompassing game that did not invoke email, text messaging, instant messaging, Facebook profiles, or web interaction. To be surrounded by media allows us to forget that media are there at all, which is why demediation becomes an important concept for media literacy as well as for media studies.

In *Digital Fandom*, I limited my explanation of demediation to alternate reality games. In this chapter, I offer some additional discussion of ARGS and also expand my discussion of demediation to look not just at the way digital media and technology have become ubiquitous, but also how that has affected the fan experience in the digital age. In a paradoxical turn, digital fandom is often just as much about the analogue experience as it is about technology: Fans use digital technology in order to augment and develop their offline experiences.[2] To this end, I will examine not just the spaces of fan demediation, but also some practices that highlight a demediated *experience*, focusing specifically on the interactive live-action game *MagiQuest* as an example of this form of play between the digital and the tangible in fandom.

MagiQuest

The interactive role-playing adventure game *MagiQuest*, a sort of LARP (Live Action Role Play) experience, offers an opportunity to explore the intersection of fans, audiences, and mediation in a digital age; at the same time, it opens up new avenues of fan studies theoretical analysis within a digital media environment. *MagiQuest* is a live-action *Harry Potter*, a consumer-driven media spectacle that exists in an über-mediated environment of computers, digital graphics, on-screen animations, and interactive production. Players embark on quests to earn mystic runes, each rune a step on a larger path towards Magi-hood. The eventual attainment of the rank Master Magi comes through perseverance, attention to detail, and a strong sense of moral fortitude. Each Master Magi possesses a stout heart, an honorable moral code, and a knowledge of magic unrivaled throughout the kingdom.

To become a Magi, one must learn to *think* like a Magi; in this way, *MagiQuest* is a pedagogical game. The quests presented by *MagiQuest* challenge and thus teach participants; the lessons learned are often more ideological in nature than they are practical. By investigating *MagiQuest* through demediation, I explore a media literacy framework as a mode of analysis. The method that *MagiQuest* uses to interpellate its players is fannish in nature, and relies on a particular set of fan-based skills to inculcate the values and ideas of the game. *MagiQuest* makes use of a concept I've defined as "playing fandom":

> As consumers of media, we play with the texts, meanings, and values created by media industries. But playing fandom isn't just what we do with our everyday media; it's also what our media do with us. As media producers ourselves, we are played by those same industries. Our creative work is used to sell products and services. Our clicks become capital. We are commoditized from and marketed to.

In other words, "Media fandom is best understood as a continual, shifting negotiation and dialogue within already-extant industrial relations."[3] Media corporations—like Creative Kingdoms, the corporation that owns and developed *MagiQuest*—utilize fannish attributes and activities to appeal to a fannish sensibility. Some of these attributes and activities include the playful use of demediation. But rather than embrace the freedom and openness of fan creativity, these corporate parodies of fandom commoditize and foster the experience of fandom.[4] They do this by integrating digital attributes of the fan experience into the everyday lives of fans: in other words, they embrace what I'm calling in this chapter a "hybrid fandom."

Hybrid fandom is an extension of demediation, wherein a fictional diegetic world will use an obvious focus on mediation (or a hypermediation) in order to immerse viewers in a more naturalistic world. Hybrid fandom, as exemplified by *MagiQuest*, is also an interaction between offline and digital, between the playfulness of fandom and the seriousness of corporate culture. *MagiQuest* illustrates what Lincoln Geraghty calls a "convergence of popular fandom, new media, nostalgia and contemporary toy culture"; as he suggests, this type of hybridity implies "that the lines between past and present, technology and culture, childhood and adulthood are increasingly porous."[5] *MagiQuest* is not the only hybrid fan experience. Indeed, I've previously written about hybrid spaces of fandom—e.g., the *Doctor Who Experience* in Cardiff, UK as a specific fan destination—and hybrid fan activities—e.g., creating *Inspector Spacetime*, and *SuperWhoLock* original fan texts.[6] Or, fans often use digital technology to augment their experiences with the show's real-world production, visiting filming locations and sets of particularly favored shows and tweeting or writing about the experience. Matt Hills calls this "textual-enunciative productivity" a form of digital transformation, where fans upload "pictures and videos of TV filming … along with fans' tweets reporting convention news," multiplying "the possibility (and opportunity) for fans to shift textual-enunciative productivity into textual productivity, pretty much in real time." A good example of this type of textual-enunciative productivity is #Setlock, the hashtag used by fans who seek out the filming of *Sherlock* and tweet out pictures of the filming, or #dwsr—Doctor Who Set Reports.[7] Fans' interaction in the digital world augments and deepens the fan experience in the offline world.

Hybrid fandom can also be seen in 21st century convention culture. Fans have been meeting since the beginnings of organized fandom, like the Sherlock Holmes fan communities from the time of Conan Doyle. But the 1930s, according to Francesca Coppa, saw the first organized science-fiction fan conventions, and of course today fan conventions are some of the most visible places for fan audiences today.[8] As I write this, the most recent San Diego Comic-Con just ended, with over 130,000 fans attending and creating a "communion" of fan culture.[9] Many conventions like Comic-Con (although most with fewer attendees) appear all over the country: In my own city of Chicago, for example, two large conventions (Wizard World and Chicago Comic & Entertainment Expo [C2E2]) each greet 50–70,000 attendees each year, and scores of smaller, more specific conventions populate each year in the city. Jennifer Porter views fan conventions as a type of pilgrimage, marked by both religious zeal and fan enthusiasm.[10] Beyond meeting in person, however, fan conventions also facilitate digital interactions. Peter Kelly and I interviewed fans and discussed the way that fans at fan conventions engage more in their fandom offline than they do online, but "their offline practices are still facilitated through and influenced by their online interactions." That is, "fans are still meeting in person to screen episodes, to game, to chat, and to participate in conventions (among other things), but they are organizing, structuring, and facilitating these interactions via online and digital technology."[11] Paul

Théberge argues that interaction between offline fan groups and online "official" fan clubs are facilitated by digital technology, and are "marked by the appearance of a more direct form of dialogue between artist and fans and a more regular, even daily, ability to connect both artists with fans and fans with one another."[12] While the "digital space has clearly augmented the practices and methods of interpersonal connection amongst fans … many fans continue to adhere to offline practices" such as convention attendance, set location visiting, museum encounters, games, and fan clubs.[13] In other words, digital fandom augments offline interaction.

In many ways, fandom is itself demediation. Fans today do not operate in a media vacuum; rather, they embrace a multitude of mediations and technologies to augment their fan practices, and if anything, such interactions reveal the ubiquity of technology and the ease with which fans embrace digital interactions, supplementing their offline lives. This online/offline interaction, a hybrid fandom, can be witnessed in how *MagiQuest* functions both as a game and an immersive environment, how *MagiQuest* can be analyzed using a *demediated* look at the game environment, and how *MagiQuest* demonstrates a rough dialogue between fans and producers. I argue that *MagiQuest* integrates a productive user experience into a pedagogically enticing environment. The *MagiQuest* environment provides a glimpse into a world of ubiquitous mediation, and the über-mediation of the game teaches the fan how to function within our mediated world.

MagiQuest and Hybrid Fandom

So let's use a Time Rune to reverse the path of this chapter a bit and see from where we started. To non-Magi, *MagiQuest* is a game conceived by child psychologist Denise Weston and played in an immersive environment by those enamored of the classic Campbellian Hero myth. Players create their own hero character, attach the character to a clan, and purchase a magic wand (an electronic, infrared device) that they can use to activate particular aspects of the gameplay through a unique identifying code tied to their character. On top of each wand sits a topper, a purchasable customization that can change players' abilities. These toppers are merchandise that players, like fans, can collect to form "solid signifiers of the historical significance" of their trip to *MagiQuest*.[14] They can also collect these toppers in order to augment their experience by changing the abilities of their wizard character. Abilities are earned by completing quests and gaining runes, mystic skills for the players' characters to manifest. Sample runes include the Reveal Rune, which allows characters to "see what would be invisible," or the Ice Arrow Rune, which can "penetrate the thick red scales of the Red Dragon."[15] Runes also allow players to duel, or battle another player in a videogame-like environment.

To earn runes, players embark on quests, which always begin in front of a computer monitor. Waving the wand in front of the monitor activates a sensor that reads the wand's signal. The computer recognizes the player-character's unique code, and offers a range of quests to play. Each quest sends players searching for particular markers or landmarks in the *MagiQuest* space in order to earn runes. Players record their finds by waving their wands toward the markers. The wand picks up the signal, records the player's status, and downloads that personal information to a database. Upon finding all markers for a particular quest, the player returns to a central computer to collect the rune and begin another quest. While *MagiQuest* is largely aimed at children, many parents have enjoyed the game as well—and while I played the game in researching this chapter, I noted just as many adults as children playing along.

A personal example may elucidate this process. To earn the Lightning Rune, or the ability to use an attack spell (one of the first spells one should learn in *MagiQuest*), I stood (in the guise of my character, Sir Paul the Wise of the Trixter Clan) at the central core computer area and waved my wand topped with the Master Magi Topper ("The Master Magi Topper holds the power of the Learned Owl"). An animated Learned Owl appeared on the screen, flapped its wings, and then dissipated to reveal a number of options. I touched "Quest" and then selected the one I wanted—Lightning Rune. A live-action wizard appeared on the screen and introduced my quest by speaking a verse that posed a simple riddle: e.g., "If you be the type/to seek out a **good book**/near the quest stones, by the entry/I'd urge you to look." I turned around and saw a large tome near me. I shook my wand, a *woosh* sound was emitted, and the book lit up in sparkling lights. I had found the first clue. Three more clues pointed to a sword in the stone, a shield, and a suit of armor, each of which lit up when I shook my wand near them. Upon finding the last clue, I returned to my initial computer, waved my wand, and the wizard appeared to congratulate me and offer me the Lightning Rune (I later used the Lightning Rune to challenge my wife to a duel, which I lost). I touched the screen to accept the rune and began my next quest (the Enchant Creature Rune). Each subsequent quest asks new riddles, some of which are quite challenging—one quest (Master Magi Rune Quest) took me the better part of 45 minutes. My location (Lombard, Illinois) offered 12 quests and three longer, more complex adventures; different locations offer alternative quests, different runes, and separate adventures.

Key to any adventure or quest in *MagiQuest* is the immersion of the participants; the game asks us to *become* Magi in mind and body. *MagiQuest*, described online as "an interactive live-action, role playing game," is played by individuals in an immersive environment. In the location that I visited, for example, the *MagiQuest* environment was housed in the lower level of a mall, and took up the space of roughly four stores. Split into nine areas—it stretched from the Twisted Woods near Pixie's Perch to the Dungeon at the Dragon's Lair—the room housed a number of stanchions and scaffolds to create a labyrinthine interior. To play a quest, one must traverse across the maze a number of times, creating not just a simulation of a hero's quest, but also a remarkably good workout. Not all *MagiQuest* locations are the same; for example, the multiple locations housed in Great Wolf Lodges take participants outside of the central *MagiQuest* zone to interact with objects in the hotel itself. Similarly, my location has quests that sent me out to find musical instruments scattered across the mall. The immersion of *MagiQuest*, therefore, goes beyond a closed stage for playing out medieval fantasies to spill out into malls, hotels, and theme parks, demonstrating the spread and pervasiveness of demediation in our everyday lives as well.

Soon after this chapter was originally written, the location at Lombard shuttered its doors; however, a new location close to it had opened and also features a *MagiQuest* themed laser tag with the wands. Despite the economic downturn of the early 2010s, *MagiQuest* and its parent company Creative Kingdom have maintained locations across the United States and in Japan. In addition, they've added new quests and experiences. The "Runes of the Quest Master" quest that I described above is not just one of three "Chronicles" that players can engage in—the others, "Guardians of the Realm Portals" and "Journey to Save the Light" follow similar procedures but different plots, as the second quotation at the top of this chapter reveals.

In a hyperreal Baudrilliardian sense, then, the bizarre falseness of these not-so "real world" locations such as the mall, hotel, and theme park lends demediation to the *MagiQuest* world. In what follows I show how a consumer's immersion in the *MagiQuest* world leads to a greater *sense* of active production (whether or not that sense is accurate). *MagiQuest* uses the dichotomy between production and consumption to teach ideological lessons to the players. Questing becomes a metaphor for life, and naturalizes the concepts of commodification and conflict for the players.

ARGs, *MagiQuest*, and Demediation

MagiQuest links the mediated world with the non-mediated world, but in doing so it creates a demediated space where technology exists seamlessly with the audience production of the game. Demediation mirrors the way our lives today display multiple media as a whole environment. With demediation, the ubiquitous multitude of mediations in the digital environment envelops the user. Yet, importantly, the media don't disappear from the user's view, but remain still *knowable* as media. In the conclusion to *Digital Fandom*, I defined demediation as an intermedial position between hypermediation and immediacy—two terms from Bolter and Grusin.[16]

I've also examined demediation in relation to the *Doctor Who Experience* in Cardiff. At the *Doctor Who Experience*, the use of technology works relatively seamlessly with the multimedia environment portrayed from the television show *Doctor Who*. That is, the *Doctor Who Experience* is a wholly created environment, much like *MagiQuest*, that uses mediation in order to create immersion within fans' personal experiences with *Doctor Who*: "The hypermediation of the experience becomes an immediate experience through immersive, interactive viewership."[17] The sensory experience of the *Experience* immerses the fan but does so only through ubiquitous technology.

In contrast, Bolter and Grusin dealt almost exclusively with singular instances of mediation—the television, the photograph, the computer, and so forth—as separate entities. When they do discuss mediated spaces like Disneyland, they focuses on how digital technology augments "our contemporary world … constituted through a series of remediations," highlighting the obviousness ("hypermediacy") of media within these spaces.[18] But mediation in our digital environment is rarely experienced in a media vacuum. That is, our mediated environment contains multiple media technologies, surrounding us and creating vast arrays of many different types of hypermediation as well as many different levels of immersion all at once. We see through some, we look at others. Disneyland is completely remediated, but it is also mediated anew by the multiple personal technologies we carry everyday—and by the experience of willfully forgetting that mediation exists in Disneyland at all.

As I described earlier, Alternate Reality Games (ARGs) are one of the most obvious examples of demediation—and in many ways, *MagiQuest* is similar to these real-world, immersive, game-like experiences. An ARG is a game played in the real world, and are, as game researcher Jane McGonigal described, "antiescapist."[19] That is, ARGs use everyday technology and real-world interactions to engage reality with game-like results. Indeed, that sense of engagement complements what games researcher Yam San Chee defined as the three most important components of immersive gaming—embodiment, embeddedness, and experience—all of which engage players in "learning by doing, observing the outcomes of [their] actions, testing [their] hypotheses about the world, and reflecting further on [their] own understanding."[20] Similarly, Schroeder argued that all play engages players and "operates within its own space."[21] This space, called at times the "Magic Circle," a "playspace," a "third space," or the "game space" exists as part of our real world, but with separate rules, activities, and actions.[22] For example, when one enters the game world of *tag*, the rules state that one should slap opponents to tag them out of that round. In the real world, tagging someone runs the risk of assault and battery charges.

Like large-scale, internet-based scavenger hunts, ARGs send players shuffling between the real world and the game world to find clues and solve puzzles.[23] ARGs rely on the distinction between real and non-real, and attempt to hide the Magic Circle of the game world within the environment of the real world. It's important to note that ARGs use everyday technology to straddle this divide between the real and the gameworld—and because the technology *is* everyday, it disappears through its very ubiquity. ARGs *reinforce* the boundaries between the ludic and the non-ludic. Using everyday media to hide the game serves a dual function that reverses common assumptions about games: They both immerse the player in the game world and also make obvious the mechanics of the real world.

In the digital fandom sense, and for the fans/players of ARGs, the Alternate Reality Game emphasizes the same hybrid economic interaction as I described in earlier chapters about the Digi-gratis economy. ARGs integrate both market and gift economies in a fannish mirror. Namely, ARGs use the participation of the audience as a form of viral marketing. Henrik Örnebring notices this as an absence in critical scholarship of ARGs: "[T]here is relatively little academic concern with how ARGs function as *marketing tools*, and critical acknowledgment of the fact that most high-profile ARGs are produced for marketing purposes is limited at best."[24] Viral marketing describes a form of advertising that emerges as the participation of the audience distributes content through word-of-mouth, viral videos, e-mail forwards, or other online user-generated content. Also called "immersive marketing," this form of advertisement relies on the activity of consumers for the promotion of a product. John Gosney, the author of the first textbook about ARGs, interprets this strategy as one of informed consent: "I don't think it belittles or otherwise reduces your enjoyment of the ARG experience by keeping in mind that what you are playing … has been developed, in many cases, to market a product."[25] As Mark Andrejevic notes, "the promise of virtual participation in the production process … facilitates the conversion of viewer feedback into potentially productive marketing and demographic information."[26] For example, *The Art of the Heist* was designed as promotional material for the car company Audi. As players progressed through the game, they had to explore websites devoted to the car maker, discuss Audi products with other players, and travel to an actual (non-ludic) Audi showroom. Further, the ARG *ilovebees* was designed to promote the release of the Xbox game *Halo 2*, while *The Beast* began life as a large marketing campaign for the Steven Spielberg film *A.I.* Other ARGs, however, are designed as both promotions and supplements to other products. *The Lost Experience* both promotes the television show *Lost*, through commercials and word-of-mouth advertising, and also supplements the backstory of the *Lost* narrative. *The Lost Experience* found players not only scouring webpages for information, but making phone calls, writing e-mails, putting together physical jigsaw puzzles, and even reading a 200-page novel, in order to uncover clues related to the larger mysteries in the show. Dena shows that this form of narrative marketing is designed specifically for an active audience, one supposed to fill in the gaps left both intentionally and unintentionally by the producers.[27] For example, Christopher Nolan, director of the Batman trilogy (*Batman Begins*, *The Dark Knight*, and *The Dark Knight Rises*) designed an ARG called *Why So Serious* that asked audiences to fill in the gaps between the first two films. By filling in these gaps, the players of the ARG function in a shared network, where the discussion of the narrative of the game operates as a dialogue about the brand the ARG advertises; and also act as fans do through the creation of fan fiction and other forms of hyperdiegesis.

Conversely, ARGs also utilize characteristics of the gift economy, which include sharing, reciprocation, and communalization. Players must communicate with each other and share information in order to continue. Most saliently, this is visible in *ilovebees*, where players from around the county were given different pieces of information that they then combined into a workable whole. By working together and contributing unique information, players successfully combined interactivity with play. They contributed information freely, as gifts to benefit the whole of the community; but they use market economics to buy the components of the game—the individual texts that make up the ARG.

Like an ARG, the magical kingdom of *MagiQuest* is composed almost entirely of elements of mediation, from the IR wands wielded by the Magi to the computer terminals scattered throughout the environment. Wave your wand at the wall and any number of exciting sound effects, images, lights, or movement may occur. Computer terminals are posted every few feet, waiting for a flick of the wand to activate and display whatever creature lives in them. In a truly demediated example, the computers sit among representations of trees and forests, streams and mountains. The computer has become part of the natural environment, as an emphasis on the ordinary and accepted inculcation of technology within our everyday lives. For example, to complete the Freezing Rune quest, one must collect a number of items around the kingdom and then find the computer terminal in which the Old Man in the Stump lives. As the player approaches the computer, a monitor reveals images of the Old Man napping; when the wand is waved in front of the terminal, he awakens and demands to know who disturbed him. Terminals such as these are scattered around the environment, and even if it is the end of their quest, users can always shake a wand at a computer to get hints, see their current standing, or discover what else they have left to do. Even beyond the immediate spaces within *MagiQuest*, the game is replete with mediation. Computer monitors posted in the Market Place (the large area at the front of any *MagiQuest* kingdom in which players can purchase costumes, wand accessories, games, models, and any manner of other accouterments for their game play) display current and all-time high standings of the players. There is even an online version of the game, *MQOnline*, which displays a player-designed avatar in the magical kingdom, facing various online quests, a la *World of Warcraft* (minus the fighting). Crucially, some of the quests that one can complete at *MagiQuest* locations requires *MQ Online* to function, further cementing the hybrid fandom between digital existence and physical interaction.

Demediation thus exists as an integral part of the mediated immersion of *MagiQuest*, and reveals the fannish roots of this game-like experience. For, like fans and the hybrid interaction of digital and physical manifestation of the fan experience, players of *MagiQuest* must navigate the use of digital within the physical, and the application of the physical within the digital.

Neither completely overtakes the other and both work together to create the contemporary fan experience.

Moments of demediation bring to mind another theory of ubiquitous mediation, Baudrillard's conception of hyperreality; however, unlike hyperreality, demediation invokes a practical and applicable theory within the game. For Baudrillard, hyperreality exists as a state of culture completely enveloped in mediation, so much so that reality itself disappears. Mediation overtakes all else. All that's left is a sense of mediated truth with no basis in what Baudrillard called "first-order reality."[28] There is no antecedent to the simulation; it is all that exists. So, a mediated event takes the place of the real event, and becomes more real than that event. We learn more about the experiences of war from war *films* than we do from actual combat. Baudrillard famously used the example of Disneyland to say that the Magic Kingdom exists to disguise the fact that the rest of the world is just as simulated as Disneyland is. We see that Disneyland is false (and enjoy it for its falseness), and then don't notice that everything else around us is equally false.

Schroeder, however, has admitted that "using Baudrillard (and Baudrillard-style theories of simulation) as a theoretical tool is a problem"—and one that demediation sidesteps. Indeed, we enter into this problematic interaction with the media environment when we examine *MagiQuest* from this Baudrillardian standpoint. Just as "the boundaries of the playspace begin to disappear," when games become immersive, so too does any theoretical analysis of that game, from a "hyperreal" standpoint.[29] *MagiQuest* is an obvious hyperreal event, but because it is played in hotels, malls, and theme parks (which can also be analyzed as hyperreal locations), the hyperreality becomes subsumed by more hyperreality. One reveals the other, in an ouroboros of mediation. If we are in the age of simulation, as Sean Cubitt noted, "we can no longer distinguish between play and work."[30] Given this shift, a hyperreal analysis would lose all defining characteristics, revealing not the loss of the real, but rather the requisite loss of the hyperreal. This effectively effaces hyperreality through its own hyperreal analysis.

Instead of venturing down this path of theoretical tautology, I think it's more explanatory to focus on the interaction of the media technology interface and the user.[31] *MagiQuest* is demediate in that the ubiquity of the mediation within the game teaches players how to use technology in an intuitive and instrumental way. The interface disappears through its obviousness, and the wand—much like the Nintendo Wii controller, the Xbox Kinect, or the PlayStation Move—becomes the McLuhanesque mediated extension of the human body. Media are internalized; reality is not sublimated. For digital fandom, *MagiQuest* serves as a heuristic for the way fans interact with fan objects in the digital environment; fans may not be waving wands and going on quests, but are instead using digital technology to change the physical world in which they are fans. The digital augments the physical just as much as the physical affects the digital.

Players become one aspect of the mediated environment, a technology in and of themselves, focusing the gameplay away from a purely deterministic dichotomy between the interface and the player, and toward an environment in which the player becomes part of the mediation. The hypermediation of the game *becomes* an immediate experience through immersive gameplay. Ultimately, *MagiQuest* highlights a shift in media theory, because the mediation of the game stands separated not from other mediation, or from the player, but rather joins the non-mediated world as merely one component of co-created player activity.

Interpellation within *MagiQuest*

As a demediated text, *MagiQuest* highlights aspects of digital fandom within contemporary media theory. But what is the practical result of this demediation of *MagiQuest*? How does it manifest in ways that apply to fan audiences? Perhaps most obviously, *MagiQuest* can perhaps best be described as a *game*—one in which participants become fully immersed as role players.[32] The use of digital technology as a manifestation of the philosophy of playfulness relates directly to the changes in the media environment described in this book, as well as the media industry's focus on fans' play.[33] To play is key to both fandom and to *MagiQuest*, and players must role-play as characters in order for the game to be successful, much like the fans I identified in MySpace profile interreality.[34] Like improvised theater, *MagiQuest* creates a story as it is being played—players help construct the series of events as events are encountered.[35] One can't play if one is self-conscious about shaking a plastic wand at a painted picture of slime on a plywood tree.[36] Fandom is similarly constructed by the activities and experiences generated by the fan: Fandom does not exist without the consciousness of fans themselves (this is not to say that every fan is self-aware. But as I discuss in chapter 8, a move to the mainstream has created a particular brand of self-awareness within fandom and within fan play). We must be aware of the media to play, but we must ignore it to be immersed.

MagiQuest also illustrates important pedagogical ramifications of game-based learning. Through its demediation, *MagiQuest* interpellates key ideological lessons within its game, encouraging *teamwork*, *imagination*, and *play*. According to a press release from its creator (Creative Kingdoms), *MagiQuest*, was designed to "get kids off their couches ... moving and interacting together." Through its interactive mechanisms—the IR wands, the touch screens, and the immersive environment—*MagiQuest* stimulates the same learning capabilities in children as do games.[37] Fandom is similarly pedagogical, is similarly a "classroom" of "thinkers and responsible media citizens."[38] Fandom is a type of pedagogy for media, using play as a mode of interaction.

Indeed, the effectiveness of play as a pedagogical tool has been well-documented. For example, Chee describes how game-based learning helps students learn more information, retain that information longer, and apply that information in different settings. Of today's generation of students, he wrote, "school is only one source of learning; out-of-school learning and informal learning increasingly account for more of a millennial's overall education."[39] Video game researcher James Paul Gee argues that new times call for new literacies: the hypermediated environment of video games, cell phones, PDAs, iPads, and TiVo demands a new type of learning, one predicated on new "semiotic domains" made up of "images, sounds, gestures, movements, graphs, diagrams, equations, objects, even people"—all aspects of *MagiQuest*, and all aspects of fandom as well.[40] Focusing on play as an integrated method of pedagogical practice fosters these new literacies, forging new students who grow up knowing how to act, interact, and react in a 21st century media environment, and "both media fans and the media industries must continually negotiate, navigate, and adjust to the presence of each other in tandem with [these] changing paradigms of technological discourse in our digital society."[41] Furthermore, James et al. have shown that many leisure activities in "the new media" online, including "gaming [and] instant messaging, social networking on Facebook and MySpace, participation in fan fiction groups, blogging, and content creation" can all be subsumed under the category "play," contributing to a new understanding of online activity.[42] Jenkins et al. found that this type of play, as a concept situated within teaching and learning practices, is key to "shaping children's relationships to their bodies, tools, communities, surroundings, and knowledge." Play is crucial to learning because "this activity is deeply motivated. The individual is willing to go through the grind because there is a goal or purpose that matters to the person. When that happens, individuals are engaged."[43] In fact, using games as pedagogical models has been shown to increase retention in schools.[44] Play allows children (of all ages) to learn in a safe environment. Failure in a playspace merely results in sitting out the next round, whereas failure in the real world can entail punitive action, castigation, or punishment. Playspaces allow players to be stimulated, to experiment, to be social. These informal, third spaces allow "conduct that is both meaningful and engaging to the participant and responsible to others."[45]

Playspaces are effective not only because they offer safe zones, however, but also because they are put into practice through players' interactive efforts. According to play researchers Shanley Dixon and Sandra Weber, there is a "fascinating continuity, flow, and ambiguity between various forms of play."[46] The spaces where play happens are the ultimate demediation: They exist as created by the people who participate in them, and are both obvious and immersive. Thus playspaces, and the games residing within them, exemplify the confluence of demediation that commands attention in today's convergence culture, where users both produce and consume media on the same device, and "participation is understood as part of the normal ways that media operate."[47] *MagiQuest* relies on this co-creation: Creative Kingdoms produces an environment and interactive media; players of *MagiQuest* produce their own adventures through role-play, interaction, and imagination within the environmental realm.[48] Fans rely on this as well: Fandom creates a community and within that community, fans must engage their own interaction within the environment. Indeed, important non-scholastic skills are also built, as Adrian noted of *MagiQuest* that "*recognition*, a much sought after societal quest in itself, is instantly accomplished."[49] That Creative Kingdoms makes money from players' own creativity invites additional critique, as part of digital fandom's Digi-gratis economy.

Playspaces also offer a unique opportunity for safe and effective pedagogical practices. As I've talked about in previous chapters, we have entered an era defined by a particular "philosophy of playfulness," where media is complex and fun to engage with, and encourages fans' and viewers' dependence on a culture of ludism. As an exemplar of how play is becoming a ubiquitous marker of this cultural ludism, *MagiQuest*, I argue, enacts a further refinement in the nature of these new literacies, and provides a deepening understanding of changes to the media environment. Whereas a video game literacy might provide a solid background on new movements in media studies, the type of literacy engendered by *MagiQuest* offers a more encompassing venture into a larger digital environment.

Specifically, by straddling the audience/producer chasm, *MagiQuest* embodies a playful learning environment that teaches children not only to be active *producers* of information and play, but also to become passive *consumers* of media information. As Jenkins has warned us, as freeing as new technology may seem for democratic and consumer-driven media (and, indeed, *MagiQuest* seems the ultimate in liberating technology), these technologies are still constructed and run by large corporations.[50] Indeed, "the more audiences 'play' at fandom, the more fandom can be commoditized" itself.[51] To be fully engaged in the material, one must embody the true Magi, a role Creative Kingdoms is only too happy to illustrate and present as the norm. That is, *MagiQuest*'s veneer of pedagogy actually ends up teaching two different types of activities. On the surface, the player's active, imaginative engagement with the material does provide a safe space for creative immersion; underneath that immersion, however, the type of play engendered by *MagiQuest* interpellates particular norms of behavior.

MagiQuest offers a relatively easy vantage point from which to observe some of the interpellation that occurs with digital fandom. From the obvious gender norms that are displayed with the characters—the Princess is traditionally beautiful, tall, blonde, and thin, and the Lady in the Leaves (the keeper of the Dazzle Rune) is highly sexualized—to the lack of any ethnic or racial diversity among the onscreen actors, *MagiQuest* seems to standardize the magical kingdom for beautiful, white people. Participants are told that they are Magi, true and virtuous adventurers in a magical land. But looking around that land, players are hailed as Magi that fit a particular vision, a particular image, and a particular act. The sneaky part of this hailing occurs when participants aren't just told that this is how to look in the kingdom, but rather actively co-create this image, both by immersing themselves into the demediate world of *MagiQuest* and by purchasing *MagiQuest* merchandise. The interactivity of the game helps foster the implied messages of the game.[52] Players both implicitly and explicitly buy into the ideological standardization portrayed through the game.

In a similar way, media corporations are creating images of fans that lead viewers to interpret fandom in particular ways. For some, the version of fandom that is represented teaches a particular lesson to fan viewers: Fans are often portrayed in gendered ways that presented "good" fans (the ones that aren't ridiculed) as more consumptive of authorized texts. Kristina Busse argues that "the constructed fan combines all the positive fan qualities such as sustained viewer interest and commercial viability while engaging fannishly in ways preferred and controlled by the studios."[53] For Busse, the negative fan representations are highly gendered, with mainly feminine activities—writing fan fiction, creating communities, sexualizing (male) stars—dismissed and mocked. This has the effect of constructing fan audiences as particular ways of being: "[T]hrough hyperstylized representations of fans, fan audiences are disciplined into particular identities and behaviors."[54] In order to become part of the "appropriate" fan audiences, fans have to "buy into" the particular image they are shown.

And, in fact, this "buying into" leads to a more complex understanding of the interpellation of the media text and of *MagiQuest*. Beyond the superficial aspects of the game, players also encounter a general metaphor for ways to live their lives. Players' engagement with the immersive aspects of the game, through their own co-creation of their game experience, helps drive their embrace of the game's ulterior mythology. As Rosenbloom described it, this very immersion helps solidify the teachings of any game in the minds of the players: "[T]he driving force in these immersive, realistic environments is the user's experience."[55] *MagiQuest* uses the game environment to normalize and facilitate a standard narrative that players take as given in a world created as fiction. It's not hard to see the truth of this fictionality, especially when that fiction is presented in a participant-focused environment. Jesper Juul, for instance, has shown how fictional worlds can help shape and influence a player's understanding of the real world. In his *Half-Real*, he showed how "the fictional world cues the players into making assumptions about the real world in which the player plays the game."[56] So, for example, whereas we all know that there aren't wizards, magical owls, and dragons in the real world, the fact that they are real in the *MagiQuest* game as we play it gives them an air of authenticity, which means that their lessons are taken more to the heart than if we were simply told about them outside that environment. In digital fandom, the representation of particular types of fans in media texts becomes a "hyperfandom," where overt representations hide the more insidious.[57]

Perhaps the most potent message *MagiQuest* offers is that consumption and competition are normal, if not crucial, aspects of modern-day cultural practices. Players are hailed as Magi before they even enter the kingdom, and are then told that Magi fit into specific, heroic stereotypes that can be commoditized. For example, when a Magi enters the kingdom, the first thing he/she does is fill out a card with his/her moniker, vital statistics, and choice of clan: Majestic, Shadow, Trixter, Warrior, or Woodsy. As shown on the *MagiQuest* website, each clan is represented by its own wand topper, with special powers granted only to those who spend 15 dollars to attach it to their wand. Magi are never told that they *must* purchase the topper; merely that the topper is a chance to "express [their] game character" and "personalize [their] MagiQuest wand." Other toppers invoke different abilities, and are available in the Market Place.

I certainly don't wish to make it sound as though this is a negative or even an unwelcome aspect of the game; speaking personally, having different wand toppers varies and gives a certain flair to the gameplay (I particularly like my "Trixter" wand topper, which looks like a gyroscope, lights up orange when flicked, and makes a magical noise). Indeed, great amusement was had by all my companions one day when I purchased the "wand extender," a topper that exists solely as a Freudian symbol. But the *normalization* of the process, the fact that consumption is a matter-of-fact aspect of the game, and is co-created by the players just as much as it is offered by Creative Kingdoms (no one *has* to purchase a topper, but I have yet to see someone who hasn't), makes this an inarguably central metaphor of the game. Indeed, *MagiQuest* was originally conceived to be centered on commodification; creator Weston stated that "There are multiple sources of revenue coming in. … 'We get you in the very beginning' when you buy the wand."[58] Fandom for the industry is a consumptive practice as well.

The consumption of *MagiQuest* extends into the game world, supporting Juul's point that game experiences mirror real-world ones. Within the game, players accumulate two things: Experience Points (XP), the accrual of which leads to promotion from mere Magi to Master Magi, and gold. Gold is earned by flicking one's wand toward magic chests of money, which open, sparkle, and offer their contents to the player. "You've found 200 gold!" they exclaim. The increase in wealth is recorded via the player's wand and is factored into the scores at the end of the game: The more wealth (and XP) one accrues, the higher one's social standing—a clear metaphor for climbing the socioeconomic ladder.

It is impossible to transfer gold from one person to another, so sharing is hardly an option. Indeed, the initial separation into clans represents another social interpellation on the part of *MagiQuest*—the clear differentiation and segmentation into social groups. Although no one group is seen more positively than another, the distinctions between groups highlight a deterministic view of social relations (mirrored in *MagiQuest*'s antecedent, *Harry Potter*, in which the Hogwarts' matriculates are sorted into four houses, each with clear-cut characteristics). Differences are connoted by the language describing each clan, and rarely affect actual gameplay; they simply help structure and identify the players. The Woodsy clan, for example, is peopled with "creatures who can sometimes be a little mischievous." Trixters are "wise-cracking jokers and jesters who use music as part of their practical joking." Members of the Shadow clan "tend to be sneaky Magi who are usually up to no good." As Gee described, the three identities we create when we play games—the virtual identity of the character, the real identity of the player, and the "projective identity" that describes the "interface between … the real-world person and the virtual character"—all interact as a "larger whole."[59] *MagiQuest* players can envision themselves in this projective identity through contrasting the same characteristic against others with that same trait—the negative connotation of the Shadow clan's "sneaky" mentality is similar to the more positively weighted "mischievous" tendency of the Woodsy clan. This articulates one particular virtual identity, affecting the "real-world" identity of the player. Identity formation in games is crucial, especially for children.[60] Narrowly defining these characteristics creates associations that could have repercussions in the real world as well.

Indeed, what the clan segmentation and the game commodification have in common is the fostering of a sense of competition in the game. Although players are not deliberately pitted against each other (except in the separate dueling part of the game), they are often implicitly set in contrast, from the competing clans to the accumulation of XP and gold. The only cooperation truly experienced in the game is at the end, when more than one person can take on the final boss, the dragon. Even that level of teamwork is optional.

In this way, the quest of *MagiQuest* becomes a metaphor for the trappings of a constituted, capitalistic life. Players engender the characters within the game, and are led through a series of quests that inexhaustibly lead to active consumption and implicit competition. Yet, players get in on the act too, becoming so immersed in the game that the interpellative forces at work become naturalized. But perhaps the most important ideological lesson imparted by *MagiQuest* is the influence of the pervasiveness and ubiquity of the contemporary media environment on the players themselves. Articles and reports about *MagiQuest* continually highlight the virtual reality aspects of the game and its similarity to video games.[61] *MagiQuest* inherently relies on digital technology to foster the immersion of the players in the game and then attempts to hide this mediation from the participants through intuitive and seamless interfaces. The demediation of the game highlights how *MagiQuest* teaches us how to live/work/play/live/be a mediated life.

The End of the Quest

MagiQuest offers players a number of things: quests and adventures, wands and magic, interaction and immersion. But *MagiQuest* also gives its players certain rules and lessons by which it instructs everyday interactions. What happens when questing is taken as a metaphor for life? What happens when overt mediation becomes ubiquitous? What happens, then, when the game ends and Magi move on to other aspects of their life? As Chee has noted, any discussion of games must include a discussion of the "set of values and beliefs" that "are manifested through what the game seeks to convey via its design." In other words, "games are often a rhetorical medium."[62] If that is the case, then *MagiQuest* tells us about life in a post-industrial, digital society where mediation is central to contemporary existence.

All quests end; the classic hero's quest could end in victory, redemption, or death. But when *MagiQuest* ends, players take away important metaphors that help structure and explain today's mediated world. That players become co-creators of this world through their gameplay fosters a greater persuasive function within the game mechanics. As the gameplay becomes normalized, as players scribe their own meaning into the game, it becomes more immersive and influential. This is not true of all contemporary games, most of which exist as mere products to purchase.[63] Games are a singular entity existing as a purchasable piece of media—the Bolter and Grusin-esque epitome of remediation. *MagiQuest*, although often described using video game terminology, represents a more complex mediated experience. How best, then, to describe a game that is more a *practice*, and less a *product*? Rather than focus on how *MagiQuest* harnesses virtual reality, simulation, and hyperreality to set it apart from the "real-world," it is more instructive, and more accurate, to show the ways *MagiQuest* demediates, invoking characteristics that highlight its connection to our everyday life. *MagiQuest* may be a representation of a magical, mythological kingdom, replete with princesses, goblins, dragons, and pixies. But in many ways, it is also an extension and remarkable facsimile of our own digital world.

As digital fandom becomes more ubiquitous in all our lives, we will encounter demediation as a type of hybrid fandom in a multitude of ways. Perhaps we will see more digital/physical interactions like *MagiQuest* develop. One such example is *WizardQuest*, a themed story environment wherein players solve puzzles and riddles in order to complete a single challenge. Like *MagiQuest*, players enter a storybook-type world and interact with digital technology, but unlike *MagiQuest*, *WizardQuest* invites more puzzle solving and physical interaction (there is a mirror maze, a slide, and a ball pit). Perhaps more themed/

interactive style museums will appropriate popular culture: exhibits like the *Star Wars* exhibits at the Seattle EMP Museum and at the Odysseum in Germany use a hybrid of fandom, popular culture, and digital technology to teach, in this case, costuming and identities. Or, perhaps Alternate Reality Games will become more mainstream.[64] Either way, fans will still be playing.

Notes

1. See Booth, *Game Play;* Montola, Stenros, and Waern, *Pervasive Games*.
2. See Wood and Baughman, "*Glee* Fandom and Twitter," 329.
3. Booth, *Playing Fans*, 1, emphasis in original.
4. Booth, *Game Play*.
5. Geraghty, *Cult Collectors*, 9.
6. Booth, *Playing Fans*; Booth, *Crossing Fandom*.
7. Hills, "Fiske's 'Textual Productivity' and Digital Fandom," 135–6.
8. Coppa, "A Brief History of Media Fandom."
9. Scott, "At Comic-Con, Bring out Your Fantasy and Fuel the Culture."
10. Porter, "Pilgrimage and the IDIC Ethic."
11. Booth and Kelly, "The Changing Faces of *Doctor Who* Fandom," 65.
12. Théberge, "Everyday Fandom."
13. Booth and Kelly, "The Changing Faces of *Doctor Who* Fandom," 57.
14. Geraghty, *Cult Collectors*, 2.
15. Unless otherwise stated, all quotations about the *MagiQuest* game mechanics and shop come from the *MagiQuest* website, www.magiquest.com.
16. Bolter and Grusin, *Remediation*.
17. Booth, *Playing Fans*, 119.
18. Bolter and Grusin, *Remediation*, 182–3.
19. McGonigal, *Reality Is Broken*, 125.
20. Chee, "Embodiment, Embeddedness, and Experience," 15.
21. Schroeder, "Playspace Invaders," 147, citing Huizinga, *Homo Ludens*.
22. Huizinga, *Homo Ludens*; Salen and Zimmerman, *Rules of Play*; Castronova, "Virtual World Economy"; Schroeder, "Playspace Invaders"; James et al., *Young People, Ethics, and the New Digital Media*; Juul, *Half-real*.
23. Sparks, "Will Your Next Job Be in an Alternate Reality?"
24. Örnebring, "Alternate Reality Gaming and Convergence Culture," 449.
25. Gosney, *Beyond Reality*, 22.
26. Andrejevic, "Watching Television without Pity," 27.
27. Dena, "Player-Created Tiers," 41.
28. Baudrillard, *Simulacra and Simulation*, 6.
29. Schroeder, "Playspace Invaders," 144–8.
30. Cubitt, *Simulation and Social Theory*, 133.
31. Stein, "This Dratted Thing."
32. Grant, "At Great Wolf Lodge, Some Grown-up Time"; Hutchens, "Great Wolf Lodge Resort Much More Than Just a Water Park"; Salas, "Enter the Dragon."

33. Booth, *Playing Fans*.
34. Salas, "Enter the Dragon."
35. Uren, "Finding the Game in Improvised Theater."
36. Kushner, "Escape to Middle-Earth."
37. Schoolfield, "Quest Masters."
38. Alvermann and Hagood, "Fandom and Critical Media Literacy"; Booth; "Fandom," ¶1.1; Black, *Adolescents and Online Fan Fiction*.
39. Chee, "Embodiment, Embeddedness, and Experience," 5.
40. Gee, *What Video Games Have to Teach Us about Learning and Literacy*, 17.
41. Booth, "Fandom," ¶2.2.
42. James et al., *Young People, Ethics, and the New Digital Media*, 11.
43. Jenkins et al., *Confronting the Challenges*, 35–7.
44. Hernandez, "Gaming+Autonomy=Academic Leadership."
45. James et al., *Young People, Ethics, and the New Digital Media*, 15.
46. Dixon and Weber, "Playspaces, Childhood, and Video Games," 15.
47. Jenkins, *Convergence Culture*, 246.
48. Schoolfield, "Quest Masters."
49. Adrian, "Families Fall under the Spell," 46, emphasis added.
50. Jenkins, "Interactive Audiences?"
51. Booth, *Playing Fans*, 1.
52. Brookey and Booth, "Restricted Play"; Brookey, *Hollywood Gamers*.
53. Busse, "Geek Hierarchies," 78.
54. Booth, *Playing Fans*, 76.
55. Rosenbloom, "A Game Experience," 31.
56. Juul, *Half-real*, 168.
57. Booth, *Playing Fans*.
58. As cited in Fleisher, "MagiQuest's Creator Franchises MB-first Game in U.S., Japan."
59. Gee, *What Video Games Have to Teach Us about Learning and Literacy*, 55–6.
60. Chee, "Embodiment, Embeddedness, and Experience."
61. Salas, "Enter the Dragon"; Szadkowski, "Resorts' MagiQuest Actives Slick Fantasy"; Kushner, "Escape to Middle-Earth."
62. Chee "Embodiment, Embeddedness, and Experience," 26.
63. Kerr, *The Business and the Culture of Digital Games*.
64. See Chess and Booth, "Lessons Down the Rabbit Hole."

Bibliography

Adrian, Clare. "Families Fall under the Spell." *Tourist Attractions and Parks* 37, no. 5 (2007): 44–6.

Alvermann, Donna E., and Margaret C. Hagood. "Fandom and Critical Media Literacy." In *Re/Mediating Adolescent Literacies*, edited by John Elkins and Allan Luke, 36–46. Newark, DE: International Reading Association, 2000.

Andrejevic, Mark. "Watching Television without Pity: The Productivity of Online Fans." *Television and New Media* 9, no. 1 (2008): 24–46.

Baudrillard, Jean. *Simulacra and Simulation*, translated by Sheila Faria Glaser. Ann Arbor: University of Michigan Press, 1994.

Black, Rebecca. *Adolescents and Online Fan Fiction*. New York: Peter Lang, 2008.

Bolter, Jay David, and Richard Grusin. *Remediation*. Cambridge, MA: The MIT Press, 1999.

Booth, Paul. *Crossing Fandoms: SuperWhoLock and the Contemporary Fan Audience*. Houndmills, UK: Palgrave Pivot, 2016.

Booth, Paul. "Fandom: The Classroom of the Future." In "European Fans and European Fan Objects: Localization and Translation," edited by Anne Kustritz, special issue, *Transformative Works and Cultures* 19 (2015). http://dx.doi.org/10.3983/twc.2015.0650.

Booth, Paul. *Game Play: Paratextuality in Contemporary Board Games*. New York: Bloomsbury, 2015.

Booth, Paul. *Playing Fans: Negotiating Fandom and Media in the Digital Age*. Iowa City: University of Iowa Press, 2015.

Booth, Paul, and Peter Kelly. "The Changing Faces of *Doctor Who* Fandom: New Fans, New Technologies, Old Practices?" *Participations* 10, no. 1 (2013). http://www.participations.org/Volume%2010/Issue%201/5%20Booth%20&%20Kelly%2010.1.pdf.

Brookey, Robert A. *Hollywood Gamers: Digital Convergence in the Film and Video Game Industries*. Bloomington: Indiana University Press, 2010.

Brookey, Robert, and Paul Booth. "Restricted Play: Synergy and the Limits of Interactivity in *The Lord of the Rings: The Return of the King* Video Game." *Games and Culture* 1, no. 3 (2006): 214–30.

Busse, Kristina. "Geek Hierarchies, Boundary Policing, and the Gendering of the Good Fan." *Participations* 10, no 1 (2013): 73–91. http://www.participations.org/Volume%2010/ Issue%201/6%20Busse%2010.1.pdf.

Castronova, Edward. "Virtual World Economy: It's Namibia, Basically." 03 August 2004. Accessed 23 March 2016. http://terranova.blogs.com/terra_nova/2004/08/virtual_world_e.html.

Chee, Yam San. "Embodiment, Embeddedness, and Experience: Game-Based Learning and the Construction of Identity." *Research and Practice in Technology Enhanced Learning* 2, no. 1 (2007): 3–30.

Chess, Shira, and Paul Booth. "Lessons Down the Rabbit Hole: Alternate Reality Gaming in the Classroom." *New Media and Society* 16, no. 6 (2015): 1002–17.

Coppa, Francesca. "A Brief History of Media Fandom." In *Fan Fiction and Fan Communities in the Age of the Internet*, edited by Karen Hellekson and Kristina Busse, 41–59. Jefferson, NC: McFarland, 2006.

Cubitt, Sean. *Simulation and Social Theory*. London: Sage, 2001.

Dena, Christy. "Emerging Participatory Culture Practices: Player-Created Tiers in Alternate Reality Games." *Convergence: The International Journal of Research into New Media Technologies* 14, no. 1 (2008): 41–57.

Dixon, Shanley, and Sandra Weber. "Playspaces, Childhood, and Video Games." In *Growing Up Online: Young People and Digital Technologies*, edited by Sandra Weber and Shanley Dixon, 15–35. New York: Palgrave, 2007.

Fleisher, Lisa. "MagiQuest's Creator Franchises MB-First Game in U.S., Japan." *Sun News*, 14 November 2007. Accessed 26 Feb 2011 from EBSCOhost. Link no longer available.

Gee, James Paul. *What Video Games Have to Teach Us about Learning and Literacy*. New York: Palgrave Macmillan, 2003.

Geraghty, Lincoln. *Cult Collectors*. London: Routledge, 2014.

Gosney, John W. *Beyond Reality: A Guide to Alternate Reality Gaming*. Boston: Thompson, 2005.

Grant, Tracy. "At Great Wolf Lodge, Some Grown-Up Time." *The Washington Post*, 24 December 2010. Accessed 23 March 2016. http://www.washingtonpost.com/wp-dyn/ content/article/2010/12/21/AR2010122105597.html.

Hernandez, Dan. "Gaming+Autonomy=Academic Leadership." *Principal Leadership* 10, no. 4 (2009): 44–8.

Hills, Matt. "Fiske's 'Textual Productivity' and Digital Fandom: Web 2.0 Democratization Versus Fan Distinction?" *Participations: The Journal of Audience and Reception Studies* 10, no. 1 (2013): 130–53. http://www.participations.org/Volume%2010/Issue%201/9%20Hills%2010.1.pdf.

Huizinga, Johan. *Homo Ludens: A Study of Play Element in Culture*. Boston: Beacon Press, 1955.

Hutchens, B. "Great Wolf Lodge Resort Much More Than Just a Water Park." *News Tribune*, 2008. Accessed 26 February 2011. Link no longer available.

James, Carrie, with Katie Davis, Andrea Flores, John M. Francis, Lindsay Pettingill, Margaret Rundle, and Howard Gardner. *Young People, Ethics, and the New Digital Media: A Synthesis from the GoodPlay Project*. Cambridge, MA: The MIT Press, 2009.

Jenkins, Henry. *Convergence Culture: Where Old and New Media Collide*. New York: New York University Press, 2006.

Jenkins, Henry. "Interactive Audiences? The 'Collective Intelligence' of Media Fans." In *Fans, Bloggers, and Gamers: Exploring Participatory Culture*, 134–52. New York: New York University Press, 2006.

Jenkins, Henry, with Ravi Purushotma, Margaret Weigel, Katie Clinton, and Alice J. Robison. *Confronting the Challenges of Participatory Culture: Media Education for the 21st Century*. Cambridge, MA: The MIT Press, 2009.

Juul, Jesper. *Half-Real: Video Games between Real Rules and Fictional Worlds*. Cambridge, MA: The MIT Press, 2005.

Kerr, Aphra. *The Business and the Culture of Digital Games: Gamework/Gameplay*. Thousand Oaks, CA: Sage, 2006.

Kushner, David. "Escape to Middle-Earth." *Wired* 17, no. 9 (2009).

McGonigal, Jane. *Reality Is Broken: Why Games Make Us Better and How They Can Change the World*. New York: Penguin, 2011.

Montola, Marcus, Jaakko Stenros, and Annika Waern, eds. *Pervasive Games: Theory and Design*. Burlington, MA: Morgan Kaufmann, 2009.

Örnebring, Henrik. "Alternate Reality Gaming and Convergence Culture: The Case of *Alias*." *International Journal of Cultural Studies* 10, no. 4 (2007): 445–62.

Porter, Jennifer E. "Pilgrimage and the IDIC Ethic: Exploring *Star Trek* Convention Attendance as Pilgrimage." In *Intersecting Journeys: The Anthropology of Pilgrimage and Tourism*, edited by Ellen Badone and Sharon R. Roseman, 160–79. Champaign: University of Illinois Press, 2004.

Rosenbloom, A. "A Game Experience–in Every Application." *Communications of the ACH*, 46, no. 7 (2003): 28–31.

Salas, Randy. "Enter the Dragon: Grab a Wand and Hunt All Manner of Fantasy Creatures at MagiQuest, a New Interactive Adventure at the Mall of America." *Star Tribune*, 13 October 2009. Accessed 23 March 2016. https://www.highbeam.com/doc/1G1- 209623490.html.

Salen, Katie, and Eric Zimmerman. *Rules of Play: Game Design Fundamentals*. Cambridge, MA: The MIT Press, 2004.

Schoolfield, J. "Quest Masters." *Funworld* 22, no. 10 (2006): 42–51.

Schroeder, Randy. "Playspace Invaders: Huizinga, Baudrillard and Video Game Violence." *Journal of Popular Culture* 30, no. 3 (1996): 143–53.

Scott, A. O. "At Comic-Con, Bring out Your Fantasy and Fuel the Culture." *New York Times*, 15 July 2015. Accessed 23 March 2016. http://www.nytimes.com/2015/07/19/movies/at-comic-con-bring-out-your-fantasy-and-fuel-the-culture.html.

Sparks, Kimberly A. "Will Your Next Job Be in an Alternate Reality?" *Canadian Screenwriter* 11, no. 2 (2009).

Stein, Louisa Ellen. "'This Dratted Thing': Fannish Storytelling through New Media." In *Fan Fiction and Fan Communities in the Age of the Internet*, edited by Karen Hellekson and Kristina Busse, 245–60. Jefferson, NC: McFarland, 2006.

Szadkowski, J. "Resorts' MagiQuest Actives Slick Fantasy." *Washington Times*, 25 August 2007. Accessed 23 March 2016.

Théberge, Paul. "Everyday Fandom: Fan Clubs, Blogging, and the Quotidian Rhythms of the Internet." *Canadian Journal of Communication* 30, no. 4 (2005). Accessed 23 March 2016. http://cjc-online.ca/index.php/journal/article/viewArticle/1673/1810.

Uren, Tim. "Finding the Game in Improvised Theater." In *Second Person: Role-Playing and Story in Games and Playable Media*, edited by Pat Harrigan and Noah Wardrip-Fruin, 279–84. Cambridge, MA: The MIT Press, 2007.

Wood, Megan M., and Linda Baughman. "*Glee* Fandom and Twitter: Something New, or More of the Same Old Thing?" *Communication Studies* 63, no. 3 (2012): 328–44.

· 7 ·

THE DIGI-GRATIS AND CROWDFUNDING

> *Money is the key that opens the door to man's dark secrets. … . And grants me the only thing in the world I respect: these books.*
>
> —Pérez-Reverte, *The Club Dumas* (p. 60)

> *Hi. LeVar Burton here. You may know me as Kunta Kinte, from ROOTS, or Geordi La Forge, from Star Trek: The Next Generation. You also may have grown up with me on Reading Rainbow. It was my mother who taught me that, by picking up a book, I could "go anywhere" and "be anything." … Now, I am hoping you will join me on my mission: to bring Reading Rainbow back for every child, everywhere.*
>
> —LeVar Burton, *Bring Reading Rainbow Back for Every Child, Everywhere!*

When *Digital Fandom* was first written, crowdfunding was in its infancy. Although now a major part of the digital fan experience, crowdfunding is barely over a decade old. The first crowdfunding website, according to Sara Bannerman, was Kiva, a microfinance agency started in 2005 that used crowdfunding to finance small loans to entrepreneurs.[1] Users of Kiva can ask for loans for any number of reasons; as of 2016, these loans were being paid back at a 98% repayment rate.[2] Kiva is still in use, but as the popularity of crowdfunding has grown, the emphasis has shifted from loans to donations.[3] Perhaps the most well-known crowdfunding site—and the focus of this chapter—is

Kickstarter, which was founded in 2009. The fact that over 34 billion dollars was crowdfunded in 2015—a rise from 1.5 billion in 2011—indicates not only that individual donations are on the rise, but that it's big business as well.[4]

For Corso in *The Club Dumas*, money is a means to an end: It is a way not only to have power over others, but also to enable a particular type of fandom. As I argue in this chapter, crowdfunding sites like Kickstarter rely on harnessing affective economies—the state wherein "emotions do things, and they align individuals with communities … through the very intensity of their attachments"—in order to generate fan-like engagement with the campaigns.[5] This engagement relies on and exemplifies the Digi-gratis economy. Henry Jenkins has used the term "affective economy" to describe a new paradigm of television branding's engagement with consumer emotions, when a television series like *American Idol* "emphasizes the emotional commitments consumers make in brands as a central motivation for their purchasing decisions."[6] As Matt Hills has noted, the concept of affective economies can also be applied to crowdfunding.[7] Crowdfunding campaigns for media texts have not just managed to tap into a vein of motivated, affective, and financially stable fan audiences in order to fund their projects, but also to utilize the same Digi-gratis economic system as fans do within fan communities. In this chapter, I examine three cinematic campaigns and one television/mobile app campaign as examples of entertainers drawing on their own fan audience for financial support. These campaigns are key examples of 21st century digital fandom.

But looking at just the Kickstarter campaign out of its fannish context only tells part of the story. As Liza Potts describes in her analysis of how musician Amanda Palmer uses fan financing:

> Palmer's fans are not passive; they are her contributors, co-conspirators, and supporters within a much larger do it yourself (DIY), participatory culture forming between artist and fan … the ability to engage with a fan base, co-produce with them, and support each other is a major factor in finding … success.

Potts calls this an "ecosystem" of fans, producers, and record labels, and maintains the importance of fan engagement as "a domain for participation."[8] To see fan participation in this system requires an exploration some of the economic tensions that emerge from crowdfunding campaigns. Relating crowdfunding to fandom more specifically, Mel Stanfill argues that "we can't assume fans are blindly throwing money. This does have structural similarities to other forms of fan microcosm. Fans aren't duped."[9] Ultimately, I hope that by reflecting on the crowdfunding phenomenon as a symptom of this larger economic and cultural transition, fan scholars can make significant and thoughtful interventions on the monetization of affect in today's Digi-gratis economy.

Digi-Gratis Economy

In the first edition of *Digital Fandom*, I introduced the concept of the Digi-gratis economy, a mash-up of the gift and the market economies (see chapter 4). My argument was that in the digital world, viewing the economic interaction between the fan and the media producer as simply one or the other did not tell the whole story; rather, the digital environment reveals a more complex interaction wherein "gifts" can be monetary, "replies" can be non-reciprocal, and monies become aspects of affection and emotion. This type of "mixed economy" revalues valuation and ties fannish enthusiasm directly to financial incentivism.[10] In the Digi-gratis economy, as Henry Jenkins, Sam Ford, and Joshua Green have demonstrated, "value gets produced through the reappraisal and recirculation of … the residual."[11] Because digital goods and services are non-tangible (and therefore non-"consumable" in the literal sense of the term), the exchange of information and goods takes on a different tenor than it would in the tangible world: Giving bytes is different than giving things, and asks for a different response in return.

In the time since *Digital Fandom*'s publication, the concept of the Digi-gratis has been used in a variety of contexts. Kevin Veale expands the discussion of the Digi-gratis to examine the affective dimension of fan experiences. For Veale, the "affective dimension of something refers to a spectrum of subjective things that we feel but that we are less conscious of than our emotions. It is possible to name an emotion and pin it down, but the affective tone of an experience is harder to label and is more amorphous." Through an analysis of the *My Little Pony: Friendship Is Magic* fandom, Veale reveals the community engagement between the creators of the show and the fan community as a type of Alternate Reality Game, wherein the Digi-gratis economy functions to give fans "pleasure … and part of that pleasure is grounded in the social capital it gains within the community."[12] His point about ARGs has been taken up by Burcu S. Bakioglu as well, writing about the economic interaction between the fans and creators of LonelyGirl15.[13] In a fan-meta piece called "The Multiverse as Asylum," (see chapter 8), fan-scholar e_cademia, drawing on Angela Ndalianis's concept of the neo-baroque, describes the result of the Digi-gratis economy: "Fandom does not exist unless multiple people produce fan works and exchange them. This model is integral to fandom."[14] Contemporary media practices—such as taking (freely written) fan fiction and commoditizing it (e.g., *Fifty Shades of Grey*)—highlight the way the Digi-gratis economy functions in both fannish and industrial models.[15]

Indeed, in her analysis of the development of copyright laws specifically in relation to *Harry Potter* fan fiction, Liberty Kohn has used the Digi-gratis to define this historical moment to argue that there has been a "greater neoliberal market intrusions into all social structures" and that "academia should not only expect, but must analyze intrusions by traditional markets and spheres into the new historical texturing of the [Web] 2.0 era."[16] Applying the Digi-gratis model to crowdfunding reveals a similar neoliberal trend: Mark Duffett argues that "as the electronic media shifted they facilitated the co-optation of fandom as a marketing device, and business became increasingly interested in fan phenomena."[17] Engaging a type of fan audience in this manner ignores many unique fannish communities and instead focuses on, as Kristina Busse argues, "a shift toward industry and production." She goes on to note that "as different business models vie for supremacy, there exist many contradictory approaches to fan labor. They all share the desire to capitalize on fannish love and its ability to generate and disseminate interest.[18] Yet, the Digi-gratis economy also reveals the emotional and affective satisfaction generated via fan communities engaged in donations and funding campaigns. Just as the economic systems are mashed up, so too are the relationships between fan and producer, between funder and fundee.

Crowdfunding and the Digi-Gratis

The space for fans to finance aspects of the media process opens up new models for understanding fan participation, redefined by the underlying social structure of digital fandom's interaction with the media. This Digi-gratis view of the economics of fandom highlights changing paradigms of fan viewership as a mutable form of participatory culture. Crowdfunding campaigns that successfully engage their fans in a more participatory manner—acknowledging previous fan work, noting the saliency of fan activities in the past, appealing to fan attention in the future—highlight fandom as a consequential role in the production of the media object.[19] This occurs whether a specific media object (e.g., the *Veronica Mars* Kickstarter campaign, the *Reading Rainbow* campaign) or a media celebrity (e.g., Amanda Palmer's successful use of fan funding; Zach Braff's maligned effort). But such reliance on fandom as a temporal structure is predicated on those fans existing in the first place. Engaging a fandom that doesn't exist may spell failure for the campaign (e.g., *Darci's Walk of Shame*).

Each of these Kickstarter campaigns drew on support from fan communities to generate funds, highlighting the power of fandom to enact media change but also problematizing fandom's transformative potential.[20] This economically networked participation—the Digi-gratis—is itself engendered by particular technological processes. Scholars have argued that participatory online culture is exploitative of fan money, time, and work;[21] in the case of *Veronica Mars* specifically, Luke Pebler notes that "Under current rules, it is possible for media entities to exploit their fan bases in a way that is pernicious and ultimately unsustainable."[22] Other scholars demonstrate how fans in the case of the *Veronica Mars* Kickstarter are well aware of their role in the financing of a film, and are not exploited at all.[23] For example, Bertha Chin argues that fan agency is crucial for understanding a desire/willingness to donate to a Kickstarter campaign: "Frustratingly, fan agency always gets left out in arguments which purports concern that fans are being duped by studios and networks."[24] Chin, like Stanfill, demonstrates that fans are not "dupes" but rather are aware of the industrial situation. Indeed, crowdfunding exists as a tension between the neoliberal emphasis on the role of individual, and a more open, participatory model of networked action.

Crowdfunding then becomes a way of employing fan social capital in a market context.[25] In some respects, fandom would seem to exemplify participation; but as Jenkins notes, "there is an urgent need to develop a more nuanced account of what is participatory about fandom."[26] As fan researchers we need to be cautious to avoid making "fandom" and "participatory culture" synonymous. As Leora Hadas has noted:

> while the logic of participation might seem to mirror the logic of fandom … they are not one and the same; … even an interpretive fannish community cannot be seen apart from its own norms and ideals; and … the loose and open nature of participatory culture as idealized in the Web 2.0 model might, in fact, clash with these ideals as much as it might clash with the wider cultural model of production it is threatening to replace.[27]

As she goes on to show, while some fans may embrace the emancipatory potential of Web 2.0 participation in digital spaces, other fans embrace the strict hierarchical functionality, neoliberal notions of capital, and regulation of fannish content engendered by Web 2.0 digital networks (for instance, Ayhan Aytes writes about the neoliberal harnessing of cognitive labor online, and Abigail De Kosnik describes fans' labor as generative of capital).[28] Fan funding seems more tied to this neoliberal style of fandom than to a transformative one, illustrating that fandom is not a homogenous grouping and the boundaries between fannish groups often remain entrenched.[29] Instead, different fans embrace different aspects of the web for different purposes.[30] Such is the power of fandom, to resist easy typification.

The problem with linking fans to participatory culture is that it simplifies both fan activity and audience participation. As I and others have noted, there is an "over-emphasis on participatory fans in the academy."[31] Fan activity can be incredibly transformative, but it can also, for example, be what Hills calls "mimetic," or the "desire … to replicate what's seen on screen."[32] This focus on the non-transformative fan activities is important for fan studies to remain grounded in all aspects of fan activity. Although crowdfunding appears to harness one particular type of fandom to generate funds for a media project, ignoring the importance of fans' financial participation risks eliding these fans from fan studies generally.

Crowdfunding emphasizes the monetization of fans' emotional connection to an object. Incentivizing donations blurs social incentives with economic ones as Talia Leibovitz, Antoni Roig Telo, and Jordi Sánchez-Navarro discuss.[33] Yet, as Venkat Kuppuswamy and Barry Bayus suggest, "social information (i.e., others' funding decisions) will play an important role in the ultimate success of a project."[34] When this support is monetized rather than reliant on social relationships, however, "[m]any potential backers do not contribute … because they assume that others will provide the necessary funding." There is also evidence that even when economic incentives are at play in crowdfunding campaigns, or even when profit-sharing is an option for crowdfunding, as Patryk Galuszka and Victor Bystrov describe, "the withdrawing of profits from the platform is sporadic. In most cases 'earning money' means reinvesting the resources in successive projects."[35] In other words, for many people, donating to a campaign is itself an affective decision, not an economic one. This is supported by Julie Hui, Elizabeth Gerber, and Mike Greenberg, who have noted that the work that goes into creating a crowdfunding campaign is often more time-consuming and requires a greater variety of skills than more traditional funding; it stands to reason that running a campaign thus stems from multiple incentives.[36]

Digi-Gratis and Fandom

There are many economic models used to develop theories of crowdfunding.[37] Much like with fandom and described earlier in this book, two of the more common are the commons-based peer production model, in which the creative energy of communities is coordinated into meaningful projects, and the model of collective intelligence, in which individuals are empowered as part of a community to contribute meaningful value to information transactions.[38] Partly, this focus may be a reflection of the very social technology that enables crowdfunding to function. In comparison to a more "slacktivist" involvement in a campaign, donating money appears to be more involved and more helpful. Slacktivism describes a type of online participation wherein individuals post support for an activity—"liking" a cause on Facebook, for example. Donating money can appear more worthwhile than simply "liking" a media object on Facebook. It can even, as described by Ethan Tussey, be a way for fans to critically engage with a broken media system.[39] But the terms of the discussion take monetization as given; other options or ideas are cut off and an economic thought determines the discourse and argument about the campaign. As Anne Kustritz argues, monetization becomes a default setting.[40]

However, in the context of the Digi-gratis economy, the use of crowdfunding is one way that fans can find meaning out of giving money to a media company. In addition to creative practices, fan studies examines how the products of professional media industries, like action figures, toys, games, and collectibles, can reflect different fan attributes. In *Cult Collectors*, Lincoln Geraghty focuses on collecting merchandise, and examines the way that fans' "relationships [are] created and embodied" by these collections. For the fans in Geraghty's study, fan identity forms through the "convergence of popular fandom, new media, nostalgia and contemporary toy culture"; for the fans that donate to crowdfunding, such nostalgia is meaningful in the way the campaigns they donate to draw on the affective memory of the fans' interactions with the original media entity.[41] Fans may give money to buy memorabilia, donate time to staff conventions, or spend energy creating fan fiction, vids, or other cultural products: Each element becomes one node in a network of value and exchange within an affective media environment. As Jenkins, Ford, and Green assert, this "reappraisal and recirculation" of value portends a new form of digital economy.[42] There are different levels and types of participation within a project, and for many fans, donating money can offer participatory pleasures.

By more fully integrating the concept of the Digi-gratis into fan research, the perception of consumption and production as *practices* shifts to see them as one cogent and chronological *process* instead. Applying the concept of the Digi-gratis economy more fully to fan crowdfunding illustrates how consumption and production are uniquely positioned as nodes in a network of fan activity. New technology shapes fannish work—the specific affordances of social media influence the development of fan communities, especially in the way those communities interact online. But to analyze each node of this network as separate ignores the larger temporal processes that develop through fan interaction, sometimes market, sometimes gift, always Digi-gratis. If each fan's individual economic interaction with a media object emerges from both gift and market elements—from the first viewing of an episode, to sharing with friends and fans, to purchasing paratextual products, to the final click of the pause button—then the diverse array of digital fan interactions generates a temporal exploration of crowdfunding.

As I've previously discussed, specifically looking at *Inspector Spacetime* fandom, fan audiences can develop and focus crowdfunding projects as a site of pleasure.[43] Giving money to a project can be enjoyable in and of itself; although the "power" of creation may still lie within the media producer, donating to a crowdfunding campaign provides the specific pleasures of connection with an artist and a feeling of community.[44] For the fans of *Inspector Spacetime*, a parody of *Doctor Who* as seen on the NBC sitcom *Community*, participation in a Kickstarter campaign to generate an *Inspector Spacetime* webseries (separate from *Community*) not only gave them the pleasure of creating new fan work, but also provided a tangible (if momentary) sense of *being a producer themselves*. Providing capital generates notions of participation. At the same time, these notions did not end up turning fans *into* producers; instead, they reified the boundary between fan and producer by establishing a clear demarcation among participants on the Kickstarter campaign. The actual producers of *Inspector Spacetime* (NBC-Universal and Sony Pictures) threatened to sue if that name was used, so the creators of the Kickstarter were forced to change the name (and identity) of the series to *Untitled Web Series about a Space Traveler Who Can Also Travel through Time*, alienating the fans that contributed to an "*Inspector Spacetime*" Kickstarter. The assertion of authority by the media industry reflects the power differential that still exists between media creators and media consumers. Yet, the fact that *Inspector Spacetime* fans still flocked to the Kickstarter even after the name change, and even donated over $25,000 to an IndieGoGo campaign for a film version of *Untitled Web Series about a Space Traveler Who Can Also Travel through Time*, indicates that fan participation doesn't always require their productivity being sanctioned by the industries.[45]

Crowdfunding campaigns designed by media corporations occupy a similar conflicted stance between production and consumption. But digital fandom is always about this conflict, of noting the saliency of fan work within the context of corporate media. By following a crowdfunding campaign from opening video to funded project, one could theoretically see the birth of a new form of creative product—one formed not just from corporate control and organization, but also from crowdsourcing and fan involvement, a hybrid system of production and consumption.[46] Fans, engaged in crowdfunding endeavors, not only personify the campaign, but also personify the production process of the campaign, from inception to reception. Rather than seeing the production process in stages—pre-production/production/post-production, e.g.—a look at the Digi-gratis economy within neoliberal fannish experiences portends a more fluid process of participation and observation.

Film Campaigns

Digital fandom illustrates the procedural logic behind fan participation in crowdfunding. By analyzing three film campaigns on Kickstarter—and then comparing them to a successful app campaign—I illustrate the applicability of the Digi-gratis economy to contemporary digital fandom. Fans on crowdfunding campaigns may generate a feeling of participation, but it is a participation harnessed by the media corporations as an appropriation of affective economies.

In March 2013, Kristen Bell and Rob Thomas, of *Veronica Mars* fame, quickly became Kickstarter superstars by completely funding a *Veronica Mars* movie within 8 hours of opening the project. Popular press and academic blogs touted the new era of film funding this heralded, while others illustrated antecedents for this type of film funding.[47] A few weeks later, Zach Braff, from the sitcom *Scrubs*, took to Kickstarter to fund his own indie film called *Wish I Was Here*, although this was met with more ire than Bell and Thomas's endeavor.[48] And a month after Braff's Kickstarter was fully funded, Melissa Joan Hart (the lead in *Clarissa Explains It All*) attempted to use Kickstarter to fund her romantic comedy *Darci's Walk of Shame*. Unlike Bell and Thomas's and Braff's campaigns, Hart's campaign failed to earn the 2 million dollars she had set as her goal.[49]

The next year, on May 29, 2014, social media exploded with news that LeVar Burton's Kickstarter campaign for a classroom app, based on his *Reading Rainbow* children's television series, had made its 1 million dollar goal in less than 24 hours. *Reading Rainbow* had been a children's public television series focusing on the importance of reading. The campaign, which aimed to bring an educational app based on the show into classrooms across the world, continued to generate momentum over the next month until it finally made over five million dollars by July 2.

The four crowdfunded campaigns analyzed here demonstrate different levels of fan engagement. While the final existence of the *object* (as a demonstrable *product*) is easily established (*Veronica Mars*, *Reading Rainbow*, and *Wish I Was Here* were funded; *Darci's Walk of Shame* was not), the continued existence of each of the campaigns allows for a more concrete exemplar of the crowdfunded campaign's temporal actuality. Each project hinges on fan nostalgia by acknowledging the presence of fandom at multiple nodes in the production process. Nostalgia plays a crucial role in understanding contemporary fan interests, and Geraghty writes that "notions of nostalgia and memory are bound up in the recreation of a contemporary fan identity rather than a recreation of the past," highlighting the temporal and historical awareness of fandom within and for a media project.[50] Examining the Digi-gratis within the economics of fandom deepens the investigation of this nostalgia as it applies to what Jenkins, Ford, and Green describe as a social transformation of technology. With access to more and more information, individual consumers are beginning to "spread" media content beyond the boundaries it was originally intended for. Spreadability—the potential for "audiences to share content for their own purposes"—is a paradigm shift in the way analysts can think about media distribution: If the broadcast paradigm saw media circulation limited to an elite few, spreadability sees more distribution power to the masses.[51]

Indeed, Kickstarter itself has spreadable (and Digi-gratis) characteristics. Specific campaigns often draw on word-of-mouth advertising by individuals spreading information about each project. And once a donation is given, Kickstarter encourages the donator to post news of the donation to social media like Twitter and Facebook. For example, I first heard about the *Veronica Mars* Kickstarter on my Facebook wall when two colleagues and a friend each separately posted that they had contributed to the campaign, and my wife mentioned the *Reading Rainbow* campaign to me. Fans are here spreading not only their own generosity, but also news of the project itself. As Kuppuswamy and Bayus note, this spreadability increases as projects near completion, both because project creators make final pleas for more donations (if not funded) and because more people wait to donate until the end of the campaign.[52]

To examine each Kickstarter project on its own, however, one risks inducing specific characteristics from particular campaigns. As Jonathan Gray suggests, individual Kickstarter campaigns appear to harness an "amorphous" crowd that does not aggregate into a community.[53] However, seeing multiple projects manifest similar characteristics allows us to see the "community of participants"—the crowd itself becomes participatory, engaging in "the production and distribution" of media products themselves.[54] In other words, crowdfunding fandom alters the experience of online participatory culture. Each of these Kickstarters reveals that different digital fan processes help shape the eventual outcomes of the campaigns.

This participatory process holds true whether or not the crowdfunding campaign could be considered successful. For example, the *Veronica Mars* Kickstarter opened on March 13, 2013 and within one day had surpassed its 2 million dollar goal. When it closed on April 12, 2013, the campaign had over 91,500 backers who invested a total of 5.7 million dollars toward the making of the film. Warner Brothers became involved in the project soon afterward, offering to distribute the film and as of June 2013, the movie had started filming, with a 2014 release date. By all accounts, this particular Kickstarter campaign was a financial success, and the thousands of people who invested will know that their money has helped bring this fan favorite show back on the screen.

This strong fan base is the reason for the highly visible success of the *Veronica Mars* Kickstarter.[55] In the three years the show was on the air (2004–2007), it was critically acclaimed but had a relatively small audience. However, the audience was loyal and dedicated, as San Diego reporter Karen Peterson describes.[56] Along with other fan campaigns, the *Veronica Mars* fandom became highly invested in restarting the show—the *Veronica Mars* "Cloud Watchers" organized viewers to garner higher ratings during the show's airing, and later sent over 10,000 Mars candy bars to the CW to try and save the show from cancellation.[57] This parallels similar efforts by other fan groups to get their favored shows back on the air—*Star Trek* fans famously writing letters to CBS, *Jericho* fans sending peanuts to CBS, *Roswell* fans mailing Tabasco sauce to the WB, and *Chuck* fans buying Subway.

Traditionally, fan studies scholars might view the activity of all these different fan communities through a reception studies lens. That is, each of these fan groups was active and participatory in the media environment, but only *after* they had already been consumers of the show. *Jericho* fans and *Roswell* fans sent food items because characters on those shows mentioned that food—it was a direct reference to what had been outlined on air. Taking quite literally Jenkins's "poaching" metaphor from *Textual Poachers*, these fans took their "save-our-show" campaigns directly from the text itself. In no way is this intended to demean the act of these fans—they are using the resources at their disposal to interact as best they are able given the media environment.

What the *Veronica Mars* Kickstarter campaign illustrates, however, is the presence of fans' active receptive practices within multiple aspects of the production process. The receptive practices of the filmmakers toward the fan campaign via acknowledgment of fans' work may have helped spur fans to want to donate to the Kickstarter. And in turn, the reward structure of Kickstarter augments fannish enthusiasm: The rewards—from the digital downloads of the film to a guest role onscreen—reflect fan interest in the future of the project (i.e., as Mittell states, with "fans as funders, we're basically just pre-buying merchandise, DVDs, or experiences").[58] Disregarding the artistic issues that might arise when a fan feels like he/she has a financial incentive to define the narrative, in a transmediated text like *Veronica Mars*, specific identifiable moments of "production" and "reception" are difficult to articulate and differentiate.[59] When *Veronica Mars* fans attempted to get the show back on the air in 2007, their efforts were highly visible to the *Veronica Mars* production team. According to an article at the time, Rob Thomas even noted fans' efforts to save the show.[60] It's possible that the post-reception influence of fans on the show helped to convince Rob Thomas and Kristen Bell to crowdfund the film.[61] In the video promoting the campaign (itself a "spreadable" component of Kickstarter projects), both Bell and Thomas mention the *Veronica Mars* fandom by name—calling them an affectionate "Marshmallows"—and use community-based language like "this is *our* chance to make the *Veronica Mars* movie happen," fully involving the fan community in the project. When the film was released a year after the Kickstarter, Thomas referenced fan work again: "Today, *Veronica Mars* returned, and it's all because of you," he wrote on the Kickstarter page, acknowledging the fandom throughout *Veronica Mars*'s existence.

That the Digi-gratis economy emerges in Thomas's engagement with fandom happened long before the Kickstarter, as Matt Hills argues: "Thomas' engagement with 'unauthorized' [*Veronica Mars*] material, his strategic use of Television without Pity and his sanctioning of self-reflexive anti-industry discourses within [*Veronica Mars*] all offered precedents for the … campaign." The interactive potential of fans and producers becomes fluid in the Digi-gratis economy; crowdfunding presents a useful metaphor for seeing this procedural relationship grow and develop. Similarly, fans of *Veronica Mars* could look back on the Kickstarter campaign as a way of "tactically poach[ing] from the Kickstarter crowdfunding process by negotiating how this is meaningful to them."[62] If the entirety of the *Veronica Mars* fandom can be identified as a process—from the moment the show was conceived of by Rob Thomas until its final download decades from now—then fans and producers both play a participatory role in this construction. Despite the unequal *financial* or *creative* role in a neoliberal media environment, digital fandom is co-created. The influence of one upon the other cannot be measured simply in dollar amounts or volumes of fan fiction produced. Rather, the *collaborative potential* of both groups portends the appearance of the Digi-gratis economy.

Although not as successful as the *Veronica Mars* Kickstarter, Zach Braff's *Wish I Was Here* Kickstarter campaign also demonstrates the participatory potential of fandom. Braff started his campaign on April 24, 2013 and as of May 24 had raised over 3.1 million dollars with 46,500 backers. This total was over a million dollars more than he had originally planned. The reward structure for Braff's campaign, similar to the *Veronica Mars* campaign, offered some tangible rewards like tee-shirts and soundtracks, and also intangible rewards like being able to attend a special screening with him in various cities around the world.

Braff's film, a follow-up to his successful indie hit *Garden State* (2004), met with ire from many online. The reason for this is aptly summed up by NBC's Maria Fernandez: "someone with plenty of money of his own and direct access to Hollywood resources has no business playing in the kiddie pool that is supposed to be Kickstarter."[63] Myles McNutt describes how the reaction considered it a "bastardization" of the Kickstarter process.[64] Even *The New York Times* questions the ethicality of his funding the film through Kickstarter when he has plenty of money himself.[65] Notably, such ethical discussion didn't appear with the *Veronica Mars* campaign, even though Kristen Bell is also a high profile actor who also has a large bank account (although she did take a pay cut to appear in the film). And, Salon.com's Erin Keane defends the project: Anyone who thinks "Kickstarter [is] an amazing resource for cash-strapped innovators who might otherwise not have access to more than their own savings accounts," she argues, is "thinking like a production company, not like a fan who doesn't mind buying into a glorified fan club to show her support for an artist she likes."[66] In other words, Braff's campaign succeeded because of his own personal star power and fannish influence. Although *Wish I Was Here* is not part of a franchise, Braff's indie film *Garden State* and his work on the TV show *Scrubs* gives his fans the power to articulate a previously established relationship with him.

Braff establishes a connection with his fans via the video on his Kickstarter page. The video opens with him articulating his knowledge of his fan base, by joking about his most commonly asked question—"Can I mount you and yell 'Eagle!'?"—and sitting next to a poster for *Garden State*. The irreverent joke is similar to many from *Scrubs*. He then appeals to his fans' appreciation of his previous work (Donald Faison, co-star of *Scrubs*, makes a cameo in the video) and explains that he would like final cut/casting decisions on his movie. Throughout the video, Braff clearly situates his audience as *his* fans—making sure they are included in the process. "Most importantly," he says, "[*Garden State*] became a very special film *to you*, who did understand what I was trying to say." However, this articulation maintains a separation between Braff as producer and his fans as viewers:

> I want to bring you, my fans, the truest representation of what I have in my brain. That means I'll have the final cut over what ends up in the movie and I get to only cast the actors I think are perfect for the roles. Please *help me* make another movie *for you* like *Garden State*. (emphasis added)

Braff distances himself from the fans, using less-inclusive language and singular personal pronouns. Braff's ownership of his own production may have caused some of the negative backlash from fans. Throughout the video, Braff aims to connect with his fan base, but also holds them at a distance. Yet, as an example of the Digi-gratis economy, the fans are spurred to donate by the previous receptive and productive practices they've engaged in in the past.

Former child star Melissa Joan Hart's Kickstarter campaign for her romantic comedy *Darci's Walk of Shame* utilized a video like Braff and Bell did, but failed to make the 2 million dollar minimum to fund the project. The rewards for her project were similar to Braff's: tee-shirts, digital downloads, and options to attend screenings. However, few of these rewards were given out. She cancelled the campaign two weeks early (although, given the statistic that most Kickstarters get funded in the last few days, that may have been a mistake).[67] What is intriguing about Hart's failed Kickstarter project is not just in the ways it failed to attract an audience, but also in the way it continues to unfold as a *process*. The participatory culture of the project suffered by neglecting digital fandom and creating a distinction between fan participation and fandom as a process.

Most saliently, Hart did not appeal to her fan base in the same way as Bell and Braff did. If fandom is itself always a process, then alienating a section of fandom through temporal negation may have hurt her chances for financial donation. Hart, made famous in the mid-1990s for staring in Nickelodeon's *Clarissa Explains It All*, has a fan base aged in their twenties and early thirties. Since *Clarissa*, Hart has appeared in another children's show, *Sabrina the*

Teenage Witch, and in a sitcom aimed at families on the ABC Family network, *Melissa and Joey*. Rather than reaching out to her fan base via direct address, Hart's video on her Kickstarter page stars her and her mother, reading through scripts that take familiar elements from her previous works and mocking them. "I think you need to do something different," her mother says. They note how Hart has moved past her roles on Nickelodeon and in *Sabrina*, and her mother suggests she takes on a new role, as an adult. Hart admits "I don't know anyone who's going to put me in a movie like this." When she directly addresses the audience, Hart doesn't appeal to her fan base's early years—she says "I need your help to prove to people that *I'm more than* just Clarissa or Sabrina" (emphasis added). Any fans attached to her previous work are shut out by her direct address. Literally talking to the fans, she tells them she does not want to appeal to them. She distances herself from her fan base, and her fans did not get behind the film, as they were alienated from the production process.

Because fans were not involved at different times, nor invoked at the "proper" times, the crowdfunding campaign for *Darci's Walk of Shame* could not invoke digital fandom into the participatory paradigm. Hart's video had the opposite effect than was intended, according to the *LA Times*: "Hart's project lacked a nostalgia factor."[68] But more so than nostalgia, it also lacked a specific acknowledgment of any Melissa Joan Hart fan's work. Hart failed to recognize her fans as a unique group; crowdfunding through fandom works through recognition of participation.

Kickstarting the Digi-Gratis with *Reading Rainbow*

Reading Rainbow was an American television series that ran on public broadcasting service (PBS) television from 1983 to 2006. In the US, public broadcasting is commercial free, and supported by individual contributions, donations, and government financing. The show featured LeVar Burton and various celebrity guests reading children's books and then being magically transported to the narratives, lessons, or experiences contained within those books. For hundreds of thousands of children, the show introduced and validated reading as a worthwhile activity. The campaign on Kickstarter wasn't to restart the television show, but rather to develop a classroom-version of *RRKidz*, an already-existing commercial mobile app based on *Reading Rainbow*, as well as teacher development materials to be given to disadvantaged classrooms and sold to others. Within this lofty-sounding goal, however, lies a particular economic tension that the larger concept of crowdfunding reveals:

the industrial commoditization of affect. By linking the gift-giving attributes of fandom to the market-driven economic actualities of the industry, *RRKidz* develops from the same Digi-gratis economy that also brought back, for instance, *Doctor Who*.[69]

It is hard to argue with a campaign that purports to bring reading to disadvantaged children. But some, including Molly Hensley-Clancy and Caitlin Dewey, were critical of the move to commercialize of this act: "The campaign borrows from the language of charities to present Reading Rainbow as a solution to the American education crisis."[70] Why are individuals donating to a *Reading Rainbow* campaign to encourage reading in schools, when we have multiple governmental and civic bodies expected to do the same? The controversy stems from a previously non-profit television series being turned into a for-profit educational app. *RRKidz* is the commercial wing of *Reading Rainbow*, the company that produces and distributes the app. This commercialism is not unique to the *Reading Rainbow* campaign (although the language is vague on the site, the campaign never claims to be a charity).

The issue here is not that one particular "good" campaign is commercialized, but rather that this critical emphasis on economics becomes the *de facto* discourse guiding crowdfunding campaigns in general. Suzanne Scott has argued that "By engaging in a manner that media industries can quantify and comprehend, fans are endeavoring to reimagine the production process, while acknowledging the realities of that process." Through the Digi-gratis economy, crowdfunding can become a tool that "relies on an embrace of fannish notions of transformative engagement with media culture, and the industry's capacity to imagine alternatives" to that transformative nature.[71] At the same time, Ethan Tussey focuses on the "long history of fan activism against the television ratings system and network executives" as the lesson of fan-ancing—helping to finance a film or app represents "an indictment of the television industry."[72]

Focusing on the commercial nature of the product on any crowdfunded campaign reveals the underlying neoliberal forces at work in contemporary society. As Robert McChesney defines the term,

> Neoliberalism is the defining political economic paradigm of our time—it refers to the policies and processes whereby a relative handful of private interests are permitted to control as much as possible of social life in order to maximize their personal profit.[73]

A political and economic shift toward neoliberalism focuses on establishing a laissez-faire market (free from governmental oversight), consumerist culture, and libertarian ideals of privatization of goods and services. A neoliberal culture harnesses the pleasures of leisure into economic profit for others—browsing social media becomes a way of making money for media corporations through advertising, for example—as "the last vestiges of private, intimate life, relationships and emotions are, often unwittingly and gradually, sacrificed to work."[74] Aspects of crowdfunding are neoliberal, not only because crowdfunding extols in the individual creator/author of the campaign, but because it harnesses and normalizes the language of economics. Crowdfunding links commercialized successes with affect and emotion, turning a charitable "feeling" into a commercialized payment—the flip side of the Digi-gratis.[75]

In this economic shift, individual affect becomes enmeshed within a commercial system.[76] Fans are implicitly, through representational, economic, and interpellative practices, told how to act, and what type of fandom is preferred.[77] Crowdfunding funnels affect into commercial ventures. In a neoliberal environment, education shifts from a basic right to an economic policy. And for Dewey, the question of education is central to the issue of the campaign. She notes that, "when *Reading Rainbow* began in 1983, the big question was, 'how do we get kids interested in reading?' By 2009, that question had become, 'how do we teach kids to read, period?'"[78] What was once a focus on ways to incentivize reading turns into an educational necessity. Moving education, in this case, to the private sector refocuses policy. When *Reading Rainbow* was on PBS, it survived through charitable donation, grants, and public funding. On Kickstarter, it thrives on individual payment.

Campaigns harness the most financially active of their fans in order to both make up their funding goal and also to spread the word about the campaign through social media, revealing a real economic focus on those with more money than others. Yet, new technologies like social media and crowdfunding "sometimes simply facilitate exploitation" of fan audiences.[79] Certainly, Zachary Griffin demonstrates how, if unchecked by regulation, crowdfunding can led to "massive fraud" of donators.[80] *Reading Rainbow*, as with many crowdfunded campaigns, wanted to use the language of inclusivity to direct a group of emotionally invested fans to center on a particular "artist" (in this case, Burton). This tension is illustrated in Burton's language used on the rewards for the campaign: "We can make sure that millions of kids learn to love reading, but we can't do it without you. That's why I hope *you'll join me* in making a difference!" (emphasis mine). By both referencing the affect individuals might have for the program as well as the power of the crowd to affirm the direction of a single author, the campaign reveals how emotion and affect spur donations.

Burton's campaign relies on individual donors' nostalgia for their own history with both the show *Reading Rainbow* and their past as readers. For example, as the second quotation at the start of this chapter reveals, Burton's initial introduction to the page focuses on nostalgia for past fandom:

> Hi. LeVar Burton here. You may know me as Kunta Kinte, from ROOTS, or Geordi La Forge, from Star Trek: The Next Generation. You also may have grown up with me on Reading Rainbow. It was my mother who taught me that, by picking up a book, I could "go anywhere" and "be anything." Ever since Reading Rainbow began in 1983, I have dedicated myself to fostering a love of reading in children, just as my mother did for me.

As Pam Cook has described, nostalgia is "a way of coming to terms with the past" through individual interpretive practices.[81] *Reading Rainbow* is history—that is, it resides in a collective memory of the past. Nostalgia, guided by a personal memory of the historical text, "dissolves the pain of the past in a memory of its beauty."[82] Burton cleverly links moments of fan nostalgia with family nostalgia in his opening here: there is both nostalgia for television series of the past (*ROOTS*, *Star Trek*, *Reading Rainbow*) as well as the more personal, affective description of his mother—an experience that I'm sure many of his readers shared about their own. Although each viewer will have a different sense of nostalgia for Burton and *Reading Rainbow* (and, presumably, their mothers), all that is required for this affective work in the present is to tap into any sense of nostalgia.[83] A connection develops between backer and creator. Affect becomes commoditized.

In the case of *Reading Rainbow*, for instance, the nostalgia within the reward structure creates hierarchy in fandom and reinforces a neoliberal interpretation of media affect. Burton's campaign page notes under "Rewards" that he "believes that spreading the love of reading is its own greatest reward. But if you're going to join our cause, we also have some wonderful prizes." These prizes are almost all fan-based, and for some of the smaller donation amounts (less than $100), the fan investor receives some *Reading Rainbow* swag (a bumper sticker, a magnet, a tote bag, or a mug) as well as access to the app. But especially at the higher donation amounts ($110+), this prize structure rewards fandom of *Star Trek*, of other media texts, or of Burton himself rather than *Reading Rainbow*. Here, monetizing fan affect creates a fan hierarchy: the more one donates, the more "fannish" one can be.[84] This is an authorized and affirmational fandom. For example, at the $110 level, the investor receives a signed headshot of Burton. At $350, Burton will follow the donator on Twitter—and will write a personal Tweet to him/her. At $400, Burton will record an outgoing voicemail message. At $600 Burton will video chat with the donator, and at $1500 Burton will host a group picnic for a small group of backers. Each additional donation brings the donator closer to Burton both textually and physically. Other donations include mixing and matching fandoms, with a pledge of $400 allowing the donor to attend an event in Los Angeles with actress Katee Sackhoff, and others from *Battlestar Galactica* (Edward James Olmos, Jamie Bamber, Michael Trucco, and Tricia Helfer), and $750 allowing the donator to see actors Kristen Bell, Enrico Colantoni, and Jason Dohring from *Veronica Mars* reading stories live in LA—a clear reference to the successful *Veronica Mars* campaign of the previous year.

In a neoliberal economic climate, where individualism is prized and emotional value is linked to economic value, the mainstream notion of fandom is intimately tied to commercial displays of affect. The Digi-gratis economic structure cuts both ways; just as much as it reveals about the non-monetary focus of fan activism, so too can it be pulled toward the commercialization and exploitation of fandom itself. Donating more money and reaping more rewards puts one closer to being a "better" fan, at least in this industry-tinged notion of fan affect. In many fan communities, egalitarianism and communalization continue to shape non-economic forces, but for the mainstream media industry, fandom is just another monetizable affective situation. By extolling the artist, the creator, the celebrated individual at the heart of the campaign, the rewards structure stratifies the spectrum of fan investors. One can be a "better" fan if one has the money, since money and affect are intimately tied together.

Participatory Cultures, Participatory Fandom

Media producers are only just beginning to learn the best ways to harness fandom's affective potential and economic viability for crowdfunding purposes. It may be, as Jenkins, Ford, and Green assert, that "the greatest advantage" of crowdfunding media projects "may rest with those producers whose work operates within genres with strong fan followings (animation, science fiction, horror) and who speak to well-defined populations (minority and activist groups)."[85] Certainly, this appears to be the case with the *Veronica Mars* Kickstarter campaign, which tapped into a rich vein of fannish enthusiasm for the series. Additionally, the fact that Rob Thomas and Kristen Bell had largely identified the group by a particular name, declaring it into being through nomenclature, solidified the identity of the fan group. Neither Braff nor Hart has particularly "well-defined populations" of fans following them (not to say they don't have fans, but rather than their fandoms may be more mainstream and diffuse than the idiosyncratic *Veronica Mars* fans).

However, instead of seeing fans as separated from the production process, the Digi-gratis economy reveals that they are part of the system through their economic value within the complex *process* by which texts becomes actualized. Looked at through traditional models of economic development, fans' support of *Veronica Mars* is certainly a success story and non-support of *Darci's Walk of Shame* a failure. To be fair, one produced a movie and one didn't. But using an analysis based on the Digi-gratis economy we can see the overall pattern of fandom both campaigns manifest. It's not just about the film that gets made at the end of the campaign (after all, the film could be considered a box office failure even when it does get made), but the process by which fans become integrated into the production of the film. And, the generative potential of fandom does not just happen at specific moments of reception, but rather becomes emblematic of the entire temporal process of creation. Fans are always-already placed within the digital media system, from start to finish.

As a subset of crowdsourcing, where many individuals working together can solve difficult problems or issues, crowdfunding turns social incentives into economic incentives. As Daren Brabham notes, "Crowdsourcing is a model for problem solving, not merely a model for doing business."[86] Crowdfunding represents both the monetization of affect, and also the potential power of crowds to enact society change. Crowdfunding draws ire, but often the anger stems from the fact that one is running up against the reality of the world we live in (e.g., money and capital) and the expectations for a free and salutatory fan culture rather than the specific campaign itself. Perhaps what is upsetting is not that people are donating to crowdfunding ventures, but that crowdfunding now seems to represent a fundamental part of Western culture.

And indeed, this is precisely the lesson fan scholars can take from the increased visibility and popularity of crowdfunding. Outside of commercialization, the mechanics of crowdfunding function in our professional lives. For instance, popular academia is crowdfunding: A single chapter can contribute to a wider academic conversation; a single donation can bring books to a disadvantaged classroom. Just as no chapter can hope to discuss all academic scholarship, no one campaign can hope to encompass all the possibilities of distributed and organized funding. Each article written, each paper graded, each review submitted—these are drops in a sea of (largely unpaid) and (largely underappreciated) knowledge economy. But, just as every 5 dollar donation moves a fund closer to its goal, so too does every step in academia move us closer to a greater understanding of the world around us. Perhaps crowdfunding is not a symptom of a new societal order, but rather a microcosm of a centuries-old practice.[87]

New concepts in technological (crowdfunding) and economic (Digi-gratis) development can be usefully applied to non-corporeal procedures, like media creation and idea generation. Fandom is an inherently emotional experience; digital technology can facilitate and channel that emotion into new avenues. Fandom excels at both, and applying theories of digital fandom through the model of crowdfunding portends a more emotive, affective *process* of media creation.

Today, when popularized through social media, crowdfunding is as varied as the people who use it. As Luke Pebler writes, "it's difficult to make generalizations about crowd funding ... the huge variety of projects it's being used for, and the accordingly wide range of collaborative and capital needs of those projects" makes any single analysis only part of a larger spectrum.[88] So perhaps this is one of the lasting lessons that crowdfunding teaches us. Crowdfunding is the latest manifestation of a process of support and reliance on a community. That this community is monetized today is merely a reflection of our contemporary neoliberal culture, but that is not the end of the story: Rather, the collaborative potential of crowdfunding portends a greater shift than what we have already witnessed. Given time, the future of crowdfunding may be the future of us.

Notes

1. Bannerman, "Crowdfunding Culture."
2. https://www.kiva.org/about/stats.
3. Belleflamme, Lambert, and Schwienbacher, "Individual Crowdfunding Practices."
4. Clifford, "Crowdfunding Nearly Tripled Last Year."
5. Ahmed, "Affective Economies," 119.
6. Jenkins, *Convergence Culture*, 319.
7. Hills, "*Veronica Mars*, Fandom, and the 'Affective Economics' of Crowdfunding Poachers."
8. Potts, "Amanda Palmer and the #LOFNOTC," 361–2, 377.
9. Stanfill, "The *Veronica Mars* Kickstarter, Fan-ancing, and Austerity Logics."
10. Terranova, "Free Labor," 35–6.
11. Jenkins, Ford, and Green, *Spreadable Media*, 96.
12. Veale, "Capital, Dialogue, and Community Engagement," ¶1.8, ¶3.1.
13. Bakioğlu, "Bull in a China Shop."
14. e-cademia, "The Multiverse as Asylum."
15. Jones, "Fifty Shades of Exploitation."
16. Kohn, "Harry Potter and the Neoliberal Hallows."

17. Duffett, *Understanding Fandom*, 16.
18. Busse, "Fan Labor and Feminism," 111–2.
19. Galuszka, "New Economy of Fandom."
20. Chin, Jones, McNutt, and Pebler, "*Veronica Mars* Kickstarter and Crowd Funding."
21. Andrejevic, "Watching Television without Pity"; Aytes, "Return of the Crowds"; Pebler, "My Gigantic Issue"; Terranova, "Free Labor."
22. Pebler, "My Gigantic Issue," 6.
23. Chin, "The *Veronica Mars* Movie"; Jones, "Fan Exploitation, Kickstarter and *Veronica Mars*."
24. Chin, "The *Veronica Mars* Movie," ¶6.
25. D'Amato, "Investors and Patrons, Gatekeepers and Social Capital."
26. Jenkins, "Fandom Studies," 93.
27. Hadas, "The Web Planet," ¶1.2.
28. Aytes, "Return of the Crowds"; De Kosnik, "Fandom as Free Labor."
29. Jones, "*My Little Pony*"; Hills, *Fan Cultures*; Hills, "'Twilight' Fans."
30. Booth and Kelly, "The Changing Faces of *Doctor Who* Fandom."
31. Bury, et al., "From Usenet to Tumblr: The Changing Role of Social Media," 316; Booth, "Reifying the Fan"; Hills, "As Seen on Screen?"; Hills, "From Dalek Half Balls."
32. Hills, "As Seen on Screen?"; Hills, "From Dalek Half Balls."
33. Leibovitz, Roig Telo, and Sánchez-Navarro, "Up Close and Personal."
34. Kuppuswamy and Bayus, "Crowdfunding Creative Ideas," 1.
35. Galuszka and Bystrov, "The Rise of Fanvestors," ¶6.3.
36. Hui, Gerber, and Greenberg, "Easy Money"; Kocer, "Social Business in Online Financing."
37. Agrawal, et al., "Some Simple Economics of Crowdfunding."
38. Benkler, *Wealth of Networks*; Lévy, *Collective Intelligence*.
39. Tussey, "Fixing Television by Funding a Movie."
40. Kustritz, "Exploiting Surplus Labours of Love."
41. Geraghty, *Cult Collectors*, 5, 9.
42. Jenkins, Ford, and Green, *Spreadable Media*, 96.
43. Booth, *Playing Fans*.
44. Potts, "Amanda Palmer and the #LOFNOTC"; and a feeling of community, Booth, "Reifying the Fan."
45. http://www.indiegogo.com/projects/the-inspector-chronicles-untitled-motion-picture-about-a-space-traveler.
46. Brabham, "Crowdsourcing."
47. e.g., Goodale, "Veronica Mars' Online Campaign Passes $2 Million"; McNutt, "Kickstarting Veronica Mars"; Bishop, "Could Kickstarter One Day Completely Oust Hollywood?"; McCracken, "The Future of Media Production?"
48. Fernandez, "Zach Braff Raises Money."
49. June, "Melissa Joan Hart's Kickstarter Campaign Fails."
50. Geraghty, *Cult Collectors*, 3.
51. Jenkins, Ford, and Green, *Spreadable Media*, 3.
52. Kuppuswamy and Bayus, "Crowdfunding Creative Ideas."
53. Gray, "Crowds, Words, and the Futures of Entertainment Conference."

54. Jenkins, Ford, and Green, *Spreadable Media*, 249; Aitamurto, "The Impact of Crowdfunding on Journalism," 431.
55. See Hills, "*Veronica Mars*, Fandom, and the 'Affective Economics' of Crowdfunding Poachers."
56. Peterson, "A Shout-out to Blog Buzz."
57. Empire, "Features: *Veronica Mars*."
58. Mittell, "*Veronica Mars* and Exchanges of Value Revisited," ¶2.
59. See Scott, "The Moral Economy of Crowdfunding and the Transformative Capacity of Fan-ancing"; Cochran, "Crowdfunding the Narrative."
60. Menon, "Mars Bars Can't Save Veronica."
61. Hills, "*Veronica Mars*, Fandom, and the 'Affective Economics' of Crowdfunding Poachers," 184.
62. Hills, "*Veronica Mars*, Fandom, and the 'Affective Economics' of Crowdfunding Poachers," 184–5.
63. Fernandez, "Zach Braff Raises Money."
64. McNutt (with Chin, Jones, and Pebler), "*Veronica Mars* Kickstarter and Crowd Funding."
65. Klosterman, "Was It Ethical for Zach Braff to Take to Kickstarter."
66. Keane, "Kickstarter: Stop Blaming Us."
67. See Kuppuswamy and Bayus, "Crowdfunding Creative Ideas."
68. Sperling, "Melissa Joan Hart."
69. See Jenkins, Ford, and Green, *Spreadable Media*, 96; Hills, *Triumph of a Time Lord*; Booth, *Playing Fans*.
70. Hensley-Clancy, "Reading Rainbow's Massive Kickstarter Campaign," ¶6; Dewey, "You Might Want to Reconsider."
71. Scott, "The Moral Economy of Crowdfunding and the Transformative Capacity of Fan-ancing," 179.
72. Tussey, "Fixing Television by Funding a Movie."
73. McChesney, "Introduction," 7.
74. Harvie, *Fair Play*, 53.
75. Booth, "Fandom."
76. De Kosnik, "Fandom as Free Labor."
77. Booth, *Playing Fans;* Scott, "Repackaging Fan Culture."
78. Dewey, "You Might Want to Reconsider," ¶7.
79. Bannerman, "Crowdfunding Culture," ¶14.
80. Griffin, "Crowdfunding," 45.
81. Cook, *Screening the Past*, 4.
82. Wilson, *Cultural Passions*, 49.
83. Booth, *Playing Fans*.
84. Scott, "The Moral Economy of Crowdfunding and the Transformative Capacity of Fan-ancing," 172.
85. Jenkins, Ford, and Green, *Spreadable Media*, 234.
86. Brabham, "Crowdsourcing," 120.
87. Hemer, "A Snapshot on Crowdfunding"; Bannerman, "Crowdfunding Culture."
88. Pebler (with Chin, Jones, and McNutt), "*Veronica Mars* Kickstarter and Crowd Funding."

Bibliography

Agrawal, Ajay, Christian Catalini, and Avi Goldfarb. "Some Simple Economics of Crowdfunding." *National Bureau of Economic Research* 14 (2013): 1–46. Accessed 22 March 2016. http://www.nber.org/papers/w19133.

Ahmed, Sara. "Affective Economies." *Social Text* 22, no. 2 (2004): 117–39.

Aitamurto, Tanja. "The Impact of Crowdfunding on Journalism." *Journalism Practice* 5, no. 4 (2011): 429–45.

Andrejevic, Mark. "Watching Television without Pity: The Productivity of Online Fans." *Television and New Media* 9, no. 1 (2008): 24–46.

Aytes, Ayhan. "Return of the Crowds: Mechanical Turks and Neoliberal States of Exception." In *Digital Labor: The Internet as Playground and Factory*, edited by Tim Scholtz, 79–97. New York: Routledge, 2012.

Bakiog˘lu, Burcu S. "Bull in a China Shop: Alternate Reality Games and Transgressive Fan Play in Social Media Franchises." *Transformative Works and Cultures* 17 (2014). http://dx.doi. org/10.3983/twc.2014.0594.

Bannerman, Sara. "Crowdfunding Culture." *Wi: Journal of Mobile Culture* 7, no. 1 (2013). Accessed 22 March 2016. http://wi.mobilities.ca/crowdfunding-culture/.

Belleflamme, Paul, Thomas Lambert, and Armin Schwienbacher. "Individual Crowdfunding Practices. *Venture Capital: An International Journal of Entrepreneurial Finance* (2013). Accessed 22 March 2016. http://papers.ssrn.com/sol3/papers.cfm?abstract_id=2151179.

Benkler, Yochai. *The Wealth of Networks: How Social Production Transforms Markets and Freedom*. New Haven, CT: Yale University Press, 2006.

Bishop, Rollin. "Could Kickstarter One Day Completely Oust Hollywood?" [video]. *The Mary Sue*, 30 March 2013. Accessed 22 March 2016. http://www.themarysue.com/kickstarter-replace-hollywood.

Booth, Paul. "Fandom: The Classroom of the Future." In "European Fans and European Fan Objects: Localization and Translation," edited by Anne Kustritz, special issue, *Transformative Works and Cultures* 19 (2015). http://dx.doi.org/10.3983/twc.2015.0650.

Booth, Paul. *Playing Fans: Negotiating Fandom and Media in the Digital Age*. Iowa City: University of Iowa Press, 2015.

Booth, Paul. "Reifying the Fan: *Inspector Spacetime* as Fan Practice." *Popular Communication: The International Journal of Media and Culture* 11, no. 2 (2013): 146–59.

Booth, Paul, and Peter Kelly. "The Changing Faces of *Doctor Who* Fandom: New Fans, New Technologies, Old Practices?" *Participations* 10, no. 1 (2013). http://www.participations.org/Volume%2010/Issue%201/5%20Booth%20&%20Kelly%2010.1.pdf.

Brabham, Daren C. "Crowdsourcing: A Model for Leveraging Online Communities." In *The Participatory Cultures Handbook*, edited by Aaron Delwiche and Jennifer. J. Henderson, 120–9. New York: Routledge, 2013.

Bury, Rhiannon, Ruth Deller, Adam Greenwood, and Bethan Jones. "From Usenet to Tumblr: The Changing Role of Social Media." *Participations* 10, no. 1 (2013): 229–318. http://www.participations.org/Volume%2010/Issue%201/14%20Bury%20et%20al%2010.1.pdf.

Busse, Kristina. "Fan Labor and Feminism: Capitalizing on the Fannish Labor of Love." *Cinema Journal* 54, no. 3 (2015): 110–15.

Chin, Bertha. "The *Veronica Mars* Movie: Crowdfunding—Or Fan-Funding—at Its Best?" *On/Off Screen*, 13 March 2013. Accessed 23 March 2016. http://onoffscreen.wordpress.com/2013/03/13/the-veronica-mars-movie-crowdfunding-or-fan-funding-at-its-best/.

Chin, Bertha, Bethan Jones, Miles McNutt, and Luke Pebler. "*Veronica Mars* Kickstarter and Crowd Funding." *Transformative Works and Cultures* 15 (2014). http://dx.doi.org/10.3983/twc.2014.0519.

Clifford, Catherine. "Crowdfunding Nearly Tripled Last Year, Becoming a $16 Billion Industry." *Entrepreneur*, 31 March 2015. Accessed 9 April 2016. https://www.entrepreneur.com/article/244503.

Cochran, Tanya. "Crowdfunding the Narrative, or the High Cost of 'Fan-ancing'." In *Crowdfunding the Future: Media Industries, Ethics, and Digital Society*, edited by Lucy Bennett, Bertha Chin, and Bethan Jones, 31–46. New York: Peter Lang, 2015.

Cook, Pam. *Screening the Past: Memory and Nostalgia in Cinema*. London: Routledge, 2005.

D'Amato, Francesco. "Investors and Patrons, Gatekeepers and Social Capital: Representations and Experiences of Fans' Participation in Fan Funding." In *Ashgate Research Companion to Fan Cultures*, edited by Linda Duits, Koos Zwaan, and Stijn Reijnders, 135–48. Farnham, UK: Ashgate, 2014.

De Kosnik, Abigail. "Fandom as Free Labor." In *Digital Labor: The Internet as Playground and Factory*, edited by Tim Scholtz, 98–112. New York: Routledge, 2013.

Dewey, Caitlin. "You Might Want to Reconsider That Donation to the *Reading Rainbow* Kickstarter." *The Washington Post*, 29 May 2014. Accessed 23 March 2016. http://www.washingtonpost.com/news/the-intersect/wp/2014/05/28/you-might-want-to-reconsider-that-donation-to-the-reading-rainbow-kickstarter.

Duffett, Mark. *Understanding Fandom: An Introduction to the Study of Media Fan Culture*. London: Bloomsbury, 2013.

e-cademia, "The Multiverse as Asylum: Madness and Neo-Baroque Fandom." *LiveJournal*, 31 March 2012. Accessed 23 March 2016. http://e-cademia.livejournal.com/732.html.

Empire online. "Features: *Veronica Mars*." *Empire*, 23 March 2016. Accessed 23 March 2016. http://www.empireonline.com/50greatesttv/default.asp?tv=48.

Fernandez, Maria Elena. "Zach Braff Raises Money—and Ire—with Kickstarter Campaign for New Film. *NBC News*, 22 May 2013. Accessed 23 March 2016. entertainment.nbcnews.com/_news/2013/05/22/18401068-zach-braff-raises-money-and-ire-with-kickstarter-campaign-for-new-film.

Galuszka, Patryk. "New Economy of Fandom." *Popular Music and Society* 38, no. 1 (2015): 25–43.

Galuszka, Patryk, and Victor Bystrov. "The Rise of Fanvestors: A Study of a Crowdfunding Community." *First Monday* 19, no. 5 (2014). Accessed 23 March 2016. http://firstmonday.org/article/view/4117/4072.

Geraghty, Lincoln. *Cult Collectors*. London: Routledge, 2014.

Goodale, Gloria. "'Veronica Mars' Online Campaign Passes $2 Million: What's Hollywood Thinking?" *The Christian Science Monitor*, 14 March 2013. Accessed 23 March 2016.

www.csmonitor.com/USA/Society/2013/0314/Veronica-Mars-online-campaign-passes-2-million-What-s-Hollywood-thinking.

Gray, Jonathan. "Crowds, Words, and the Futures of Entertainment Conference." *Antenna*, 15 November 2011. Accessed 22 March 2016. http://blog.commarts.wisc.edu/2011/11/15/crowds-words-and-the-futures-of-entertainment-conference.

Griffin, Zachary. "Crowdfunding: Fleecing the American Masses." *Case Western Reserve University School of Law*, 14 March 2012. Accessed 23 March 2016. http://papers.ssrn.com/sol3/papers.cfm?abstract_id=2030001.

Hadas, Leora. "The Web Planet: How the Changing Internet Divided *Doctor Who* Fan Fiction Writers." *Transformative Works and Cultures* 3 (2009). http://dx.doi.org/10.3983/twc.2009.0129.

Harvie, Jen. *Fair Play: Art, Performance and Neoliberalism*. London: Palgrave Macmillan, 2013.

Hemer, Joachim. "A Snapshot on Crowdfunding." *Working Papers Firms and Region*, R2 (2011). Accessed 23 March 2016. http://hdl.handle.net/10419/52302.

Hensley-Clancy, Molly. "Reading Rainbow's Massive Kickstarter Campaign Doesn't Mention That Most Schools Will Have to Pay for Access." *Buzzfeed*, 30 May 2014. Accessed 23 March 2016. http://www.buzzfeed.com/mollyhensleyclancy/reading-rainbows-massive-kickstarter-campaign-doesnt-mention.

Hills, Matt. "As Seen on Screen? Mimetic SF Fandom & the Crafting of Replica(nt)s." *In Media Res*, 10 September 2010. Accessed 23 March 2016. http://mediacommons.futureofthebook.org/imr/2010/09/10/seen-screen-mimetic-sf-fandom-crafting-replicants.

Hills, Matt. *Fan Cultures*. London: Routledge, 2002.

Hills, Matt. "From Dalek Half Balls to Daft Punk Helmets: Mimetic Fandom and the Crafting of Replicas." *Transformative Works and Cultures* 16 (2014). http://dx.doi.org/10.3983/twc.2014.0531.

Hills, Matt. *Triumph of a Time Lord: Regenerating* Doctor Who *in the Twenty-First Century*. London: IB Tauris, 2010.

Hills, Matt. "'Twilight' Fans Represented in Commercial Paratexts and Inter-Fandoms: Resisting and Repurposing Negative Fan Stereotypes." In *Genre, Reception, and Adaptation in the Twilight Series*, edited by Anne More, 113–31. Farnham, UK: Ashgate, 2012.

Hills, Matt. "*Veronica Mars*, Fandom, and the 'Affective Economics' of Crowdfunding Poachers." *New Media and Society* 17, no. 2 (2015): 183–97.

Hui, Julie S., Elizabeth Gerber, and Mike Greenberg. "'Easy Money? The Demands of Crowdfunding Work." *Segal Segal Technical Report* 12, no. 4 (2012). Accessed 23 March 2016. http://egerber.mech.northwestern.edu/wp-content/uploads/2012/11/Easy-Money-_The-Demands-of-CrowdfundingWork-_2012.pdf.

Jenkins, Henry. *Convergence Culture: Where Old and New Media Collide*. New York: New York University Press, 2006.

Jenkins, Henry. "Fandom Studies as I See It." *Journal of Fandom Studies* 2, no 2. (2014): 89–109.

Jenkins, Henry, Sam Ford, and Joshua Green. *Spreadable Media: Creating Value and Meaning in a Networked Culture*. New York: New York University Press, 2013.

Jones, Bethan. "Fan Exploitation, Kickstarter and *Veronica Mars*." *Fan Academic Blogger*. 15 March 2014. Accessed 23 March 2016. http://bethanvjones.wordpress.com/2013/03/15/fan-exploitation-kickstarter-and-veronica-mars/.

Jones, Bethan. "Fifty Shades of Exploitation: Fan Labor and *Fifty Shades of Grey*." In "Fandom and/as Labor," edited by Mel Stanfill and Megan Condis, special issue, *Transformative Works and Cultures* 15 (2014). http://dx.doi.org/10.3983/twc.2014.0501.

Jones, Bethan. "*My Little Pony*, Tolerance Is Magic: Gender Policing and Brony Anti-fandom." *The Journal of Popular Television* 3, no. 1 (2015): 119–25.

June, Daniel. "Melissa Joan Hart's Kickstarter Campaign Fails Utterly to Kick Off." *JD Journal*, 16 May 2013. Accessed 23 March 2016. http://www.jdjournal.com/2013/05/16/melissa-joan-harts-kickstarter-campaign-fails-utterly-to-kick-off/#.

Keane, Erin. "Kickstarter: Stop Blaming Us for Zach Braff and 'Veronica Mars'." *Salon*. 10 May 2013. Accessed 23 March 2016. http://www.salon.com/2013/05/10/kickstarter_stop_blaming_us_for_zach_braff_and_veronica_mars/.

Klosterman, Chuck. "Was It Ethical for Zach Braff to Take to Kickstarter?" *The New York Times*, 24 May 2013. Accessed 23 March 2016. http://www.nytimes.com/2013/05/26/magazine/was-it-ethical-for-zach-braff-to-take-to-kickstarter.html?_r=0.

Kocer, Suncem. "Social Business in Online Financing: Crowdfunding Narratives of Independent Documentary Producers in Turkey." *New Media and Society* 17, no. 2 (2015): 231–48.

Kohn, Liberty. "Harry Potter and the Neoliberal Hallows: Copyright, Counterpublics and 2.0 Fan Fiction." *Technoculture* 2 (2012). Accessed 23 March 2016. http://tcjournal.org/drupal/vol2/kohn.

Kuppuswamy, Venkat, and Barry L. Bayus. "Crowdfunding Creative Ideas: The Dynamics of Project Backers in *Kickstarter*." *SSRN Working Paper* (2013). Accessed 23 March 2016. http://papers.ssrn.com/sol3/papers.cfm?abstract_id=2234765.

Kustritz, Anne. "Exploiting Surplus Labours of Love: Narrating Ownership and Theft in Crowdfunding Controversies." In *Crowdfunding the Future: Media Industries, Ethics, and Digital Society*, edited by Lucy Bennett, Bertha Chin, and Bethan Jones, 47–63. New York: Peter Lang, 2015.

Leibovitz, Talia, Antoni Roig Telo, and Jordi Sánchez-Navarro. "Up Close and Personal: Exploring the Bonds between Promoters and Backers in Audiovisual Crowdfunded Projects." In *Crowdfunding the Future: Media Industries, Ethics, and Digital Society*, edited by Lucy Bennett, Bertha Chin, and Bethan Jones, 15–30. New York: Peter Lang, 2015.

Lévy, Pierre. *Collective Intelligence: Mankind's Emerging World in Cyberspace*, translated by Robert Bononno. New York: Perseus, 1997.

McChesney, Robert. "Introduction." In *Profit over People: Neoliberalism and Global Order* by Noam Chomsky, 7–18. New York: Seven Stories Press, 1999.

McCracken, Chelsea. "The Future of Media Production?" *Antenna*, 03 April 2013. Accessed 23 March 2016. http://blog.commarts.wisc.edu/2013/04/03/the-future-of-media-production/.

McNutt, Myles. "Kickstarting *Veronica Mars*: A Moment in a Movement." *Antenna*, 15 March 2013. Accessed 23 March 2016. http://blog.commarts.wisc.edu/2013/03/15/kickstarting-veronica-mars-a-moment-in-a-movement/.

Menon, Vinay. "Mars Bars Can't Save Veronica." *Toronto Star*, 12 June 2007. Accessed 23 March 2016. http://www.thestar.com/opinion/columnists/2007/06/12/mars_bars_cant_save_veronica.html.

Mittell, Jason. "*Veronica Mars* and Exchanges of Value Revisited." *Just TV*, 15 March 2013. Accessed 23 March 2016. http://justtv.wordpress.com/2013/03/15/veronica-mars-and-exchanges-of-value-revisited.

Pebler, Luke. "My Gigantic Issue with the *Veronica Mars* Kickstarter." *Revenge of the Fans* [Suzanne Scott], 13 March 2013. Accessed 23 March 2016. http://www.suzanne-scott.com/2013/03/15/guest-post-my-gigantic-issue-with-the-veronica-mars-kickstarter/.

Peterson, Karla. "A Shout-out to Blog Buzz: TV Shows Embrace the Passion (and Value) of Internet Fan Sites." *Union-Tribune*, 10 April 2006. Accessed 23 March 2016. http://legacy.sandiegouniontribune.com/news/features/20060410-9999-1c10karla.html.

Potts, Liza. "Amanda Palmer and the #LOFNOTC: How Online Fan Participation Is Rewriting Music Labels." *Participations* 9, no. 2 (2012). http://www.participations.org/Volume%209/Issue%202/20%20Potts.pdf.

Scott, Suzanne. "The Moral Economy of Crowdfunding and the Transformative Capacity of Fan-ancing." *New Media and Society* 17, no. 2 (2015): 167–82.

Scott, Suzanne. "Repackaging Fan Culture: The Regifting Economy of Ancillary Content Models." *Transformative Works and Cultures* 3 (2014). http://doi:10.3983/twc.2009.0150.

Sperling, Nicole. "Melissa Joan Hart Says Kickstarter Failure 'Started out Wrong'." *LA Times*, 20 May 2013. Accessed 23 March 2016. http://articles.latimes.com/2013/may/20/entertainment/la-et-mn-melissa-joan-hart-kickstarter-failure-20130520.

Stanfill, Mel. "The Veronica Mars Kickstarter, Fan-ancing, and Austerity Logics." *Mel Stanfill*, 25 March 2013. Accessed 23 March 2016. http://www.melstanfill.com/ the-veronica-mars-kickstarter-fan-ancing-and-austerity-logics/.

Terranova, Tiziana. "Free Labor: Producing Culture for the Digital Economy." *Social Text* 63 18, no. 2 (2000): 33–58.

Tussey, Ethan. "Fixing Television by Funding a Movie: The Crowdfunding of *Veronica Mars*." In *Crowdfunding the Future: Media Industries, Ethics, and Digital Society*, edited by Lucy Bennett, Bertha Chin, and Bethan Jones, 157–73. New York: Peter Lang, 2015.

Veale, Kevin. "Capital, Dialogue, and Community Engagement—*My Little Pony: Friendship Is Magic* Understood as an Alternate Reality Game." *Transformative Works and Cultures* 14 (2013). http://journal.transformativeworks.org/index.php/twc/article/view/510/405.

Wilson, Elizabeth. *Cultural Passions: Fans, Aesthetes and Tarot Readers*. London: IB Tauris, 2013.

· 8 ·

FAN STUDIES AS/AND TUMBLR

It was early evening. There was very little room at the bar and they were pressed shoulder to shoulder, surrounded by cigarette smoke and the murmur of conversation …

—Pérez-Reverte, *The Club Dumas* (p. 18)

To no small degree, it was my fannish enthusiasm and not my academic curiosity that led me to consider an advanced degree in media studies and to spend all of the time required to gain competency in that discipline (though by that point, I would be hard pressed to separate out the two). I have found approaching popular culture as a fan gives me new insights into the media by releasing me from the narrowly circumscribed categories and assumptions of academic criticism and allowing me to play with textual materials.

—Henry Jenkins, *Textual Poachers* (p. 5)

The social media site Tumblr has changed the face of fandom from both a technological and a cultural standpoint. As Alena Karkanias notes, its design has made it "conducive to fan interactions and so [it] is one of the most popular spaces for fandoms."[1] With its image-focused community and its rebloggable remit, Tumblr offers fans new ways of experiencing, creating, and sharing texts. The medium of Tumblr is the message of fandom; that is, Tumblr makes participation look like fandom whether or not the users necessarily consider themselves "fans." Tumblr crowds everyone together, flattening participation into Pérez-Reverte's crowded bar of fandom. This interpretation of and within the Tumblr technology is a prime example of *Digital Fandom*'s "philosophy of playfulness" and the development of a culture of play. By emulating characteristics of playfulness—as I discuss in this chapter, characteristics like community-minded posting, emotional motivation for use, and engagement—Tumblr engenders in its users a playful attitude and creates a culture where play becomes the default setting; where individuals don't just accept their community and culture but continually adjust their own positioning within that culture.

In this chapter, I want to use Tumblr not to talk about fandom *per se*, but rather as a way of exploring and analyzing the state of fan studies. As fan studies becomes more formalized through conventions, journals, collegiate degree programs, and academic societies, by definition it becomes more *disciplined*—that is, it becomes structured and formalized around particular ways of being.[2] These traditional academic settings may bring with them prestige, community, attention, and growth; but these settings also run the risk of creating silos around the traditionally interdisciplinary field of fan studies, of formalizing that which has previously been free-flowing, of alienating instead of embracing new ways of interacting with academia. I believe we need to keep fan studies playful, to retain the sense that fans—and fan scholars—are pushing against the boundaries of formality by developing alternative, non-mainstream modes of investigation and appreciation. If fan studies embodies the philosophy of playfulness, then Tumblr represents one site and symbol for continuing this philosophy within an era of disciplinary growth.

At the same time, this formalization of fan studies is happening at a moment when fandom itself is becoming mainstream—an aspect of the Tumblr effect. As fans can use digital technology like Tumblr to reach others, create new works, and develop new ways of interacting with their media texts and communities, their work becomes more visible and more approachable. Fandom, as we are reminded in fan studies literature over the past decade, has become mainstream. Yet, in the introduction to their *Ashgate Research Companion to Fan Studies*, Koos Zwaan, Linda Duits, and Stijn Reijnders ask:

> If the paradigm of fan studies has become mainstream, is there still a need for a separate field for the study of fan cultures? How distinctive are its concepts and methodologies? These developments thus call for a critical rethinking of the function and significance of fandom, fan practices and fan studies.[3]

This chapter is an attempt to answer this question; a step toward critically rethinking fan studies as a discipline; and an application of digital methodologies toward fan studies itself.

Tumblr provides a model of collective authorship that becomes fruitful for exploring the development of fan studies. More specifically, I think that fan studies can learn from Tumblr, and that our engagement as fan scholars can and should be informed by the lessons of fandom within the Tumblr environment. As Alexander Cho has described, rather than focusing on one specific mode of engagement, "Tumblr seem[s] like a terrain of affinities speaking at thousand miles a minute, one that regard[s] written language as a simple, runty cousin. … Tumblr trade[s] in affect."[4] Similarly, Mark Duffett, in his textbook for fandom studies, argues that fan studies offers "a plausible language through which to talk about emotion."[5] Emotionality and affect lie at the heart of both Tumblr and fan studies, and fan studies can model its own modes of collectivity—its own nexus of voices—from the multiplicity on Tumblr.

Tumblr's active fan bases also engender the type of media literacy that is espoused by fan studies. Tumblr provides a "glimpse inside" fans' thinking, as it represents not a finished product, but what Busse and Hellekson call a "work in progress" for fan writing.[6] This chapter is also a type of *work in progress* that emulates Tumblr discourse. Academics aren't supposed to talk about the process of writing within the text (barring, of course, books about writing). A completed book represents *knowledge* and *completeness*. What is missing—what is unsaid—is the hundreds of thousands of words that get left out, the conversations that shaped the development of the chapter, the wrong turns and broken links. What's missing is the *trouble*, the *excess*, the *hyperdiegesis* of academia. But as with Tumblr, that's often the most instructive part—and it's the part that took over as I was writing to become this final chapter's central principle. Let me back up a bit with a bit of revision history (again, not something that normally happens in an academic book, but something that does happen in digital environments). Importantly, the revision is the point.

I knew Tumblr had to feature in *Digital Fandom 2.0*. As I sat down to write this chapter, I realized that I had written about Tumblr before—analyzing *SuperWhoLock* and *Inspector Spacetime* as key Tumblr texts that demonstrated not just particular media fandoms, but also how fandom itself was becoming an object of fannish attention.[7] A particular "fandom of fandom" manifests on Tumblr in both the way fan discussion moved toward *being* fans, and the turn to "meta" in fan Tumblr posts. But for this chapter I didn't know what I specifically wanted to write about. I didn't know how to grasp the *enormity* of Tumblr, the *diversity* in Tumblr, or the *multivocality* of Tumblr. Until I realized that *that* should be the focus. It is impossible to talk about Tumblr's influence on fandom because *every Tumblr blog is a cacophony of voices;* every reblog is a vocal reminder of key themes in the Tumblr conversation. To give just a few statistics: As of October 8, 2015, Tumblr users have posted over 257.8 million blogs and 121.5 billion blog posts.[8] Almost 60 million blog posts are uploaded each day. This increases exponentially each year; according to Chang, Tang, Inagaki, and Liu, in January 2014 Tumblr had "166.4 millions of users and 73.4 billions of posts."[9] According to Alexa.com, Tumblr is in the top 20 websites in the United States (#19) and in the top 50 worldwide (#42). More important than mere numbers, however, Tumblr also has an incredibly *engaged* audience, as these users create and post content both fannish and not in a multi-vocal and polyphonous environment.

Tumblr is a dialogue, not an argument, and it is this dialogue that we must be vigilant to maintain within a rapidly disciplining fan studies.

Fast forward to the 2015 Fan Studies Network conference. I was giving a paper on *SuperWhoLock* as a metaphor for fan studies.[10] From the audience, Kristina Busse quite aptly questioned this assertion—why not Tumblr itself as the exemplar of fan studies? Busse's contribution to my thinking on the topic had enormous consequences, for as it turns out, that was the keystone for this conclusion to *Digital Fandom 2.0*.

As a discipline, fan studies is in its infancy. But to call fan studies a discipline is to limit it; to relate it to the academy is to deny it the very power that makes it pertinent to millions of people outside the academy. For Jenkins writing in 2014, fandom studies is a "dispersed multidisciplinary" field.[11] An academic discipline comes with academic baggage. Fan studies *needs* to avoid this baggage in order to remain relevant and meaningful in the digital age.

As with any new discipline, a settling of sorts occurs as the discipline fits into already established contours of the academy. According to Grindstaff, writing about the use of popular culture studies in the well-established social science of sociology, disciplines struggle to find legitimacy when emerging.[12] New disciplines have growing pains, as Christine Geraghty describes with the emergence of media studies as a popular major:

> Many media and communications degrees would be seen by those organizing them as, what we might call, lower-case degrees inevitably drawing on a wide range of disciplinary histories and methods. Even within the narrower confines of specifically upper-case subject areas, there are debates about how to define the area of study and what precisely different disciplinary contributions should be.[13]

In a discussion of the disciplining of Comics Studies as a field, Henry Jenkins asks important questions undergirding any disciplinary formation: "Who participates in the activity … (intellectuals, academics, artists, fans, reformers)? What questions (aesthetic, cultural, ideological, formal, historical) get asked and through what methods do they get answered? What audiences (fans, the general public, would-be artists [and authors], students) get addressed and through what channels?"[14] New disciplines always find a niche in which to sit, but then that niche becomes a larger paradigm against which other, newer disciplines become formalized.

Fan studies' strength lies in its applicability outside the academy, in its focus on interdisciplinary scholarship, in its varied multi-vocality. Despite becoming incorporated into departments and colleges with majors and degree programs, fan studies needs to be an outsider to continue to maintain its conversational and dialogic properties and its fluidity. Fan studies can do something that fans themselves are doing; it can learn from Tumblr and apply fandom to the field. On the surface, there are many similarities between fans and academics: an expertise in a particular, obscure field, a desire to share that expertise with others, a need to explore an object in depth.[15] Even the commonalities between fan conventions and scholarly conferences signal this doubling: panels allow the exchange of ideas while keynote speakers attract large crowds (the use of hashtags to follow events further exemplifies the similarities), question and answer sessions reveal disparate levels of knowledge, exhibit halls at conferences display books while dealer's alleys at conventions display wares, etc.

Tumblr is a metaphor for what fan studies can be—an undisciplined discipline, a chaotic system. Sam Ford has noted in his discussion of the state of the field in 2014, "The strength of fan studies and its rich body of research stems in part from the interdisciplinarity that has allowed scholars from varied fields to bring their research methods and theoretical constructs to bear on the relationships active audiences have with and around media texts." The key aspect of fan studies moving forward lies in understanding how to "maintain a cohesive community around the concept of fan studies while not losing its porous boundaries that are essential to facilitate the continued incorporation of new methodological approaches, new types of fandoms and media industries, and new issues/angles to be tackled."[16] Tumblr offers fan scholars and fans themselves the opportunity to see a chaotic, undisciplined system at work; to focus on open-endedness, playfulness, and liveness within an evolving structure.

In this chapter, therefore, I offer some ideas about how fan studies is beginning to become disciplined, some advantages to this disciplinary attention, and then some ways to resist this typification through the metaphor of Tumblr. (However, I don't want to make this a "Tumblr is like this, therefore fan studies is like that" analysis. Such comparative work not only essentializes one particular type of fan studies, but also runs the risk of *a priori* analyzing before the analysis is complete. But relating the two helps us shape the way the discipline is forming.) Hills writes that "it is hugely unlikely that there is going to be one theoretical framework through which we can best understand media fandom."[17] Similarly, I don't think there's one way—one particular methodological framework or disciplinary action—that helps us understand fan studies, other than in its very undisciplined nature. Thus, I look at formal and informal characteristics of a discipline and the "imagined subjectivity" of fans studies.[18] Following this, I offer insights from fan studies literature itself about why fan studies should *not* be disciplined, and focus on the ways that fandom and fan studies resists this formalization.[19]

Disciplining Fan Studies

Throughout these next few pages, I'll make a case for how fan studies has become *legitimized* and *formalized* as a discipline. In 2005 for C. Lee Harrington and Denise Bielby's special issue of *American Behavioral Scientist* focused on fan studies, the topic of the discipline had

> exploded as a vibrant, interdisciplinary, and cross-national focus of inquiry. Intersecting audience studies, media studies, and consumption studies in ways that advance all three areas of inquiry (as well as others), recent studies of fans and fandom force us to rethink key questions of identity, performance, practice, genre, gender, sexuality, self, affect, race, ethnicity, and nationalism.[20]

As Lucy Bennett has argued, in an article charting the history of fan studies:

> Fan cultures and its scholarship is a vibrant, rich and colourful discipline that is almost impossible to capture completely within one article, when considering all the different forms and areas that being a media fan, and a researcher, can encompass. In this sense, this perceived impossibility I have faced is a testament to the field and its current shape and flux.[21]

Bennett indicates that the variety of fan work and fan cultures makes capturing them in one discipline difficult. Louisa Stein, at the same time, notes that formalizing any sort of discipline comes with inherent political and ideological complications. Here, comparing fan culture and digital culture, she argues that:

> both of these categories … exist as discursive constructions with specific and ideologically loaded histories. Fan culture and millennial mainstream digital cultures are, to a degree, discursive fictions that in any given moment break down into a diversity of specific experiences[.] We must be wary of equating the two or assuming causal relationships of influence between the two.[22]

To extrapolate from her discussion of fandom, we can say that fan studies is also a discursive construction with political implications.

With that in mind, and certainly with an awareness of the political implications of constructing fan studies as a discipline, I'd like to explore some of the ways that we can conceptualize fan studies as a formal discipline (before arguing against this in the conclusion). The term "discipline" means multiple things in a variety of contexts. The formalization of fan studies as a discipline brings worthwhile and necessary opportunities for the field. However, I do think that there are consequences that fan scholars need to be mindful of moving forward, and we can learn from Tumblr how to navigate some of these consequences. As Jenkins notes, here writing about Comics Studies but that can be easily applied to Fan Studies as well:

> Make no mistake about it: a discipline disciplines. Disciplines define borders and set priorities. Disciplines decide what counts and on what terms. These decisions are, to be sure, made collectively and they evolve over time, but the starting premises for a field matters. Disciplines are defined as much by what they exclude as they are by what they include … So, the decisions we make right now … matter because they represent the road map for a field that is still trying to grab its turf, define its boundaries, and find its way.[23]

A. S. Oldnall notes that "The concept of an academic discipline is ambiguous and not easily defined because, as with many concepts, it is open to interpretation, according to the context in which it is being used." That is to say, the formation of a discipline is always political—disciplinary functionality is deployed for particular reasons, in particular spaces, and with particular incentives. Oldnall goes on to argue that the most practical way of "defining a discipline is through the structural framework and organization of academic institutions."[24] Becher concurs.[25] In an organizational sense, then, the discipline of fan studies has become formalized through courses taught at universities, advanced coursework, and majors and minors in the subject. To date and to the best of my knowledge there are only very few "fan studies" -centric programs at institutions of higher education;[26] however, there certainly are courses being taught around the country (see http://fanlore.org/wiki/Syllabi for some examples), and a growing number of related courses.

Approaching the academic foundation of a discipline is problematic, however, especially in the neoliberal economy of education. As more universities find themselves caught in the "'student as consumer' model of education," universities are becoming "the bastion of the administrator and the accountant, not the instructor and the student. … Students see critical thinking and liberal arts as less valuable than more 'practical' fields." As a field focused in affect, trading in popular culture, and dependent upon "engaged, passionate, and critical" works from fans, fan studies does not immediately present the type of practical know-how that a neoliberal educational system might engender.[27] Although teaching fan studies is about teaching critical reading strategies (certainly an important topic and something all students need) it is a hard sell when compared with a major or program with a quantitative and definitive outcome, especially for students (and parents) who link a university degree to career outcomes. Fan studies is a hard sell to administrators worried about the bottom line and to parents concerned about their child's career prospects. Yet, as Becher notes, "even in the most crassly utilitarian social climate, the external standing of a discipline is at least in some measure independent of its proven economic worth." Thus, we need to look both beyond "the attitudes, activities and cognitive styles of groups of academics representing a particular discipline" and toward "the characteristics and structures of the knowledge domains with which such groups are professionally concerned" in order to develop a more thorough understanding of the disciplinary function of fan studies.[28]

In an early study, King and Brownell give an authoritative analysis of academic disciplines, which I quote at length:

> The disciplines of knowledge are not clearly described as areas of study or of knowledge, but metaphorically as **communities of scholars** who share a domain of intellectual inquiry or discourse. In essence these societies of specialists are engaged in a variety of styles of human imagination in which the spirit of inquiry is applied to **defined domains** of human concern. Each group of intellectual discoursers has **a heritage** and is striving to bring the development of its domain or field to a continually higher and more fruitful state of knowing and meaning. The body of intellectual discourses in a field has one or more **characteristic ways of knowing**—of warranting knowledge—or it may share modes of inquiry with other disciplines. The group shares the precious resource of a **specialized language** or other system of symbols which makes precision of definition and inquiry possible. The body of discoursers has a set of more or less well-related concepts. The community has an **inheritance**

> **of books, articles, and research reports**, and a system for communication among the membership. **Members of the community share affective as well as cognitive links**, with the excitement of discovery and the pleasure of sharing with colleagues as common characteristics. The work of the community emphasizes style and the **search for truth**. Further, the community has either an explicit or a **tacit conception of [hu] man**. Finally, the discipline is an **instructive** community.[29]

These formal characteristics help to define a discipline as specifically separate from other types of communities. And not to make too simplistic a comparison, I think each one of King and Brownell's concerns are met by fan studies: Communities of scholars like the Fan Studies Network (formed in 2012) and the International Association of Audience and Fan Studies (formed in 2008 and apparently dormant as of 2016) reveal an international and interdisciplinary group of scholars practicing fan studies work.[30] Robin Reid in *The Routledge Companion to Science Fiction* has written about the development of fan studies as a field, arguing that "international scholarship began to develop" in the early 1990s as scholars were publishing in a variety of fields.[31] Fan studies' domain explores the way individuals form attachment to media, music, sports, celebrity, and other texts. It has a clear heritage: 1992 is a pivotal year,[32] although Rebecca Wanzo has recently revisited the canonization of Fiske, Jenkins, Bacon-Smith, and Lewis, ed. as eliding the large contribution of African American fans.[33] A key element of disciplining fan studies is through a canonization of key texts—in this case, *Textual Poachers*. Mark Duffett, in *Understanding Fandom*, argues that "*Textual Poachers* … marked a new era in which reception studies merged with fandom research to create *fan studies*."[34] Gray, Sandvoss, and Harrington convincingly (and influentially) break the field of fan studies into three "waves" that resonate with different emphases: The early days focused on fandom as a "beautiful" reading strategy, the second wave examined the construction of fans' own identities within new media, and the third wave expands to investigate "fandom as part of the fabric of our everyday lives" in which fan studies "aims to capture fundamental insights into modern life."[35]

Jenkins's *Textual Poachers* is only canonical in hindsight, although his identity as a scholar-fan has been influential for fan studies scholars moving forward, as the second quotation at the start of this chapter reveals. As Sam Ford describes, *Textual Poachers* "made no call for creating a new field."[36] In contrast, Matt Hills's *Fan Cultures* has been hugely influential in the field—one might argue it has become even more relevant to fan studies as a discipline than Jenkins's *Textual Poachers*—as it presents a theoretical basis for understanding the relationship between fans *and academics* most directly and arguing the study of fans needs a unique theoretical framework.[37] Indeed, Jenkins himself argues that "while I have drawn on theory as a tool for understanding fandom as a set of cultural, social, and interpretive practices, I have not drawn upon fandom as a means of developing a new theory of media consumption."[38] Hills makes the case for how fandom has been perceived and studied through different disciplinary approaches in the academy. At the same time, Hills argues that fan studies is less disciplined within the academy and the distinctions between fan groups "are often produced through the institutional agendas of university departments, such as the need to establish 'film studies' in the 1970s, and the attempt to counter this with 'TV studies' as a rival field." That is, "theorists follow their own institutional or theoretical agendas, and use fandom within these theory wars and territorial skirmishes." Fandom is a concept used by multiple disciplines, at least in 2002, as a "rhetorical tailoring … in order to act upon particular academic institutional spaces and agendas … fandom is *used*" not discovered.[39] Jenkins's view of fandom, according to Hills, is *formalized* and therefore replicates already existent value systems of "good" and "bad" subjectivity, of "critical" vs. "affective" interpretation, etc. To make the connection to Tumblr, it is also used by multiple groups that are labeled "fannish" even if they might not self-identify that way, and an ethos of Tumblr is discussed even if the people using Tumblr use it differently from each other.

In addition to a shared history, having a similar emphasis on methodology gives fan studies a disciplinary structure. Fan studies tends to focus on qualitative research, based in textual analysis, ethnography, interviews, and extended digital analysis. Importantly, understanding emotional reaction and affect is key to developing many fan studies and to understanding Tumblr. Researcher Katharina Freund, in an article written with Dianna Fielding, described how "in ethnographic research it is precisely these emotional, subjective, and personal aspects that are the most valuable from an anthropological perspective. The social justifications provided by my participants in interviews revealed key tensions and gaps in their community."[40] There is a particular language to fan studies, as scholars in the field speak of transformative work, affective play, and aca-fandom (among many other) as unique terms to the discipline. It is worth noting that Tumblr fits many of these disciplinary constructs as well: Users on Tumblr "have appropriated and created terminology that fits the types of activities they participate in."[41]

Another method of determining disciplinary status is through the existence of content-specific journals and conferences. The aforementioned Fan

Studies Network has organized an annual fan studies conference for the past four years. Even well-established scholarly groups like the Society for Cinema and Media Studies (SCMS) has created the Fan and Audience Studies Special Interest Group (although the creation of this group was also fraught, as the inclusion of "audience studies" with fan studies was seen by some to be a unfocused disciplinary boundary). Previously, panels and workshops had been organized at SCMS focusing on fan studies, most notably one about the term "aca-fandom" organized by Louisa Stein. Ford also argues that the moment when Kristina Busse and Henry Jenkins brought male and female scholars together to discuss gender, in 2007 on Jenkins's blog:

> was a pivotal point in the development of fan studies as a discipline of sorts. It played a significant role in helping bridge divides between male and female fan studies scholars—and between approaches to fan studies that had been perceived as predominantly masculine or feminine, or focused on predominantly male or female fan bases. But, beyond this primary goal, the series helped create a common dialogue among fan studies scholars.[42]

Journals like *Transformative Works and Cultures* and the *Journal of Fandom Studies* are both specifically about fan work and publish studies and scholarship on fandom.

Finding common sources of material—fan work—is also elemental in establishing a fan studies discipline, so a good portion of the disciplinary functioning of fan studies has focused on understanding what a fan actually *is*. Cornel Sandvoss and Laura Kearns relate the academic analysis of fans to "two dimensions: creativity and collectivity. These characteristics of fandom in turn appear to set the practices of fans apart from other less productive and tightly knit audience groups."[43] Of course, not all fans are collective, and fan studies finds fan antagonism relevant as well.[44] Some of this definitional research finds a home in the spaces of research—fans are self-defined as those who frequent fan conventions (ethnographic research) or write fan fiction (textual analysis on Archive of Our Own or FanFiction.net), or make vids. But the development of fan studies has led to numerous comparisons between fans and other groups of people, like gamers, costumers, and artists. Megan Condis, for example, explores "who is able to lay claim to titles like 'fan' or 'gamer,' how those titles are being contested along gendered, racialized, sexualized, and classed lines, what happens when new users lay claim to those titles, and how some fans are reacting to the loss of their privileged relationships with content producers."[45] Fandom is a multi-vocal term, a prism of identities and practices.[46]

Perhaps most importantly, as Becher notes of academic disciplines, fan studies "provide[s] a clear example of the interaction between academia and the world which lies beyond its confines."[47] That is, fan studies *can make a difference* to the community about which it investigates. As I wrote in 2015:

> First, I believe our educational system can benefit from more fandom enthusiasm, both from us as scholars and from our students, fans themselves. Second, as fan-scholars, we have a momentous opportunity and responsibility to learn from and to teach fannish ardor in the classroom, and to contribute to it as both fans and scholars. And third, new generations of fans will become future teachers, thinkers, and responsible media citizens. It is our responsibility, as today's media scholars and teachers, to help train them in these critical skills of fandom by exposing them to thoughtful, engaged fan work.[48]

Fan studies is instructive, in King and Brownell's terms, but it is also an exploration of what it means to be human—to have feelings and emotions, to express them in a community, to be angry and sad and loving and passionate—in an era of digital commercialization and a neoliberal focus on assertive individualism and isolationism. Through an organized scholarly community, some key canonized texts, conceptual methodologies, and a detailed roadmap for its creation, fan studies has become a destination rather than a journey.

And yet—there are consequences of disciplining fandom; of concretizing the affective. Indeed, I think we need to be critical of disciplinary work, as formulating a *beginning* merely concretizes (and privileges) one narrative over others. Wanzo's re-examination of the genealogy of fan studies reveals, specifically, the important (and often ignored) contributions of African American audiences toward fandom, but also, more generally, how an alternate genealogy can generate multiple (parallel) pathways through a discipline.[49] The "imagined subjectivity" of academia reveals how different disciplinary mechanisms actually constrain rather than open up avenues of research. Being wedded to one particular methodology, one particular corpus of texts, one particular group will always delimit and define the subject of study, effectively muting the field.[50] As Katherine Morrissey argues in *Transformative Works and Cultures*:

> A focus on fandom from multiple perspectives is critical, given ongoing challenges in conceptualizing what it is to be a fan. How do we attempt to process a concept that is simultaneously claimed as an activity, an identity, and a connection to others? Rather than seeing this confusion as a problem, perhaps it is more useful to see it as

> precisely the point. In trying to understand an aspect of media culture that we all, to some degree, engage in, the field of fan studies needs to approach fans and fandom in a variety of ways: at the level of the individual, at the level of practices, and as a framework in which the self encounters media culture. In our current moment, the media environment is undergoing dramatic changes. It is critical that fan studies continues to question the control of cultural production and consider the ways that today's media industries are working to accommodate both fans and fan practices.[51]

Indeed, Hills echoes Morrissey's sentiments in an interview with Clarice Greco in *Matrizes*:

> The idea that one can just come up with a definition of fandom I think is problematized through the fact that it's performed in so many different ways and in so many varied contexts. Rather than focusing on defining fandom we need to think about how fandom is performed and for whom and in what context, and try to think, really, about what kind of subset of this vast overarching diffused category of fandom we might be dealing with.[52]

Both Morrissey and Hills highlight a key element that has strengthened fan studies throughout its development: the flexibility in articulating a specific focus. Fan studies needs to remain engaged in the fan communities, which themselves are ever shifting and changing. To define fan studies is to (artificially) define the fan; a limitation of disciplining fan studies would therefore be imposing an (already hypothesized) identity of the fan onto the investigative subject.

In comparison, Tumblr offers a site of ever-changing dynamics wherein fan audiences can be encountered but also where the multi-vocality of the audiences on the site means multiple discourses can be followed at once. There is no one "ethos" of Tumblr. Hills relates fans to academics and academics to fans. Today, as fans use Tumblr, so too do scholars-fans use fan studies. Hills writes, "just as academia cannot be thought of as a singular entity, fandom also cannot be depicted as a univocal site or as one 'thing.'"[53] To remain constant—to remain a present force within academic scholarship—fan studies must navigate both sides of its remit. It must be both fannish and scholarly, both emotional and critical. Fan studies is rapidly becoming formalized and while that brings with it prestige and validation from the academy, it also makes rigid the boundaries and concretizes the subjects at the heart of the field. Following Tumblr as a symbol—with all its unruly messiness and uncharacterized affect—is one way fan studies will be assured of remaining relevant.

Tumbling through Fandom; Tumbling through Fan Studies

In this section I outline how fans are using Tumblr and in what ways Tumblr provides a social experience unobtainable through other social media sites. Then, I tie Tumblr's motifs to the possibilities of an undisciplined fan studies discipline. Specifically, I want to discuss the way that Tumblr as a technology can symbolize the nature of fan studies, especially as fan studies appears to be moving toward a particular concretization. I want to use this comparison as a caution against formalization and toward an understanding that, as Becher points out, "The concept of an academic discipline is not altogether straightforward, in that, as is true of many concepts, it allows room for some uncertainties of application."[54] Tumblr represents this uncertainty—an ambiguity that I believe strengthens rather than diminishes the functionality of fan studies as a field of study. As Petersen notes, "Tumblr is organized in a certain way and operates by a particular grammar, and this enforces a particular manner of communication that differs from other mediated and non-mediated contexts."[55] In other words, Tumblr creates a unique space for user interaction but also prescribes a particular way of being and identity through Tumblr community(ies). Tumblr disciplines through its undisciplined nature—and becomes a model for fan studies scholarship moving forward.

Tumblr is a quest, not a destination. Like MySpace before it (see chapter 4), the user-generated content within the specific space of Tumblr articulates a diversity of characteristics even within the standard infrastructure of its system. Most Tumblr sites fit a basic profile—there are posts, reblogs, likes, images, GIFs, and other content presented in a chronological fashion. Unlike Facebook, though (and perhaps most dramatically different from other social sites), Tumblr utilizes a system of user-developed "themes" that can radically change the appearance of each person's site. Changing themes has become a substantial part of the new web—a web that has really only thrived over the past half-decade (since the first edition of *Digital Fandom* was released). DIY and drag-and-drop website services like Weebly and Wix flourish as web space becomes ever more available. Theme-creation allows for a safe middle ground between the cacophony of MySpace's original flexible interface—what David Gauntlet argues was "originally abused by people who just seemed to want to make a mess" but was also "how individual *identity* is expressed online"[56]—and the static, rigid Facebook monolith. In an email conversation with me, Tumblr Content and Community Associate Amanda

Brennan noted that its popularity is often attributable to this flexibility: "As a medium, Tumblr is unique in our post types and the ability to truly present what you want to in a highly customizable way you want to. People can be and do anything on the site."

Each Tumblr page, then, is both partly user-created and partly community-developed. Even at its base level, before any content has been added, a Tumblr page sits at the intersection of user creativity and professional context—a nexus that has proven fruitful in academic scholarship of fans as well, for this mix of bottom-up and top-down creation symbolizes a strength of fan studies, stronger through its combination of academic voices and fan voices.[57] The appeal of Tumblr for fan communities is not necessarily just the flexibility of theme; it is also the community feeling that develops through the interaction between participants. Brennan went on in her email to point out that:

> people can separate out parts of their personality easily as well. … Additionally, the community is layers of different communities bonded by passions. Everyone can experience Tumblr in a different way based on the things they really care about, and the ways people participate is truly based on that. (Brennan email 2015)

Despite the veneer of transparency, however, there are some elements of the Tumblr user base that are rarely revealed; I was denied specific stats of numbers of users or percentage of users that use particular tags, and was told that Tumblr does not reveal any statistical data regarding users, content, or tagging.

Yet, the notion of "community" is uncertain on Tumblr.[58] Hillman, Procyk, and Neustaedter note that "the concept of belonging to a fandom in Tumblr is fuzzy. Unlike Facebook, you do not get accepted to groups. You are part of the fandom when you think you are." Rather than forming social groups and interactive communities, users on Tumblr "follow" other people's blogs. Communities of interest form, but these communities are not formalized through the participants themselves, but rather organized by the topics generated by the community: "Some users joined Tumblr not specifically to participate in fandoms, yet as they began following users, they learned about fandoms through their dashboard."[59] As pointed out by *Washington Post* columnist Caitlin Dewey, Tumblr is "a vast, unmapped network of niche communities."[60] This is not unlike the way academic communities emerge: Just to draw from my own experience, which cannot be unique, I wrote a dissertation on a particularly narrow topic and as I have grown as an academic I have both followed other academic's work and expanded my own interests through this following of others. In this way, both Tumblr users and academics moderate their own journey through a fandom/discipline within the Tumblr/academic space: It is not simply that one follows and becomes part of the community, but rather that one must necessarily *choose* the connections one wants to have. Both Tumblr and academics are fandoms of choice and selection.

This selection becomes key to what Alexander Cho calls the "authorial locus" on Tumblr—that is, it is not that participation on Tumblr is an act of creation, but rather "it is the act of *curation*."[61] One curates their own personal journey through their fandoms. According to Jen Kelly, "this focus on curation, literally built into Tumblr's infrastructure, is one of the central ways that the site facilitates the cultivation of analytical skills"—a crucial function of Tumblr for fannish media literacy.[62] There are advantages and disadvantages to this style of fannish interaction—while one may be able to drill down deeply into a particular fan identity, one also runs the risk of becoming too inured to one particular viewpoint. Diversity is easy to eliminate simply by deciding not to follow people who disagree with your particular style of fandom or activity. For example, you might be a *Doctor Who* fan on Tumblr and only follow blogs that highlight the relationship between the Tenth Doctor and Rose because you really dig that relationship. However, following those blogs reinforces your particular viewpoint at the expense of a variety of other types of *Doctor Who* Tumblr blogs (e.g., a Classic series blog, a blog that focuses on the Twelfth Doctor and Clara, etc.). Thus, "users do not experience 'noise' in conversations, but non-reflectively engage in conversations based on the norm for *loosened relevance*. This works particularly well with humorous or playful [or fannish] conversations."[63] This type of fandom "filter bubble"[64] or "individual echo chamber"[65] reproduces the existing affective meanings through social media curation, rather than developing new meanings. Fan studies should avoid the filter bubble of academia by remaining open to multiple interdisciplinary conversations with other academic areas and developing connections with areas that might appear disparate; for instance, links between fan studies and political science might reveal civic commonalities, or between fan studies and history could help nuance discussions of fans through time.

Kelly uses this type of curation to position Tumblr within a personal public sphere, and considers "Tumblr [as], at once, both the space structured through the technological platform as well as the imagined community of people who use it."[66] Petersen argues that the "affordances of Tumblr establish specific boundaries within which its users operate and use them in order to serve their own purposes. Tumblr users interpret the affordances of Tumblr in a certain way, and respond by applying a specific cultural logic through their fan-centric activities."[67] Fan studies can develop an understanding of contemporary fandom by studying and using Tumblr as a mechanism for viewing fan interaction as structured into non-mutually exclusive groups, for a type of "trans-fandom" experience.[68]

This logic can be most obviously actualized through the use of tagging, or adding hashtagged text to an image in order to both label it and organize it. In my interview with Brennan, she noted that "people tag for a myriad of reasons," including: "personal organization on their own blog, to get their content found, [and] to add additional context to the post at hand." Fan studies can learn from tagging: Tagging allows fans to participate in a fandom or community by finding likeminded individuals sharing content on a certain topic (artists, fitness enthusiasts, librarians, medical students and professionals, people who love organization, photographers, etc.). But as discussed in the Organization for Transformative Works' "Spotlight on Tag Wrangling," tagging is also inherently messy: As a "curated folksonomy" the organization is both bottom-up (e.g., fans creating their own tags) and top-down (e.g., curators monitoring and systematizing tags).[69] Negotiating the use of tags allows a freedom of characterization to appear but a consistency of tone.[70] Fan studies as an academic field uses a form of tagging (e.g., keywords in an article) but a more comprehensive and associative form of tagging might enable implicit connections between work or authors. Louisa Stein describes how "emotion-oriented tags often intertwine with tags that function as analysis or interpretation to swamp organizational schemas"—a type of tagging mash-up that presents a multivocal theme to Tumblr. She goes on to describe how tags can also become the "site of fan fiction authorship in what is known as tagfic," when tags become a means for "transformative authorship" rather than (in)formal organizational style.[71] Similarly, CommentFic, or fiction written in the comments on blogs (see chapter 2) sees fans using the technological affordances to create original, creative work. In this way, Tumblr might best be perceived as a delimited quest for a particular brand of fandom: a quest for community, but a community founded on individualized preferences.

One of the clearest ways to see this individualized sense of community is through Tumblr's arguably most original feature, the "reblog" function. Reblogging is when a user finds a particular post they like and clicks a button to post it on their own blog. For Nistasha Perez, reblogging relates to the fan practice of

> sharing what they love. Sharing as a fan practice has increased the need for a simple and easy way to spread fannish content to friends and followers. Instead of dealing

> with photo hosting sites, html code, and formatting, Tumblr users are able to click on button and share a post with their followers.[72]

Reblogging, as Petersen notes, "enforces this sense of fan identity and belonging to a fan community. … In users who primarily participate through reblogging, there remains a level of performance and identity-making, as they choose what type of fan blog their blog is through the posts they reblog."[73] In other words, reblogging becomes a marker of identity, another way of curating one's own digital identity. Fan studies can speak to the ideas of reblogging: as the repetition of content across multiple avenues of scholarship matches the idea of the affirmational interaction but also builds on work through repetition. Lousia Stein argues that reblogging speaks to the affirmational nature of the fan community: reblogging "hails the viewer/reader directly; it requests (or perhaps demands) *individual collective affirmation*."[74]

Another element of Tumblr that speaks to its individualized nature is its emphasis on the image rather than the written word. Previous fannish blogs like LiveJournal presented a new way of organizing fans into communities (see the first edition of *Digital Fandom*). On Tumblr, the use of images and the reblog function change the focus on the fan communities. Using "a river of images," users on Tumblr create "their own units of meaning which circulate in a contemporary quotation culture, particularly on websites of the sharing internet. We can read them, and the audiences who reproduce them, alongside the [text], in relation to it, sometimes in tension with it."[75] This "quotation culture" finds repetition rather than content generation to be indicative of community interaction. Indeed, as a style of play, repetition becomes the sincerest form of community. Along with static images, GIFs—Graphics Interchange Format—are an essential tool for Tumblr users. I've previously discussed GIFs as a type of fannish activity, as a way of creating blended modes of fan creativity, and as a unique characteristic of the web.[76] GIFs are moving images—in reality, animations where one image is rapidly overlaid on another—and as such represent an important form of "play" in Tumblr culture.[77] GIFs can be quotations from the original text, can have fan-added content (e.g., fan's deliberately changing the text to make it seem as though the images reference different aspects), or can be wholly original pieces of art. GIFs represent the combination of transformative and affirmational fandoms at work on Tumblr.[78] Whereas affirmational fandom tends to uphold the text and its creator, transformative fandom tends to take the text and mold it, to create something new with it. GIFs allow us to see beyond that binary, revealing its artificiality—using actual images from the text it is affirmational but, as fans have to transform those texts to fit into the GIF format, there must necessarily be transformative aspects as well.[79] Straddling the divide between affirmational and transformational scholarship, fan studies highlights aspects of fandom that depend on affective interaction with the text.

I certainly don't want to make it sound like use of images, GIFs, and reblogging are the only ways that Tumblr differentiates itself from other social media sites, but as key elements of the Tumblr fan communities—importantly, the elements that most directly speak to the fan experience on Tumblr—they highlight the shared experiences of these communities and build on the concept of the philosophy of playfulness. Importantly, then, rather than focusing on these few elements as being the quintessential aspects of Tumblr, I want to broaden this discussion onto what these all *mean*—why images, reblogging, GIFs, and tagging are so relevant not just for Tumblr but for fan studies academics as well.

Tumblr communities, as Cho discusses, are communities formed through *affect*. The "shared social experiences … [create] a shared sense of purpose and understanding that [unite] fandom users and [allow] them to relate to one another."[80] As Stein notes, "Step by step, we are moving to a millennial media landscape no longer dominated by fears of the excesses of the unruly fan, one that instead embraces personal investment, performativity, emotion, and excess within the content of shared digital creativity."[81] *Emotionality and affect* lie at the heart of Tumblr communities. Fan communities are also focused on emotion—fandom is about shared passion (which can be directed positively or negatively), but Tumblr revalues emotional connectivity. Whereas fan communities in the past were often devalued for their "excessive" emotions or their "unruly" behavior, and female fandoms especially have been undervalued (and pathologized) because of their emotional activity, Tumblr fan communities inherently rely on and embrace the charge of affect online.[82] Images and GIFs play on emotions—although there are cerebral moments—and unabashedly ask the audience to value the emotional experience of fandom. As I noted above, Petersen argues that "humorous or playful conversations" work best on Tumblr, but she goes on to state that the type of emotional engagement works "less so with serious knowledge-heavy conversations."[83] And yet, rather than seeing a "less serious knowledge-heavy" conversation as *inherently better* or *more valuable* a style of fan engagement (i.e., affirmational fandom), Tumblr fan communities regard the undisciplined, unfettered emotionality of playfulness as valuable in its own right. Such a shift in fan community tenor reveals the underlying importance of Tumblr fandom—not that fans on Tumblr are necessarily *practicing* fandom in a different way, but that the particular modes of engagement are revalued in such a way as to establish a different set of norms. The revaluation of affect on Tumblr applies both to the disciplinary functionality of fan studies and to the revaluation of affect in the academy.

There are already fannish academic activities happening on Tumblr, enacted by those Jenkins has called fan critics, Hills called fan-scholars, and today we might call meta ("meta" being a type of fan writing that dissects and discusses media texts as an academic might also).[84] Kelly describes how Tumblr blogs

> offer great examples of counterdiscourses … particularly for marginalized groups including women, people of color and LGBT and queer people. I believe that the way Tumblr functions as a counterpublic space enables and facilitates the circulation of these types of media critique on the site. … Through both the media critiques shared on Tumblr and the media curation encouraged through the use of the site, Tumblr functions as a space in which free, fairly accessible, and public education about media is occurring.

This type of fan/educational discourse is happening in many other venues online, but Tumblr forms a central nexus of fan engagement.[85]

If, as Kelly indicates, Tumblr is facilitating the development of a more media literate audience—and if fandom itself serves as a type of classroom where people learn and practice critical thinking skills, then the relationship between Tumblr and fandom as specific *sites of critical discourse* becomes clear. Both fandom and Tumblr trade in affect; both Tumblr and fandom reflect key interpretive and transformative practices.

Tumblr and Fan Studies

Upon starting this chapter, I was uncertain about comparing fan studies to Tumblr, as I didn't want to link them superficially. Tumblr is a social media site owned by Yahoo. It is popular now, but one day it will not be (like MySpace). But what Tumblr represents—a space for multi-vocality, a site for affective play, a location for fannish curation, a place for media literacy to flourish—transcends the site. For these are characteristics not of a social media space, but of a fannish mentality; not of a specific site, but of a growing trend toward understanding how emotion and identity shape the world around us. Tumblr is a blueprint for fan studies not because of what it is, but because of what it *represents*. Tumblr is a space for experimentation and exploration; fan studies needs to retain these characteristics as an ever-increasing neoliberal and administrative emphasis overtakes the academy. Despite its formalization, fan studies needs to remain informal; despite its disciplinary functions, it needs to remain undisciplined. Rather than seeing fan studies become incorporated into the system, I think the value of fan studies lies in it remaining on the periphery.

Writing a second edition of *Digital Fandom* has been a decidedly self-reflective affair. It is indulgent in a way, but also allowed me to revisit ideas I had a decade ago as I was working in a graduate program and trying to sound more confident than I actually was. There is nothing more humbling than examining (and editing) work from your past. To be sure, I'm grateful for the opportunity and I think this is a stronger book than the first edition—more varied, with a greater scope and more streamlined discussions. But it's also been a reminder of my past, and an indication of how one's ideas may grow and develop over time.

I knew I would have to integrate Tumblr into the new edition and I wanted to make that chapter meaningful not just for fans, but for the community of scholars that I've read and learned from since I started writing *Digital Fandom*. To that end I wanted to expand out from the concerns of the first edition—largely, how fan work can be read through literary theory in order to develop original characteristics and theories—toward disciplinary concerns within fan studies. As fan studies rapidly changes, it's important to recognize the importance of the variety of disciplines and interdisciplinary scholarship that first began the field. For fan studies, I agree with what Jenkins writes about comic studies: "I embrace an approach that is radically undisciplined, taking its tools and vocabulary where it can find them, expansive in its borders to allow the broadest possible range of objects of study, inclusive in who it allows to participate and in the sites where critical conversations occur."[86] A narrowly defined field ignores interdisciplinary approaches. Ford argues that

> Most in fan studies would agree that we can't have an absolutist answer to what does or does not warrant fandom, or whether certain kinds of fan activities constitute "a greater level of engagement" than others. Yet, we sometimes adopt modes of thinking that explicitly or implicitly prioritize certain types of fan activities over others. … [Scholars place] value in certain products of fandom over the process of fandom.[87]

And Bennett notes that, today:

> Just as the theory of "textual poaching" is embraced and reapplied by some scholars, it is rejected by others due to its rigidity and irrelevance in accounting for the new forms of powerful interactions within participatory culture. These paradoxes then, are quite similar to the field itself as it is operating right now.[88]

The contradictions within fan studies are powerful and, I think, are the most valuable aspects of the discipline.

Notes

1. Karkanias, "The Intra- and Inter-Sub-Community Dynamics of Fandom," 5.
2. See Jenkins, "Fandom Studies," 93.
3. Zwaan, Duits, and Reijnders, "Introduction to *The Ashgate Research Companion to Fan Cultures*," 2.
4. Cho, "Queer Reverb," 44.
5. Duffett, *Understanding Fandom*,135.
6. Busse and Hellekson, "Introduction."
7. Booth, "Reifying the Fan"; Booth, *Playing Fans*; Booth, *Crossing Fandoms*.
8. https://www.tumblr.com/about.
9. Chang, Tang, Inagaki, and Liu, "What Is Tumblr," 1.
10. Booth, *Crossing Fandoms*.
11. Jenkins, "Fandom Studies," 103.
12. Grindstaff, "Culture and Popular Culture"; see also Becher, *Academic Tribes and Territories*, 140.
13. Geraghty, "Doing Media Studies," 28.
14. Jenkins, "Introduction," 4.
15. Hills, *Fan Cultures*; Booth, "Augmenting Fan/Academic Dialogue."
16. Ford, "Fan Studies," 53–4.
17. Hills, "Fandom as an Object," 150.
18. Hills, *Fan Cultures*.
19. Geraghty, "Doing Media Studies."
20. Harrington and Bielby, "New Directions," 799, my emphasis.
21. Bennett, "Tracing *Textual Poachers*," 14.
22. Stein, *Millennial Fandom*, 10.
23. Jenkins, "Introduction," 5.
24. Oldnall, "Nursing as an Emerging Academic Discipline," 605–6.
25. Becher, *Academic Tribes and Territories*, 19.
26. Full disclosure: I helped start one at DePaul University in 2016.
27. Booth, "Fandom," ¶3.1, ¶1.5.
28. Becher, *Academic Tribes and Territories*, 142, 20.
29. King and Brownell, *The Curriculum and the Disciplines of Knowledge*, 68–95, emphases mine.

30. As discussed in Lucas, "Notes and Correspondence."
31. Reid, "Fan Studies," 204.
32. Bennett, "Tracing *Textual Poachers*."
33. Wanzo, "African American Acafandom and Other Strangers."
34. Duffett, *Understanding Fandom*, 65, 68.
35. Gray, Sandvoss, and Harrington, "Introduction," 9.
36. Ford, "Fan Studies," 55.
37. Hills, *Fan Cultures*.
38. Jenkins, *Textual Poachers*, 286.
39. Hills, *Fan Cultures*, 2, 10, Hills's emphasis.
40. Freund and Fielding, "Research Ethics in Fan Studies," 330.
41. Hillman, Procyk, and Neustaedter, "Tumblr Fandoms, Community & Culture," 287.
42. Ford, "Fan Studies," 57.
43. Sandvoss and Kearns, "From Interpretative Communities to Interpretative Fairs," 91.
44. Hills, "'Twilight' Fans"; Jones, "My *Little Pony*"; Booth, *Crossing Fandoms*.
45. Condis, "No Homosexuals in *Star Wars*?" 199.
46. Booth and Bennett, "Introduction," 10; Stein, "Sherlock, SuperWhoLock and Transfandom."
47. Becher, *Academic Tribes and Territories*, 140.
48. Booth, "Fandom," ¶1.1.
49. Wanzo, "African American Acafandom and Other Strangers."
50. Hills, *Fan Cultures*, 11.
51. Morrissey, "Fan/Dom," ¶ 2.5, emphasis mine.
52. Hills, "Fandom as an Object," 149.
53. Hills, *Fan Cultures*, 7.
54. Becher, *Academic Tribes and Territories*, 19.
55. Petersen, "*Sherlock* Fans Talk," 90.
56. Gauntlet, *Making Is Connecting*, 196.
57. Jenkins, *Convergence Culture*; Jenkins, Ford, and Green, *Spreadable Media*; Booth, "Fandom."
58. See Hillman, Procyk, and Neustaedter "'alksjdf;lksfd,'" 6.
59. Hillman, Procyk, and Neustaedter, "Tumblr Fandoms, Community & Culture," 287.
60. Dewey, "2015 Is the Year," ¶10.
61. Cho, "Queer Reverb," 46.
62. Kelly, "Tumblr."
63. Petersen, "*Sherlock* Fans Talk," 90, my emphasis.
64. Pariser, *The Filter Bubble*.
65. Pande, "Squee from the Margins."
66. Kelly, "Tumblr."
67. Petersen, "*Sherlock* Fans Talk," 91.
68. Hills, "Fandom as an Object," 158–9;
69. https://otw-news.dreamwidth.org/169697.html.
70. Although this policy of "tag wrangling" has itself been criticized: http://fanlore.org/wiki/AO3_Tagging_Policy_Debate.
71. Stein, *Millennial Fandom*, 158.
72. Perez, "Gif Fics," 151.
73. Petersen, "*Sherlock* Fans Talk," 96.

74. Stein, *Millennial Fandom*, 155, my emphasis.
75. Newman, "Say 'Pulp Fiction' One More Goddamn Time," 127.
76. Booth, "Reifying the Fan"; Booth, *Playing Fans;* Booth, *Crossing Fandoms.*
77. Hillman, Procyk, and Neustaedter, "Tumblr Fandoms, Community & Culture," 287.
78. See obsession_inc, "Affirmational Fandom vs. Transformational Fandom."
79. See Booth, *Playing Fans;* Hills, "From Dalek Half Balls."
80. Hillman, Procyk, and Neustaeder, "'alksjdf;lksfd,'" 8.
81. Stein, *Millennial Fandom,* 15.
82. Jensen, "Fandom as Pathology"; Click, "'Rabid,' 'Obsessed,' and 'Frenzied'"; Busse, "Geek Hierarchies."
83. Petersen, "*Sherlock* Fans Talk," 90.
84. Jenkins, *Textual Poachers;* Hills, *Fan Cultures.*
85. Kelly, "Tumblr"; Booth, "Fandom."
86. Jenkins, "Introduction," 6.
87. Ford, "Fan Studies," 63.
88. Bennett, "Tracing *Textual Poachers*," 14.

Bibliography

Becher, Tony. *Academic Tribes and Territories: Intellectual Enquiry and the Cultures of Disciplines.* Milton Keynes, UK: Open University Press, 1989.

Bennett, Lucy. "Tracing Textual Poachers: Reflections on the Development of Fan Studies and Digital Fandom." *The Journal of Fandom Studies* 2, no. 1 (2014): 5–20.

Booth, Paul. "Augmenting Fan/Academic Dialogue: New Directions in Fan Research." *The Journal of Fandom Studies* 1, no. 2 (2013): 119–37.

Booth, Paul. *Crossing Fandoms: SuperWhoLock and the Contemporary Fan Audience.* Houndmills, UK: Palgrave Pivot, 2016.

Booth, Paul. "Fandom: The Classroom of the Future." In "European Fans and European Fan Objects: Localization and Translation," edited by Anne Kustritz, special issue, *Transformative Works and Cultures* 19 (2015). http://dx.doi.org/10.3983/twc.2015.0650.

Booth, Paul. *Playing Fans: Negotiating Fandom and Media in the Digital Age.* Iowa City: University of Iowa Press, 2015.

Booth, Paul. "Reifying the Fan: *Inspector Spacetime* as Fan Practice." *Popular Communication: The International Journal of Media and Culture* 11, no. 2 (2013): 146–59.

Booth, Paul, and Lucy Bennett. "Introduction: Seeing Fans." In *Seeing Fans: Representations of Fandom in Media and Popular Culture*, edited by Lucy Bennett and Paul Booth, 1–10. New York: Bloomsbury, 2016.

Busse, Kristina. "Geek Hierarchies, Boundary Policing, and the Gendering of the Good Fan." *Participations* 10, no 1 (2013): 73–91. http://www.participations.org/Volume%2010/Issue%201/6%20Busse%2010.1.pdf.

Busse, Kristina, and Karen Hellekson. "Introduction: Work in Progress." In *Fan Fiction and Fan Communities in the Age of the Internet*, edited by Karen Hellekson and Kristina Busse, 5–32. Jefferson, NC: McFarland and Co., 2006.

Chang, Yi, Lei Tang, Yoshiyuki Inagaki, and Yan Liu. "What Is Tumblr: A Statistical Overview and Comparison," 2014. Accessed 23 March 2016. http://arxiv.org/abs/1403.5206.

Cho, Alexander. "Queer Reverb: Tumblr, Affect, Time." In *Networked Affect*, edited by Ken Hillis, Susanna Paasonen, and Michael Petit, 43–58. Cambridge, MA: The MIT Press, 2015.

Click, Melissa. "'Rabid,' 'Obsessed," and 'Frenzied': Understanding *Twilight* Fangirls and the Gendered Politics of Fandom." *Flow*, 18 December 2009. Accessed 23 March 2016. http://www.fl-girls-and-the-gendered-politics-of-fandom-melissa-click-university-of-missouri/.

Condis, Megan. "No Homosexuals in *Star Wars*? BioWare, 'Gamer' Identity, and the Politics of Privilege in a Convergence Culture." *Convergence: The International Journal of Research into New Media Technologies* 21, no. 2 (2015): 198–212.

Dewey, Caitlin. "2015 Is the Year That Tumblr Became the Front Page of the Internet." *The Washington Post*, 11 March 2015. Accessed 22 March 2016. https://www.washingtonpost.com/news/the-intersect/wp/2015/03/11/move-over-reddit-tumblr-is-the-new-front-page-of-the-internet/.

Duffett, Mark. *Understanding Fandom: An Introduction to the Study of Media Fan Culture*. London: Bloomsbury, 2013.

Ford, Sam. "Fan Studies: Grappling with an 'Undisciplined' Discipline." *Journal of Fandom Studies* 2, no. 1 (2014): 53–71.

Freund, Katharina, and Dianna Fielding. "Research Ethics in Fan Studies." *Participations* 10, no. 1 (2013): 329–34. http://www.participations.org/Volume%2010/Issue%201/16%20Freund%20Fielding%2010.1.pdf.

Gauntlett, David. *Making Is Connecting: The Social Meaning of Creativity, from DIY and Knitting to YouTube and Web 2.0*. Cambridge, UK: Polity Press, 2011.

Geraghty, Christine. "Doing Media Studies: Reflections on an Unruly Discipline." *Art, Design & Communication in Higher Education* 1, no. 1 (2002): 25–36.

Gray, Jonathan, Cornel Sandvoss, and C. Lee Harrington. "Introduction: Why Study Fans?" In *Fandom: Identities and Communities in a Mediated World*, edited by Jonathan Gray, Cornel Sandvoss, and C. Lee Harrington, 1–16. New York: New York University Press, 2007.

Grindstaff, Laura. "Culture and Popular Culture: A Case for Sociology." *Annals of the American Academy of Political and Social Science* 619 (2008): 206–22.

Harrington, C. Lee, and Denise D. Bielby. "New Directions in Fan Studies." *American Behavioral Scientist* 48, no. 7 (2005): 799–800.

Hillman, Serena, Jason Procyk, and Carman Neustaedter. "'alksjdf;lksfd': Tumblr and the Fandom User Experience." 2014. Accessed 23 March 2016. http://dl.acm.org/citation.cfm?id=2600887&dl=ACM&coll=DL&CFID=594181856&CFTOKEN=93137949.

Hillman, Serena, Jason Procyk, and Carman Neustaedter. "Tumblr Fandoms, Community & Culture." *CSCW' 14 Companion* (2014). Accessed 23 March 2016. http://dl.acm.org/citation.cfm?id=2557634&dl=ACM&coll=DL&CFID=594181856&CFTOKEN=93137949.

Hills, Matt. *Fan Cultures*. London: Routledge, 2002.

Hills, Matt. "Fandom as an Object and the Objects of Fandom." *Matrizes* 9, no. 1 (2015): 147–62.

Hills, Matt. "From Dalek Half Balls to Daft Punk Helmets: Mimetic Fandom and the Crafting of Replicas." *Transformative Works and Cultures* 16 (2014). http://dx.doi.org/10.3983/twc.2014.0531.

Hills, Matt. "'Twilight' Fans Represented in Commercial Paratexts and Inter-Fandoms: Resisting and Repurposing Negative Fan Stereotypes." In *Genre, Reception, and Adaptation in the Twilight Series*, edited by Anne More, 113–31. Farnham, UK: Ashgate, 2012.

Jenkins, Henry. *Convergence Culture: Where Old and New Media Collide*. New York: New York University Press, 2006.

Jenkins, Henry. "Fandom Studies as I See It." *Journal of Fandom Studies* 2, no 2. (2014): 89–109.

Jenkins, Henry. "Introduction: Should We Discipline the Reading of Comics?" In *Critical Approaches to Comics: Theories and Methods*, edited by Matthew J. Smith and Randy Duncan, 1–14. New York: Routledge, 2012.

Jenkins, Henry. *Textual Poachers: Television Fans and Participatory Culture*. New York: Routledge, 1992.

Jenkins, Henry, Sam Ford, and Joshua Green. *Spreadable Media: Creating Value and Meaning in a Networked Culture*. New York: New York University Press, 2013.

Jensen, Joli. "Fandom as Pathology: The Consequences of Characterization." In *The Adoring Audience*, edited by Lisa Lewis, 9–29. London: Routledge, 1992.

Jones, Bethan. "*My Little Pony*, Tolerance Is Magic: Gender Policing and Brony Anti-fandom." *The Journal of Popular Television* 3, no. 1 (2015): 119–25.

Karkanias, Alena. "The Intra- and Inter-Sub-Community Dynamics of Fandom." Summer Research Paper 230 (2014). Accessed 23 March 2016. http://soundideas.pugetsound.edu/summer_research/230.

Kelly, Jen. "Tumblr: A Site of Media Literacy and Public Education." Paper presented at 2015 Midwest Popular Culture Association, Cincinnati, OH, 03 October 2015.

King, Arthur A. Jr., and John A. Brownell. *The Curriculum and the Disciplines of Knowledge: A Theory of Curriculum Practice*. New York: John Wiley and Sons, 1966.

Lucas, Barbara. "Notes and Correspondence." *Science Fiction Studies* 35(2): 359–60.

Morrissey, Katherine E. "Fan/dom: People, Practices, and Networks." *Transformative Works and Cultures* 14 (2013). http://dx.doi.org/10.3983/twc.2013.0532.

Newman, Michael Z. "Say 'Pulp Fiction' One More Goddamn Time: Quotation Culture and an Internet-Age Classic." *New Review of Film and Television Studies* 12, no. 2 (2014): 125–42.

obsession_inc. "Affirmational Fandom vs. Transformational Fandom," 01 June 2009. Accessed 23 March 2016. http://obsession-inc.dreamwidth.org/82589.html.

Oldnall, Andrew S. "Nursing as an Emerging Academic Discipline." *Journal of Advanced Nursing* 21 (1995): 605–12.

Pande, Rukmini. "Squee from the Margins: Racial/Cultural/Ethnic Identity in Global Media Fandom." In *Seeing Fans: Representations of Fandom in Media and Popular Culture*, edited by Lucy Bennett and Paul Booth, 209–20. New York: Bloomsbury, 2016.

Pariser, Eli. *The Filter Bubble: What the Internet Is Hiding from You*. London: Penguin UK, 2011.

Perez, Nistasha. "GIF Fics and the Rebloggable Canon of SuperWhoLock." In *Fan Phenomena: Doctor Who*, edited by Paul Booth, 148–57. Bristol, UK: Intellect, 2013.

Petersen, Line Nybro. "*Sherlock* Fans Talk: Mediatized Talk on Tumblr." *Northern Lights* 12 (2014): 87–104.

Reid, Robin Anne. "Fan Studies." In *The Routledge Companion to Science Fiction*, edited by Mark Bould, Andrew Butler, Adam Roberts, and Sherryl Vint, 204–13. London: Routledge, 2009.

Sandvoss, Cornel. *Fans: The Mirror of Consumption*. Malden, MA: Polity Press, 2005.

Stein, Louisa. "Sherlock, SuperWhoLock and Transfandom." Presentation at the Society for Cinema and Media Studies conference, April 2016, Atlanta, GA.

Stein, Louisa Ellen. *Millennial Fandom: Television Audiences in the Transmedia Age*. Iowa City: University of Iowa Press, 2015.

Wanzo, Rebecca. "African American Acafandom and Other Strangers: New Genealogies of Fan Studies." *Transformative Works and Cultures* 20 (2015). http://dx.doi.org/10.3983/twc.2015.0699.

Zwaan, Koos, Linda Duits, and Stijn Reijnders. "Introduction to *The Ashgate Research Companion to Fan Cultures*," 1–9. Farnham, UK: Ashgate, 2014.

BIBLIOGRAPHY

Aarseth, Espen. "Nonlinearity and Literary Theory." In *Hyper/Text/Theory*, edited by George Landow, 51–86. Baltimore, MD: The Johns Hopkins University Press, 1994.

Abercrombie, Nicholas, and Brian Longhurst. *Audiences: A Sociological Theory of Performance and Imagination*. London: Sage Publications, 1997.

Adrian, Clare. "Families Fall under the Spell." *Tourist Attractions and Parks* 37, no. 5 (2007): 44–6.

Agger, Ben. *The Virtual Self*. London: Blackwell, 2007.

Agrawal, Ajay, Christian Catalini, and Avi Goldfarb. "Some Simple Economics of Crowdfunding." *National Bureau of Economic Research* 14 (2013): 1–46. Accessed 22 March 2016. http://www.nber.org/papers/w19133.

Ahmed, Sara. "Affective Economies." *Social Text* 22, no. 2 (2004): 117–39.

Aitamurto, Tanja. "The Impact of Crowdfunding on Journalism." *Journalism Practice* 5, no. 4 (2011): 429–45.

Allen, Graham. *Intertextuality*. London: Routledge, 2000.

Allen, Nancy. "Telling Our Stories in New Ways." Review of *Remediation*, by Jay David Bolter and Richard Grusin, and *The Language of New Media*, by Lev Manovich. *Computers and Composition* 18 (2001): 187–94.

"Alternate Reality Game." *Wikipedia*. Accessed 21 March 2016. http://en.wikipedia.org/wiki/Alternate_reality_game.

Alvermann, Donna E., and Margaret C. Hagood. "Fandom and Critical Media Literacy." In *Re/Mediating Adolescent Literacies*, edited by John Elkins and Allan Luke, 36–46. Newark, DE: International Reading Association, 2000.

Amesley, Cassandra. "How to Watch *Star Trek*." *Cultural Studies* 3, no. 3 (1989): 323–39.

Andrejevic, Mark. "Watching Television without Pity: The Productivity of Online Fans." *Television and New Media* 9, no. 1 (2008): 24–46.

Anelli, Melissa. *Harry, a History*. New York: Pocket Books, 2008.

Arnold, Thomas K. "'Lost 2' Finds Way to Top of DVD Sales." *Hollywood Reporter*, 14 September 2006. Accessed 01 November 2008. http://www.hollywoodreporter.com/hr/search/article_display.jsp?vnu_content_id=1003121876. Link no longer available.

Ascott, Roy. "Homo Telematicus in the Garden of A-Life - Editorial." *TightRope 1/95* (1994). Accessed 21 March 2016. http://www.geelhaar.de/TR/artikel/text/asc/index.html.

Averintsev, Sergei S. "Bakhtin and the Russian Attitude to Laughter." In *Bakhtin: Carnival and Other Subjects: Selected Papers from the Fifth International Bakhtin Conference University of Manchester, July 1991*, edited by David Shepherd, 13–19. Atlanta, GA: Rodopi, 1993.

Aytes, Ayhan. "Return of the Crowds: Mechanical Turks and Neoliberal States of Exception." In *Digital Labor: The Internet as Playground and Factory*, edited by Tim Scholtz, 79–97. New York: Routledge, 2012.

Bacon-Smith, Camile. *Enterprising Women: Television, Fandom and the Creation of Popular Myth*. Philadelphia: University of Pennsylvania Press, 1992.

Bakhtin, Mikhail M. "Discourse in the Novel." In *The Dialogic Imagination: Four Essays*, edited by Michael Holquist, translated by Caryl Emerson and Michael Holquist, 259–422. Austin: University of Texas Press, 1981.

Bakhtin, Mikhail M. *Problems of Dostoevsky's Poetics*, edited and translated by Caryl Emerson. Minneapolis: University of Minnesota Press, 1984.

Bakhtin, Mikhail M. *Rabelais and His World*, translated by Hélène Iswolsky. Bloomington: Indiana University Press, 1984.

Bakioğlu, Burcu S. "Bull in a China Shop: Alternate Reality Games and Transgressive Fan Play in Social Media Franchises." *Transformative Works and Cultures* 17 (2014). http://dx.doi.org/10.3983/twc.2014.0594.

Bal, Mieke. *Narratology: Introduction to the Theory of Narrative*, 2nd ed., translated by Christine Van Boheemen. Toronto: University of Toronto Press, 1997.

Bannerman, Sara. "Crowdfunding Culture." *Wi: Journal of Mobile Culture* 7, no. 1 (2013). Accessed 22 March 2016. http://wi.mobilities.ca/crowdfunding-culture/.

Barthes, Roland. "The Death of the Author." In *Image-Music-Text*, translated by Stephen Heath, 142–8. New York: Hill and Wang, 1977.

Barthes, Roland. "From Work to Text." In *Image-Music-Text*, translated by Stephen Heath, 155–64. New York: Hill and Wang, 1977.

Barthes, Roland. *S/Z: An Essay*, translated by Richard Miller. New York: Hill and Wang, 1974.

Barthes, Roland. "Theory of the Text." 1973. In *Untying the Text: A Post-Structuralist Reader*, edited by Robert Young, translated by Ian McLeod, 31–47. London: Routledge, 1981.

Baudrillard, Jean. "Aesthetic Illusion and Virtual Reality." In *Reading Images*, edited by Julia Thomas, 198–206. Houndsmills, UK: Palgrave, 2000.

Baudrillard, Jean. *For a Critique of the Political Economy of the Sign*, translated by Charles Levin. St. Louis, MO: Telos Press, 1981.

Baudrillard, Jean. *Simulacra and Simulation*, translated by Sheila Faria Glaser. Ann Arbor: University of Michigan Press, 1994.

Baudrillard, Jean. *The Consumer Society*, translated by Chris Turner. London: Sage, 1998.

Baym, Nancy. "The Lost Librarians of National Defense." *Online Fandom: News and Perspectives on Fan Communication and Online Life*, 30 April 2008. Accessed 22 March 2016. http://www.onlinefandom.com/archives/the-lost-librarians-of-national-defense.

Baym, Nancy. *Tune In, Log On: Soaps, Fandom, and Online Community*. London: Sage, 2000.

Becher, Tony. *Academic Tribes and Territories: Intellectual Enquiry and the Cultures of Disciplines*. Milton Keynes, UK: Open University Press, 1989.

Bell, Duran. "Modes of Exchange: Gift and Economy." *Journal of Socio-Economics* 20, no. 2 (1991): 155–67.

Belleflamme, Paul, Thomas Lambert, and Armin Schwienbacher. "Individual Crowdfunding Practices. *Venture Capital: An International Journal of Entrepreneurial Finance* (2013). Accessed 22 March 2016. http://papers.ssrn.com/sol3/papers.cfm?abstract_id=2151179.

Benedikt, Michael. "Cyberspace: First Steps." In *Cybercultures Reader*, edited by David Bell and Barbara M. Kennedy, 19–33. London: Routledge, 2000.

Benkler, Yochai. "Coase's Penguin, or, Linux and the Nature of the Firm." In *CODE: Collaborative Ownership and the Digital Economy*, edited by Rishab Aiyer Ghosh, 169–206. Cambridge, MA: The MIT Press, 2005.

Benkler, Yochai. *The Wealth of Networks: How Social Production Transforms Markets and Freedom*. New Haven, CT: Yale University Press, 2006.

Bennett, James, and Tom Brown, ed. *Film and Television after DVD*. London: Routledge, 2008.

Bennett, Lucy. "Tracing Textual Poachers: Reflections on the Development of Fan Studies and Digital Fandom." *The Journal of Fandom Studies* 2, no. 1 (2014): 5–20.

Bennett, Tony, and Janet Woollacott. *Bond and Beyond: The Political Career of a Popular Hero*. London: MacMillan, 1987.

Berners-Lee, Tim, and Mark Fischetti. *Weaving the Web: The Original Design and Ultimate Destiny of the World Wide Web*. London: Orion Business, 1999.

Birdsall, W. "Web 2.0 as Social Movement." *Webology* 4, no. 2 (2007). Accessed 22 March 2016. http://www.webology.org/2007/v4n2/a40.html.

Bishop, Rollin. "Could Kickstarter One Day Completely Oust Hollywood?" [video]. *The Mary Sue*, 30 March 2013. Accessed 22 March 2016. http://www.themarysue.com/kickstarter-replace-hollywood.

Black, Rebecca. *Adolescents and Online Fan Fiction*. New York: Peter Lang, 2008.

Blair, Elizabeth. "Interactive Games Make Museums a Place to Play." *NPR*, 12 January 2009. Accessed 21 March 2016. http://www.npr.org/templates/story/story.php?storyId=99244253.

Blood, Rebecca. "Weblogs: A History and Perspective." *Rebecca's Pocket*, 07 September 2000. Accessed 22 March 2016. http://www.rebeccablood.net/essays/weblog_history.html.

Bolter, Jay David. *Writing Space: Computers, Hypertext, and the Remediation of Print*, 2nd ed. Mahwah, NJ: Lawrence Erlbaum Associates, 2001.

Bolter, Jay David, and Richard Grusin. *Remediation*. Cambridge, MA: The MIT Press, 1999.

Boorstin, Daniel. *The Image*. New York: Random House, 1961.

Booth, Paul. "ARGS, Fandom, and the Digi-gratis Economy." Interview with Henry Jenkins. *Confessions of an Aca-Fan*, 13–18 August 2010. Accessed 22 March 2016. http://henryjenkins.org/2010/08/args_fandom_and_the_digi-grati.html.

Booth, Paul. "Augmenting Fan/Academic Dialogue: New Directions in Fan Research." *The Journal of Fandom Studies* 1, no. 2 (2013): 119–37.

Booth, Paul. *Crossing Fandoms: SuperWhoLock and the Contemporary Fan Audience*. Houndmills, UK: Palgrave Pivot, 2016.

Booth, Paul. "Fandom: The Classroom of the Future." In "European Fans and European Fan Objects: Localization and Translation," edited by Anne Kustritz, special issue, *Transformative Works and Cultures* 19 (2015). http://dx.doi.org/10.3983/twc.2015.0650.

Booth, Paul. *Game Play: Paratextuality in Contemporary Board Games*. New York: Bloomsbury, 2015.

Booth, Paul. "Intermediality in Film and Internet: *Donnie Darko* and Issues of Narrative Substantiality." *Journal of Narrative Theory* 38, no. 3 (2008): 398–415.

Booth, Paul. "Mediating New Technology: The Realization of a Digital Intellect." *Nebula* 5, no. 1–2 (2008): 28–43.

Booth, Paul. *Playing Fans: Negotiating Fandom and Media in the Digital Age*. Iowa City: University of Iowa Press, 2015.

Booth, Paul. "Reifying the Fan: *Inspector Spacetime* as Fan Practice." *Popular Communication: The International Journal of Media and Culture* 11, no. 2 (2013): 146–59.

Booth, Paul, and Brian Ekdale. "Translating the Hyperreal (Or How *The Office* Came to America, Made Us Laugh, and Tricked Us into Accepting Hegemonic Bureaucracy)." In *American Remakes of British Television: Transformations and Mistranslations*, edited by Carlen Lavigne and Heather Marcovitch, 193–210. Lanham, MD: Lexington Books, 2011.

Booth, Paul, and Lucy Bennett. "Introduction: Seeing Fans." In *Seeing Fans: Representations of Fandom in Media and Popular Culture*, edited by Lucy Bennett and Paul Booth, 1–10. New York: Bloomsbury, 2016.

Booth, Paul, and Peter Kelly. "The Changing Faces of *Doctor Who* Fandom: New Fans, New Technologies, Old Practices?" *Participations* 10, no. 1 (2013). http://www.participations.org/Volume%2010/Issue%201/5%20Booth%20&%20Kelly%2010.1.pdf.

Bourdieu, Pierre. *Distinction: A Social Critique of the Judgment of Taste*, translated by Richard Nice. Cambridge, MA: Harvard University Press, 1984.

boyd, danah. "Facebook's Little Gifts." *Apophenia*, 13 February 2007. Accessed 21 March 2016. http://www.zephoria.org/thoughts/archives/2007/02/13/facebooks_littl.html.

boyd, danah. "Why Youth (Heart) Social Network Sites: The Role of Networked Publics in Teenage Social Life." In *MacArthur Foundation Series on Digital Learning – Youth, Identity, and Digital Media Volume*, edited by David Buckingham. Cambridge, MA: The MIT Press, 2007. Accessed 22 March 2016. http://www.danah.org/papers/WhyYouthHeart.pdf.

boyd, danah, and Nicole Ellison. "Social Network Sites: Definition, History, and Scholarship." *Journal of Computer-Mediated Communication* 13, no. 1 (2007). Accessed 22 March 2016. http://onlinelibrary.wiley.com/doi/10.1111/j.1083-6101.2007.00393.x/epdf.

Brabham, Daren C. "Crowdsourcing: A Model for Leveraging Online Communities." In *The Participatory Cultures Handbook*, edited by Aaron Delwiche and Jennifer. J. Henderson, 120–9. New York: Routledge, 2013.

Britton, Piers. *TARDISBound: Navigating the Universes of Doctor Who*. London: I.B. Tauris, 2011.

Brockman, John. *Digerati: Encounters with the Cyber Elite*. San Francisco, CA: Hardwired, 1996.

Brooker, Will. *Batman Unmasked: Analyzing a Cultural Icon*. New York: Continuum, 2001.

Brooker, Will. *Using the Force: Creativity, Community, and Star Wars Fans*. London: Continuum Press, 2002.

Brookey, Robert A. *Hollywood Gamers: Digital Convergence in the Film and Video Game Industries*. Bloomington: Indiana University Press, 2010.

Brookey, Robert, and Paul Booth. "Restricted Play: Synergy and the Limits of Interactivity in *The Lord of the Rings: The Return of the King* Video Game." *Games and Culture* 1, no. 3 (2006): 214–30.

Brookey, Robert, and Robert Westerfelhaus. "Hiding Homoeroticism in Plain View: The *Fight Club* DVD as Digital Closet." *Critical Studies in Media Communication* 19, no. 1 (2002): 21–43.

Brown, Tom. "'The DVD of Attractions'? *The Lion King* and the Digital Theme Park." *Convergence: The International Journal of Research into New Media Technologies* 13, no. 2 (2007): 169–83.

Bruns, Axel. *Blogs, Wikipedia, Second Life, and Beyond*. New York: Peter Lang, 2008.

Burke, Kenneth. *A Rhetoric of Motives*. Berkeley: University of California Press, 1969.

Burke, Kenneth. "'Terministic Screens." In *Language as Symbolic Action: Essays on Life, Literature and Method*, 44–62. Berkeley: University of California Press, 1966.

Bury, Rhiannon. *Cyberspaces of Their Own: Female Fandoms Online*. New York: Peter Lang, 2005.

Bury, Rhiannon, Ruth Deller, Adam Greenwood, and Bethan Jones. "From Usenet to Tumblr: The Changing Role of Social Media." *Participations* 10, no. 1 (2013): 229–318. http://www.participations.org/Volume%2010/Issue%201/14%20Bury%20et%20al%2010.1.pdf.

Busse, Kristina. "Fan Labor and Feminism: Capitalizing on the Fannish Labor of Love." *Cinema Journal* 54, no. 3 (2015): 110–15.

Busse, Kristina. "Geek Hierarchies, Boundary Policing, and the Gendering of the Good Fan." *Participations* 10, no 1 (2013): 73–91. http://www.participations.org/Volume%2010/Issue%201/6%20Busse%2010.1.pdf.

Busse, Kristina, and Karen Hellekson. "Introduction: Work in Progress." In *Fan Fiction and Fan Communities in the Age of the Internet*, edited by Karen Hellekson and Kristina Busse, 5–32. Jefferson, NC: McFarland and Co., 2006.

Caldwell, John. "Convergence Television: Aggregating Form and Repurposing Content in the Culture of Conglomeration." In *Television after TV*, edited by Lynn Spigel and Jan Olsson, 41–74. Durham, NC: Duke University Press, 2004.

Carey, James. "A Cultural Approach to Communication." In *Communication and Culture: Essays on Media and Society*, 13–36. New York: Routledge, 1992.

Carroll, Laura. "Cruel Spoiler, That Embosom'd Foe." *The Verve: A Literary Organ*, 09 October 2005. Accessed 21 March 2016. http://www.thevalve.org/go/valve/article/cruel_spoiler_that_embosomd_foe.

Cartmel, Andrew. *Through Time: An Unauthorized and Unofficial History of Doctor Who*. New York: Continuum, 2005.

Castronova, Edward. "Virtual World Economy: It's Namibia, Basically." 03 August 2004. Accessed 23 March 2016. http://terranova.blogs.com/terra_nova/2004/08/virtual_world_e.html.

Cerdán, Raquel, and Eduardo Vidal-Abarca. "The Effects of Tasks on Integrating Information from Multiple Documents." *Journal of Educational Psychology* 100, no. 1 (2008): 209–22.

Chang, Yi, Lei Tang, Yoshiyuki Inagaki, and Yan Liu. "What Is Tumblr: A Statistical Overview and Comparison," 2014. Accessed 23 March 2016. http://arxiv.org/abs/1403.5206.

Chatman, Seymour. *Coming to Terms*. Ithaca, NY: Cornell University Press, 1990.

Chatman, Seymour. *Story and Discourse: Narrative Structure in Fiction and Film*. Ithaca, NY: Cornell University Press, 1978.

Chee, Yam San. "Embodiment, Embeddedness, and Experience: Game-Based Learning and the Construction of Identity." *Research and Practice in Technology Enhanced Learning* 2, no. 1 (2007): 3–30.

Chess, Shira, and Paul Booth. "Lessons Down the Rabbit Hole: Alternate Reality Gaming in the Classroom." *New Media and Society* 16, no. 6 (2015): 1002–17.

Chin, Bertha. "The *Veronica Mars* Movie: Crowdfunding—Or Fan-Funding—at Its Best?" *On/Off Screen*, 13 March 2013. Accessed 23 March 2016. http://onoffscreen.wordpress.com/2013/03/13/the-veronica-mars-movie-crowdfunding-or-fan-funding-at-its-best/.

Chin, Bertha, Bethan Jones, Miles McNutt, and Luke Pebler. "*Veronica Mars* Kickstarter and Crowd Funding." *Transformative Works and Cultures* 15 (2014). http://dx.doi.org/10.3983/twc.2014.0519.

Cho, Alexander. "Queer Reverb: Tumblr, Affect, Time." In *Networked Affect*, edited by Ken Hillis, Susanna Paasonen, and Michael Petit, 43–58. Cambridge, MA: The MIT Press, 2015.

Clark, Katerina, and Michael Holquist. *Mikhail Bakhtin*. Cambridge, MA: The Belknap Press of Harvard University Press, 1984.

Click, Melissa. "'Rabid,' 'Obsessed," and 'Frenzied': Understanding *Twilight* Fangirls and the Gendered Politics of Fandom." *Flow*, 18 December 2009. Accessed 23 March 2016. http://www.flowjournal.org/2009/12/rabid-obsessed-and-frenzied-understanding-twilight-fangirls-and-the-gendered-politics-of-fandom-melissa-click-university-of-missouri/.

Clifford, Catherine. "Crowdfunding Nearly Tripled Last Year, Becoming a $16 Billion Industry." *Entrepreneur*, 31 March 2015. Accessed 9 April 2016. https://www.entrepreneur.com/article/244503.

Cobley, Paul. *Narrative*. London: Routledge, 2001.

Cochran, Tanya. "Crowdfunding the Narrative, or the High Cost of 'Fan-ancing'." In *Crowdfunding the Future: Media Industries, Ethics, and Digital Society*, edited by Lucy Bennett, Bertha Chin, and Bethan Jones, 31–46. New York: Peter Lang, 2015.

Condis, Megan. "No Homosexuals in *Star Wars*? BioWare, 'Gamer' Identity, and the Politics of Privilege in a Convergence Culture." *Convergence: The International Journal of Research into New Media Technologies* 21, no. 2 (2015): 198–212.

Cook, Pam. *Screening the Past: Memory and Nostalgia in Cinema*. London: Routledge, 2005.

Cooper, Dennis. "This Is Not an Isolated Incident: An Introduction." In *Userlands: New Fiction from the Blogging Underground*, edited by Dennis Cooper, 11–13. New York: Akashic Books, 2007.

Coppa, Francesca. "A Brief History of Media Fandom." In *Fan Fiction and Fan Communities in the Age of the Internet*, edited by Karen Hellekson and Kristina Busse, 41–59. Jefferson, NC: McFarland, 2006.

Crusius, Timothy, W. "A Case for Kenneth Burke's Dialectic and Rhetoric." *Philosophy and Rhetoric* 19, no. 1 (1986): 23–37.

Cubitt, Sean. *Simulation and Social Theory*. London: Sage, 2001.

D'Amato, Francesco. "Investors and Patrons, Gatekeepers and Social Capital: Representations and Experiences of Fans' Participation in Fan Funding." In *Ashgate Research Companion to Fan Cultures*, edited by Linda Duits, Koos Zwaan, and Stijn Reijnders, 135–48. Farnham, UK: Ashgate, 2014.

de Certeau, Michel. *The Practice of Everyday Life*, translated by Steven Randall. Berkeley: University of California Press, 1984.

De Kosnik, Abigail. "Fandom as Free Labor." In *Digital Labor: The Internet as Playground and Factory*, edited by Tim Scholtz, 98–112. New York: Routledge, 2013.

de Sena Caires, Carlos Duarte. "Towards the Interactive Filmic Narrative: 'Transparency': An Experimental Approach." *Computers and Graphics* 31 (2007): 800–8.

Deery, June. "TV.com: Participatory Viewing on the Web." *The Journal of Popular Culture* 37, no. 2 (2003): 161–83.

Dena, Christy. "ARG Stats." *Universe Creation 101*, 08 December 2008. Accessed 21 March 2016. http://www.christydena.com/online-essays/arg-stats/.

Dena, Christy. "Emerging Participatory Culture Practices: Player-Created Tiers in Alternate Reality Games." *Convergence: The International Journal of Research into New Media Technologies* 14, no. 1 (2008): 41–57.

Dena, Christy. "Patterns in Cross-Media Interaction Design: It's Much More than a URL … (Part 1)." *Crossmediaentertainment*, 2007. Accessed 21 March 2016. http://www.christydena.com/wp-content/uploads/2007/11/dena_morethan_part1.pdf.

Dentith, Simon. *Bakhtinian Thought: An Introductory Reader*. London: Routledge, 1995.

Derecho, Abigail. "Archontic Literature: A Definition, a History, and Several Theories of Fan Fiction." In *Fan Fiction and Fan Communities in the Age of the Internet*, edited by Karen Hellekson and Kristina Busse, 61–78. Jefferson, NC: McFarland, 2006.

Derrida, Jacques. "Afterward: Toward an Ethic of Discussion." In *Limited Inc.*, translated by Samuel Webber, 111–54. Evanston, IL: Northwestern University Press, 1988.

Derrida, Jacques. *Archive Fever: A Freudian Impression*, translated by Eric Prenowitz. Chicago: University of Chicago Press, 1996.

Derrida, Jacques. *Given Time: I. Counterfeit Money*, translated by Peggy Kamuf. Chicago: University of Chicago Press, 1992.

Derrida, Jacques. *Positions*, translated by Alan Bass. Chicago: University of Chicago Press, 1981.

Derrida, Jacques. *Specters of Marx: The State of the Debt, the Work of Mourning, and the New International*, translated by Peggy Kamuf. London: Routledge, 1994.

Dewey, Caitlin. "2015 Is the Year That Tumblr Became the Front Page of the Internet." *The Washington Post*, 11 March 2015. Accessed 22 March 2016. https://www.washingtonpost.com/news/the-intersect/wp/2015/03/11/move-over-reddit-tumblr-is-the-new-front-page-of-the-internet/.

Dewey, Caitlin. "You Might Want to Reconsider That Donation to the *Reading Rainbow* Kickstarter." *The Washington Post*, 29 May 2014. Accessed 23 March 2016. http://www.washingtonpost.com/news/the-intersect/wp/2014/05/28/you-might-want-to-reconsider-that-donation-to-the-reading-rainbow-kickstarter.

Dixon, Shanley, and Sandra Weber. "Playspaces, Childhood, and Video Games." In *Growing Up Online: Young People and Digital Technologies*, edited by Sandra Weber and Shanley Dixon, 15–35. New York: Palgrave, 2007.

Donath, Judith. "Identity and Deception in the Virtual Community." In *Communities in Cyberspace*, edited by Marc A. Smith and Peter Kollock, 29–59. London: Routledge, 1999.

Donath, Judith, and danah boyd. "Public Displays of Connection." *BT Technology Journal* 22, no. 4 (2004): 71–82.

du Gay, Paul, Jessica Evans, and Peter Redman, ed. *Identity: A Reader*. London: Sage, 2000.

Duffett, Mark. *Understanding Fandom: An Introduction to the Study of Media Fan Culture*. London: Bloomsbury, 2013.

Durkheim, Emile. *The Elementary Forms of Religious Life*, translated by Karen Fields. New York: Free Press, 1995.

e-cademia, "The Multiverse as Asylum: Madness and Neo-Baroque Fandom." *LiveJournal*, 31 March 2012. Accessed 23 March 2016. http://e-cademia.livejournal.com/732.html.

Eagleton, Terry. "Bakhtin, Schopenhauer, Kundera." In *Bakhtin and Cultural Theory*, edited by Ken Hirschkop and David Shepherd, 178–88. Manchester, UK: Manchester University Press, 1989.

Eagleton, Terry. *Walter Benjamin, or Towards a Revolutionary Criticism*. New York: Schocken Books, 1981.

Eco, Umberto. "*Casablanca:* Cult Movies and Intertextual Collage." *SubStance* 47 (1984): 3–12.

Empire online. "Features: *Veronica Mars*." *Empire*, 23 March 2016. Accessed 23 March 2016. http://www.empireonline.com/50greatesttv/default.asp?tv=48.

Erickson, Thomas. "Social Interaction on the Net: Virtual Community as Participatory Genre." Proceedings of the Thirtieth Annual Hawaii International Conference on System Sciences, *IEEE* (1997).

Eskelinen, Markku. "The Gaming Situation." *Game Studies 0101* 1 no. 1 (2001). Accessed 23 March 2016. http://gamestudies.org/0101/eskelinen/.

Fairclough, Norman. *Language and Power*. London: Longman, 1989.

Fernandez, Maria Elena. "Zach Braff Raises Money—and Ire—with Kickstarter Campaign for New Film. *NBC News*, 22 May 2013. Accessed 23 March 2016. entertainment.nbcnews.com/_news/2013/05/22/18401068-zach-braff-raises-money-and-ire-with-kickstarter-campaign-for-new-film.

Firewolfe. "Battlestar Galactica 1980/1985 Finding Earth." *Fanfiction.net*, 20 November 2007. Accessed 23 March 2016. http://www.fanfiction.net/s/3777645/1/Battle_Star_Galactica_1980_1985 Finding Earth.

Fiske, John. "The Cultural Economy of Fandom." In *The Adoring Audience: Fan Culture and Popular Media*, edited by Lisa A. Lewis, 30–49. London: Routledge, 1992.

Fiske, John. *Reading the Popular*. New York: Routledge, 1989.

Fiske, John. *Understanding Popular Culture*. New York: Routledge, 1989.

Fleisher, Lisa. "MagiQuest's Creator Franchises MB-First Game in U.S., Japan." *Sun News*, 14 November 2007. Accessed 26 Feb 2011 from EBSCOhost. Link no longer available.

Ford, Sam. "Fan Studies: Grappling with an 'Undisciplined' Discipline." *Journal of Fandom Studies* 2, no. 1 (2014): 53–71.

Foster, Derek. "'Jump in the Pool': The Competitive Culture of *Survivor* Fan Networks." In *Understanding Reality Television*, edited by Su Holmes and Deborah Jermyn, 270–89. London: Routledge, 2004.

Foucault, Michel. "Self-Writing." In *Ethics: Subjectivity and Truth Vol 1*, edited by Paul Rabinow, translated by Robert Hurley and others, 207–22. New York: The New Press, 1997.

Foucault, Michel. "What Is an Author?" In *Language, Counter-memory, Practice*, edited and translated by Donald Bouchard and Sherry Simon, 113–38. Ithaca, NY: Cornell University Press, 1977.

Freund, Katharina, and Dianna Fielding. "Research Ethics in Fan Studies." *Participations* 10, no. 1 (2013): 329–34. http://www.participations.org/Volume%2010/Issue%201/16%20Freund%20Fielding%2010.1.pdf.

Galuszka, Patryk. "New Economy of Fandom." *Popular Music and Society* 38, no. 1 (2015): 25–43.

Galuszka, Patryk, and Victor Bystrov. "The Rise of Fanvestors: A Study of a Crowdfunding Community." *First Monday* 19, no. 5 (2014). Accessed 23 March 2016. http://firstmonday.org/article/view/4117/4072.

Gane, Nicholas, and David Beer. *New Media: The Key Concepts*. Oxford, UK: Berg, 2008.

Gauntlett, David. *Making Is Connecting: The Social Meaning of Creativity, from DIY and Knitting to YouTube and Web 2.0*. Cambridge, UK: Polity Press, 2011.

Gauntlett, David. *Media, Gender and Identity: An Introduction*. London: Routledge, 2002.

Gee, James Paul. *What Video Games Have to Teach Us about Learning and Literacy*. New York: Palgrave Macmillan, 2003.

Genette, Gérard. *Narrative Discourse: An Essay in Method*, translated by Jane Lewin. Ithaca, NY: Cornell University Press, 1980.

Genette, Gérard. *Paratexts: Thresholds of Interpretation*, translated by Jane Lewin. Cambridge, UK: Cambridge University Press, 1997.

Geraghty, Christine. "Doing Media Studies: Reflections on an Unruly Discipline." *Art, Design & Communication in Higher Education* 1, no. 1 (2002): 25–36.

Geraghty, Lincoln. "Aging Toys and Players: Fan Identity and Cultural Capital." In *Finding the Force of the Star Wars Franchise: Fans, Merchandise, and Critics*, edited by Matthew Wilhelm Kapell and John Shelton Lawrence, 209–23. New York: Peter Lang, 2006.

Geraghty, Lincoln. *Cult Collectors*. London: Routledge, 2014.

Ghosh, Rishab Aiyer. "Why Collaboration Is Important (Again)." In *CODE: Collaborative Ownership and the Digital Economy*, edited by Rishab Aiyer Ghosh, 1–6. Cambridge, MA: The MIT Press, 2005.

Ghosh, Rishab Aiyer, ed. *CODE: Collaborative Ownership in the Digital Economy*. Cambridge, MA: The MIT Press, 2005.

Giles, David C. "Parasocial Interaction: A Review of the Literature and a Model for Future Research." *MediaPsychology* 4 (2002): 279–305.

Gindin, Sergei I. "Contributions to Textlinguistics in the Soviet Union." In *Current Trends in Textlinguistics*, edited by Wolfgang U. Dressler. Berlin: de Gruyter, 1978.

Godelier, Maurice. *The Enigma of the Gift*, translated by Nora Scott. Chicago: University of Chicago Press, 1999.

Goffman, Erving. *The Presentation of Self in Everyday Life*. New York: Doubleday, 1959.

Goodale, Gloria. "'Veronica Mars' Online Campaign Passes $2 Million: What's Hollywood Thinking?" *The Christian Science Monitor*, 14 March 2013. Accessed 23 March 2016. www.csmonitor.com/USA/Society/2013/0314/Veronica-Mars-online-campaign-passes-2-million-What-s-Hollywood-thinking.

Gosney, John W. *Beyond Reality: A Guide to Alternate Reality Gaming*. Boston: Thompson, 2005.

Goss, James, and Rob Francis (producers). *Love Off-Air*. Documentary feature on *Doctor Who: The Invasion*, DVD. London: British Broadcasting Corporation, 2006.

Grafton, Anthony. *The Footnote: A Curious History*. Cambridge, MA: Harvard University Press, 1999.

Grant, Tracy. "At Great Wolf Lodge, Some Grown-Up Time." *The Washington Post*, 24 December 2010. Accessed 23 March 2016. http://www.washingtonpost.com/wp-dyn/content/article/2010/12/21/AR2010122105597.html.

Gray, Jonathan. "Crowds, Words, and the Futures of Entertainment Conference." *Antenna*, 15 November 2011. Accessed 22 March 2016. http://blog.commarts.wisc.edu/2011/11/15/crowds-words-and-the-futures-of-entertainment-conference.

Gray, Jonathan. *Show Sold Separately: Promos, Spoilers, and Other Media Paratexts*. New York: New York University Press, 2010.

Gray, Jonathan. *Television Entertainment*. New York: Routledge, 2008.

Gray, Jonathan. *Watching with The Simpsons: Television, Parody, and Intertextuality*. New York: Routledge, 2006.

Gray, Jonathan, and Jason Mittell. "Speculation on Spoilers: *Lost* Fandom, Narrative Consumption and Rethinking Textuality." *Particip@tions* 4, no. 1 (2007). http://www.participations.org/Volume%204/Issue%201/4_01_graymittell.htm.

Gray, Jonathan, Cornel Sandvoss, and C. Lee Harrington, eds. *Fandom: Identities and Communities in a Mediated World*. New York: New York University Press, 2007.

Gray, Jonathan, Cornel Sandvoss, and C. Lee Harrington. "Introduction: Why Study Fans?" In *Fandom: Identities and Communities in a Mediated World*, edited by Jonathan Gray, Cornel Sandvoss, and C. Lee Harrington, 1–16. New York: New York University Press, 2007.

Griffin, Zachary. "Crowdfunding: Fleecing the American Masses." *Case Western Reserve University School of Law*, 14 March 2012. Accessed 23 March 2016. http://papers.ssrn.com/sol3/papers.cfm?abstract_id=2030001.

Grindstaff, Laura. "Culture and Popular Culture: A Case for Sociology." *Annals of the American Academy of Political and Social Science* 619 (2008): 206–22.

Gunkel, David. *Hacking Cyberspace*. Boulder, CO: Westview Press, 2001.

Gunkel, David. "Rethinking the Digital Remix: Mash-Ups and the Metaphysics of Sound Recording." *Popular Music and Society* 31, no. 4 (2008): 489–510.

Gunkel, David. *Thinking Otherwise*. West Lafayette, IN: Purdue University Press, 2007.

Gwenllian-Jones, Sara. "Virtual Reality and Cult Television." In *Cult Television*, edited by Sara Gwenllian-Jones and Roberta E. Pearson, 83–98. Minneapolis: University of Minnesota Press, 2004.

Gwenllian-Jones, Sara, and Roberta E. Pearson, eds. *Cult Television*. Minneapolis: University of Minnesota Press, 2004.

Gwenllian-Jones, Sara, and Roberta E. Pearson. Introduction to *Cult Television*, ix–xx. Minneapolis: University of Minnesota Press, 2004.

Gwenllian-Jones, Sara. "The Sex Lives of Cult Television Characters." *Screen* 43, no. 1 (2002): 79–90.

Gwenllian-Jones, Sara. "Web Wars: Resistance, Online Fandom and Studio Censorship." In *Quality Popular Television*, edited by Mark Jancovich and James Lyons, 163–77. London: BFI, 2003.

Hadas, Leora. "The Web Planet: How the Changing Internet Divided *Doctor Who* Fan Fiction Writers." *Transformative Works and Cultures* 3 (2009). http://dx.doi.org/10.3983/twc.2009.0129.

Hall, Stuart. "Introduction: Who Needs 'Identity'?" In *Questions of Cultural Identity*, edited by Stuart Hall and Paul Du Gay, 1–17. London: Sage Publications, 1996.

Harrington, C. Lee, and Denise D. Bielby. "Global Fandom/Global Fan Studies." In *Fandom: Identities and Communities in a Mediated World*, edited by Jonathan Gray, Cornel Sandvoss, and C. Lee Harrington, 179–97. New York: New York University Press, 2007.

Harrington, C. Lee, and Denise D. Bielby. "New Directions in Fan Studies." *American Behavioral Scientist* 48, no. 7 (2005): 799–800.

Harty, E. R. "Text, Context, Intertext." *Journal of Literary Studies* 1, no. 2 (1985): 1–14.

Harvey, David. *A Brief History of Neoliberalism*. Oxford, UK: Oxford University Press, 2005.

Harvie, Jen. *Fair Play: Art, Performance and Neoliberalism*. London: Palgrave Macmillan, 2013.

Hellekson, Karen. "A Fannish Field of Value: Online Fan Gift Culture." *Cinema Journal* 48, no. 4 (2009): 113–8. http://dx.doi.org/10.1353/cj.0.0140.

Hellekson, Karen. "SF Fan Wikis: Source, Reference, World." *Res gestae—Documentary and Digital Evidence of the Trace*, 20 July 2008. Accessed 21 March 2016. http://khellekson.wordpress.com/2008/07/20/sf-fan-wikis/.

Hellekson, Karen, and Kristina Busse, eds. *Fan Fiction and Fan Communities in the Age of the Internet*. Jefferson, NC: McFarland, 2006.

Hemer, Joachim. "A Snapshot on Crowdfunding." *Working Papers Firms and Region*, R2 (2011). Accessed 23 March 2016. http://hdl.handle.net/10419/52302.

Hensley-Clancy, Molly. "Reading Rainbow's Massive Kickstarter Campaign Doesn't Mention That Most Schools Will Have to Pay for Access." *Buzzfeed*, 30 May 2014. Accessed 23 March 2016. http://www.buzzfeed.com/mollyhensleyclancy/reading-rainbows-massive-kickstarter-campaign-doesnt-mention.

Herman, David. *Story Logic*. Lincoln: University of Nebraska Press, 2002.

Hernandez, Dan. "Gaming+Autonomy=Academic Leadership." *Principal Leadership* 10, no. 4 (2009): 44–8.

Hess, Charlotte, and Elinor Ostrom. "Introduction: An Overview of the Knowledge Commons." In *Understanding Knowledge as Commons: From Theory to Practice*, edited by Charlotte Hess and Elinor Ostrom, 3–26. Cambridge, MA: The MIT Press, 2006.

Hevern, Vincent W. "Threaded Identity in Cyberspace: Weblogs & Positioning in the Dialogical Self." *Identity: An International Journal of Theory and Research* 4, no. 4 (2004): 321–35.

Hillman, Serena, Jason Procyk, and Carman Neustaedter. "'alksjdf;lksfd': Tumblr and the Fandom User Experience." 2014. Accessed 23 March 2016. http://dl.acm.org/citation.cfm?id=2600887&dl=ACM&coll=DL&CFID=594181856&CFTOKEN=93137949.

Hillman, Serena, Jason Procyk, and Carman Neustaedter. "Tumblr Fandoms, Community & Culture." *CSCW'14 Companion* (2014). Accessed 23 March 2016. http://dl.acm.org/citation.cfm?id=2557634&dl=ACM&coll=DL&CFID=594181856&CFTOKEN=93137949.

Hills, Matt. "As Seen on Screen? Mimetic SF Fandom & the Crafting of Replica(nt)s." *In Media Res*, 10 September 2010. Accessed 23 March 2016. http://mediacommons.futureofthebook.org/imr/2010/09/10/seen-screen-mimetic-sf-fandom-crafting-replicants.

Hills, Matt. "Defining Cult TV: Texts, Inter-texts, and Fan Audiences." In *The Television Studies Reader*, edited by Robert C. Allen and Annette Hill, 509–23. London: Routledge, 2004.

Hills, Matt. *Fan Cultures*. London: Routledge, 2002.

Hills, Matt. "Fandom as an Object and the Objects of Fandom." *Matrizes* 9, no. 1 (2015): 147–62.

Hills, Matt. "Fiske's 'Textual Productivity' and Digital Fandom: Web 2.0 Democratization Versus Fan Distinction?" *Participations: The Journal of Audience and Reception Studies* 10, no. 1 (2013): 130–53. http://www.participations.org/Volume%2010/Issue%201/9%20Hills%2010.1.pdf.

Hills, Matt. "From Dalek Half Balls to Daft Punk Helmets: Mimetic Fandom and the Crafting of Replicas." *Transformative Works and Cultures* 16 (2014). http://dx.doi.org/10.3983/twc.2014.0531.

Hills, Matt. "Michael Jackson Fans on Trial? 'Documenting' Emotivism and Fandom in 'Wacko about Jacko.'" *Social Semiotics* 17, no. 4 (2007): 459–77.

Hills, Matt. "Strategies, Tactics and the Question of *Un Lieu Propre*: What/Where Is 'Media Theory'?" *Social Semiotics* 14, no. 2 (2004): 133–49.

Hills, Matt. *Triumph of a Time Lord: Regenerating* Doctor Who *in the Twenty-First Century*. London: IB Tauris, 2010.

Hills, Matt. "'Twilight' Fans Represented in Commercial Paratexts and Inter-Fandoms: Resisting and Repurposing Negative Fan Stereotypes." In *Genre, Reception, and Adaptation in the Twilight Series*, edited by Anne More, 113–31. Farnham, UK: Ashgate, 2012.

Hills, Matt. "*Veronica Mars*, Fandom, and the 'Affective Economics' of Crowdfunding Poachers." *New Media and Society* 17, no. 2 (2015): 183–97.

Horton, David, and Richard Wohl. "Mass Communication and Parasocial Interaction." *Psychiatry 19* (1956): 215–29.

Huffaker, David. "Teen Blogs Exposed: The Private Lives of Teens Made Public." *American Association for Advancement of Science*. St. Louis, MO, 2006. Accessed 23 March 2016. https://

www.researchgate.net/publication/241644648_Teen_Blogs_Exposed_The_Private_Lives_of_Teens_Made_Public.

Huffaker, David A., and Sandra L. Calvert. "Gender, Identity, and Language Use in Teenage Blogs." *Journal of Computer-Mediated Communication* 10, no. 2 (2005). http://onlinelibrary.wiley.com/doi/10.1111/j.1083-6101.2005.tb00238.x/abstract.

Hui, Julie S., Elizabeth Gerber, and Mike Greenberg. "'Easy Money? The Demands of Crowdfunding Work." *Segal Segal Technical Report* 12, no. 4 (2012). Accessed 23 March 2016. http://egerber.mech.northwestern.edu/wp-content/uploads/2012/11/Easy-Money-_The-Demands-of-CrowdfundingWork-_2012.pdf.

Huizinga, Johan. *Homo Ludens: A Study of Play Element in Culture*. Boston: Beacon Press, 1955.

Hutchens, B. "Great Wolf Lodge Resort Much More Than Just a Water Park." *News Tribune*, 2008. Accessed 26 February 2011. Link no longer available.

James, Carrie, with Katie Davis, Andrea Flores, John M. Francis, Lindsay: Pettingill, Margaret Rundle, and Howard Gardner. *Young People, Ethics, and the New Digital Media: A Synthesis from the GoodPlay Project*. Cambridge, MA: The MIT Press, 2009.

Jenkins, Henry. "Afterward: The Future of Fandom." In *Fandom: Identities and Communities in a Mediated World*, edited by Jonathan Gray, Cornel Sandvoss, and C. Lee Harrington, 357–64. New York: New York University Press, 2007.

Jenkins, Henry. *Convergence Culture: Where Old and New Media Collide*. New York: New York University Press, 2006.

Jenkins, Henry. "Fandom Studies as I See It." *Journal of Fandom Studies* 2, no 2. (2014): 89–109.

Jenkins, Henry. "Interactive Audiences? The 'Collective Intelligence' of Media Fans." In *Fans, Bloggers, and Gamers: Exploring Participatory Culture*, 134–52. New York: New York University Press, 2006.

Jenkins, Henry. "Introduction: Should We Discipline the Reading of Comics?" In *Critical Approaches to Comics: Theories and Methods*, edited by Matthew J. Smith and Randy Duncan, 1–14. New York: Routledge, 2012.

Jenkins, Henry. "Lecture 06: Media Literacy as a Strategy for Combatting Moral Panic [sic]." *MIT OCW: CMS.930/21F.034 Media, Education, and the Marketplace, Fall 2001*. Video Podcast on iTunes University. Accessed 21 January 2008.

Jenkins, Henry. "Star Trek Rerun, Reread, Rewritten: Fan Writing as Textual Poaching." In *Fans, Bloggers, and Gamers: Exploring Participatory Culture*, 37–60. New York: New York University Press, 2006.

Jenkins, Henry. *Textual Poachers: Television Fans and Participatory Culture*. New York: Routledge, 1992.

Jenkins, Henry, Sam Ford, and Joshua Green. *Spreadable Media: Creating Value and Meaning in a Networked Culture*. New York: New York University Press, 2013.

Jenkins, Henry, Xiaochang Li, Ana Domb Krauskopf, with Joshua Green. "If It Doesn't Spread, It's Dead." *Confessions of an Aca-Fan*, 11 February 2009. Accessed 23 March 2016. http://henryjenkins.org/2009/02/if_it_doesnt_spread_its_dead_p.html.

Jenkins, Henry, Xiaochang Li, Ana Domb Krauskopf, with Joshua Green. "If It Doesn't Spread, It's Dead (Part Three): The Gift Economy and Commodity Culture." *Confessions of an*

Aca-Fan, 16 February 2009. Accessed 23 March 2016. http://henryjenkins.org/2009/02/if_it_doesnt_spread_its_dead_p_2.html.

Jenkins, Henry, with Ravi Purushotma, Margaret Weigel, Katie Clinton, and Alice J. Robison. *Confronting the Challenges of Participatory Culture: Media Education for the 21st Century*. Cambridge, MA: The MIT Press, 2009.

Jensen, Jeff. "'Lost' Adds Fifth New Cast Member." *Entertainment Weekly*, 29 August 2007. Accessed 23 March 2016. http://www.ew.com/ew/article/0,,20053479,00.html.

Jensen, Joli. "Fandom as Pathology: The Consequences of Characterization." In *The Adoring Audience*, edited by Lisa Lewis, 9–29. London: Routledge, 1992.

Jensen, Klaus Bruhn. "Mixed Media: From Digital Aesthetics towards General Communication Theory." *Northern Lights* 5 (2007): 7–24.

Johnson, Keith M. "'The Coolest Way to Watch Movie Trailers in the World': Trailers in the Digital Age." *Convergence* 14, no. 2 (2008): 145–60.

Johnson, Steven. *Everything Bad Is Good for You*. New York: Riverside, 2005.

Jones, Bethan. "Fan Exploitation, Kickstarter and *Veronica Mars*." *Fan Academic Blogger*. 15 March 2014. Accessed 23 March 2016. http://bethanvjones.wordpress.com/2013/03/15/fan-exploitation-kickstarter-and-veronica-mars/.

Jones, Bethan. "Fifty Shades of Exploitation: Fan Labor and *Fifty Shades of Grey*." In "Fandom and/as Labor," edited by Mel Stanfill and Megan Condis, special issue, *Transformative Works and Cultures* 15 (2014). http://dx.doi.org/10.3983/twc.2014.0501.

Jones, Bethan. "My *Little Pony*, Tolerance Is Magic: Gender Policing and Brony Anti-fandom." *The Journal of Popular Television* 3, no. 1 (2015): 119–25.

June, Daniel. "Melissa Joan Hart's Kickstarter Campaign Fails Utterly to Kick Off." *JD Journal*, 16 May 2013. Accessed 23 March 2016. http://www.jdjournal.com/2013/05/16/melissa-joan-harts-kickstarter-campaign-fails-utterly-to-kick-off/#.

Juul, Jesper. *Half-Real: Video Games between Real Rules and Fictional Worlds*. Cambridge, MA: The MIT Press, 2005.

Kaplan, Deborah. "Construction of Character through Narrative." In *Fan Fiction and Fan Communities in the Age of the Internet*, edited by Karen Hellekson and Kristina Busse, 134–52. Jefferson, NC: McFarland, 2006.

Karkanias, Alena. "The Intra- and Inter-Sub-Community Dynamics of Fandom." Summer Research Paper 230 (2014). Accessed 23 March 2016. http://soundideas.pugetsound.edu/summer_research/230.

Keane, Erin. "Kickstarter: Stop Blaming Us for Zach Braff and 'Veronica Mars'." *Salon*. 10 May 2013. Accessed 23 March 2016. http://www.salon.com/2013/05/10/kickstarter_stop_blaming_us_for_zach_braff_and_veronica_mars/.

Keen, Andrew. *The Cult of the Amateur*. New York: Random House, 2007.

Kelly, Jen. "Tumblr: A Site of Media Literacy and Public Education." Paper presented at 2015 Midwest Popular Culture Association, Cincinnati, OH, 03 October 2015.

Kendall, Kenneth E., and Allen Schmidt. "Mash-Ups: The Art of Creating New Applications by Combining Two or More Web Sites." *Decision Line*, 15–16 March 2007. Accessed 23 March 2016. https://www.researchgate.net/publication/228400062_Mashups_The_Art_of_Creating_New_Applications_by_Combining_Two_or_More_Web_Sites.

Kerr, Aphra. *The Business and the Culture of Digital Games: Gamework/Gameplay*. Thousand Oaks, CA: Sage, 2006.

King, Arthur A. Jr., and John A. Brownell. *The Curriculum and the Disciplines of Knowledge: A Theory of Curriculum Practice*. New York: John Wiley and Sons, 1966.

Klosterman, Chuck. "Was It Ethical for Zach Braff to Take to Kickstarter?" *The New York Times*, 24 May 2013. Accessed 23 March 2016. http://www.nytimes.com/2013/05/26/magazine/was-it-ethical-for-zach-braff-to-take-to-kickstarter.html?_r=0.

Kocer, Suncem. "Social Business in Online Financing: Crowdfunding Narratives of Independent Documentary Producers in Turkey." *New Media and Society* 17, no. 2 (2015): 231–48.

Kohn, Liberty. "Harry Potter and the Neoliberal Hallows: Copyright, Counterpublics and 2.0 Fan Fiction." *Technoculture* 2 (2012). Accessed 23 March 2016. http://tcjournal.org/drupal/vol2/kohn.

Kollock, Peter. "The Economics of Online Cooperation: Gifts and Public Goods in Cyberspace." In *Communities in Cyberspace*, edited by Marc A. Smith and Peter Kollock, 220–39. London: Routledge, 1999.

Kollock, Peter, and Marc A. Smith. Introduction to *Communities in Cyberspace*, edited by Marc A. Smith and Peter Kollock, 3–27. London: Routledge, 1999.

Kopytoff, Verne G. "Blogs Wane as the Young Drift to Sites Like Twitter." *New York Times*, 20 February 2011. Accessed 29 December 2015. www.nytimes.com/2011/02/21/technology/internet/21blog.html.

Kraus, Wolfgang. "The Narrative Negotiation of Identity and Belonging." *Narrative Inquiry* 16, no. 1 (2006): 103–11.

Kristeva, Julia. "The Bounded Text." In *Desire in Language: A Semiotic Approach to Literature and Art*, edited by Leon S. Roudiez, translated by Thomas Gora, Alice Jardine, and Leon S. Roudiez, 36–63. New York: Columbia University Press, 1980.

Kristeva, Julia. "Word, Dialogue, Novel." In *Desire in Language: A Semiotic Approach to Literature and Art*, edited by Leon S. Roudiez, translated by Thomas Gora, Alice Jardine, and Leon S. Roudiez, 64–91. New York: Columbia University Press, 1980.

Kuppuswamy, Venkat, and Barry L. Bayus. "Crowdfunding Creative Ideas: The Dynamics of Project Backers in *Kickstarter*." *SSRN Working Paper* (2013). Accessed 23 March 2016. http://papers.ssrn.com/sol3/papers.cfm?abstract_id=2234765.

Kushner, David. "Escape to Middle-Earth." *Wired* 17, no. 9 (2009).

Kustritz, Anne. "Exploiting Surplus Labours of Love: Narrating Ownership and Theft in Crowdfunding Controversies." In *Crowdfunding the Future: Media Industries, Ethics & Digital Society*, edited by Lucy Bennett, Bertha Chin, and Bethan Jones, 47–63. New York: Peter Lang, 2015.

Lachmann, Renate. "Bakhtin and Carnival: Culture as Counter-Culture." *Critical Critique* 11 (1988): 115–52.

Lancaster, Kurt. *Interacting with Babylon 5*. Austin: University of Texas Press, 2001.

Landow, George. *Hypertext 3.0: Critical Theory and New Media in an Era of Globalization*, rev. ed. Baltimore, MD: The Johns Hopkins University Press, 2006.

Le Guern, Philippe. "Toward a Constructivist Approach to Media Cults," translated by Richard Crangle. In *Cult Television*, edited by Sara Gwenllian-Jones and Roberta E. Pearson, 3–26. Minneapolis: University of Minnesota Press, 2004.

Leibovitz, Talia, Antoni Roig Telo, and Jordi Sánchez-Navarro. "Up Close and Personal: Exploring the Bonds between Promoters and Backers in Audiovisual Crowdfunded Projects." In *Crowdfunding the Future: Media Industries, Ethics & Digital Society*, edited by Lucy Bennett, Bertha Chin, and Bethan Jones, 15–30. New York: Peter Lang, 2015.

Lessig, Lawrence. *Code: Version 2.0*. New York: Basic Books, 2006.

Lessig, Lawrence. "The Creative Commons." *Florida Law Review* 55 (1994): 763–77.

Lessig, Lawrence. *Free Culture: The Nature and Future of Creativity*. New York: Penguin, 2004.

Lessig, Lawrence. *The Future of Ideas: The Fate of the Commons in a Connected World*. New York: Random House, 2001.

Lessig, Lawrence. *Remix: Making Art and Commerce Thrive in the Hybrid Economy*. New York: Penguin, 2008.

Leuf, Bo, and Ward Cunningham. *The Wiki Way: Quick Collaboration on the Web*. Boston: Addison-Wesley, 2001.

Lévi-Strauss, Claude. "Selections from *Introduction to the Work of Marcel Mauss*," translated by Felicity Baker. In *The Logic of the Gift: Toward an Ethic of Generosity*, edited by Alan D. Schrift, 45–69. New York: Routledge, 1997.

Levinson, Paul. *New New Media*. Boston: Allyn & Bacon, 2009.

Lévy, Pierre. *Collective Intelligence: Mankind's Emerging World in Cyberspace*, translated by Robert Bononno. New York: Perseus, 1997.

Lévy, Pierre. *Cyberculture*, translated by Robert Bononno. Minneapolis: University of Minnesota Press, 2001.

Littleton, Cynthia. "'Lost': The Weight of the Wait." *Variety.com*, 12 October 2007. Accessed 23 March2016.http://www.variety.com/article/VR1117974014.html?categoryid=2641&cs=1.

Lucas, Barbara. "Notes and Correspondence." *Science Fiction Studies* 35(2): 359–60.

Manovich, Lev. *The Language of New Media*. Cambridge, MA: The MIT Press, 2001.

Manovich, Lev. "What Comes after Remix?" 2007. Accessed 23 March 2016. http://remixtheory.net/?p=169.

Marx, Karl. *Capital: A Critique of Political Economy*, translated by Ben Fowkes. London: Penguin Classics, 1990.

Mauss, Marcel. *The Gift: The Form and Reason for Exchange in Archaic Societies*, translated by W. D. Halls. New York: Norton, 1990.

McChesney, Robert. "Introduction." In *Profit over People: Neoliberalism and Global Order* by Noam Chomsky, 7–18. New York: Seven Stories Press, 1999.

McCracken, Chelsea. "The Future of Media Production?" *Antenna*, 03 April 2013. Accessed 23 March 2016. http://blog.commarts.wisc.edu/2013/04/03/the-future-of-media-production/.

McGonigal, Jane. "I Love Bees: A Buzz Story." Paper presented at AD:TECH, San Francisco, 25–27 April 2005. Accessed 23 March 2016. http://avantgame.com/McGonigal_42%20Entertainment_ADTECH_April%202005.pdf.

McGonigal, Jane. "The Puppetmaster Problem: Design for Real-World, Mission-Based Gaming." In *Second Person: Role-Playing and Story in Games and Playable Media*, edited by Pat Harrigan and Noah Wardrip-Fruin, 251–63. Cambridge, MA: The MIT Press, 2007.

McGonigal, Jane. "A Real Little Game: The Performance of Belief in Pervasive Play." Level Up. Proceedings of DiGRA conference, Utrecht, Netherlands, November 2003.

McGonigal, Jane. *Reality Is Broken: Why Games Make Us Better and How They Can Change the World*. New York: Penguin, 2011.

McGonigal, Jane. "'This Is Not a Game': Immersive Aesthetics and Collective Play." Digital Arts & Culture Conference Proceedings, 2003. Accessed 23 March 2016. http://edgaps.org/gaps/wp-content/uploads/mcgonigal_ting_03.pdf.

McLuhan, Marshall. *The Medium Is the Message*. New York: Random House, 1994.

McNutt, Myles. "Kickstarting *Veronica Mars*: A Moment in a Movement." *Antenna*, 15 March 2013. Accessed 23 March 2016. http://blog.commarts.wisc.edu/2013/03/15/kickstarting-veronica-mars-a-moment-in-a-movement/.

Meadows, Mark Stephen. *Pause and Effect*. Indianapolis, IN: New Riders, 2003.

Menon, Vinay. "Mars Bars Can't Save Veronica." *Toronto Star*, 12 June 2007. Accessed 23 March 2016. http://www.thestar.com/opinion/columnists/2007/06/12/mars_bars_cant_save_veronica.html.

Merrin, William. *Baudrillard and the Media*. Cambridge, UK: Polity Press, 2005.

Merrin, William. *Media Studies 2.0*. London: Routledge, 2014.

Merrin, William. "Media Studies 2.0: Upgrading and Open-Sourcing the Discipline." *Interactions: Studies in Communication and Culture* 1, no 1 (2009): 17–34.

Miall, David S. "Episode Structures in Literary Narratives." *Journal of Literary Semantics* 33 (2004): 111–29.

Miller, Carolyn R., and Dawn Shepherd. "Blogging as Social Action: A Genre Analysis of the Weblog." In *Into the Blogosphere: Rhetoric, Community, and Culture of Weblogs*, edited by Laura Gurak, Smiljana Antonijevic, Laurie Johnson, Clancy Ratliff, and Jessica Reyman. Minneapolis: University of Minnesota Press, 2004. Accessed 23 March 2016. http://blog.lib.umn.edu/blogosphere/.

Miller, Paul Allen. "The Otherness of History in Rabelais' Carnival and Juvenal's Satire, or Why Bakhtin Got It Right the First Time." In *Bakhtin and the Other*, edited by Peter Barta, Paul Allen Miller, Charles Platter, and David Shepherd, 141–64. London: Routledge, 2001.

Mitchell, W. J. T. *What Do Pictures Want?: The Lives and Loves of Images*. Chicago: University of Chicago Press, 2006.

Mittell, Jason. *Complex TV: The Poetics of Contemporary Television Storytelling*. New York: New York University Press, 2015.

Mittell, Jason. "The Loss of Value (or the Value of Lost)." *Flow* 2, no. 5 (2005). Accessed 23 March 2016. http://flowtv.org/?p=165.

Mittell, Jason. "Narrative Complexity in Contemporary American Television." *The Velvet Light Trap* 58 (2006): 29–40.

Mittell, Jason. "Sites of Participation: Wiki Fandom and the Case of Lostpedia." *Transformative Works and Cultures* 3 (2009). http://dx.doi.org/10.3983/twc.2009.0118.

Mittell, Jason. "The Value of Lost, Part Two." *FlowTV* 2, no. 10 (2005). Accessed 23 March 2016. http://flowtv.org/?p=435.

Mittell, Jason. "*Veronica Mars* and Exchanges of Value Revisited." *Just TV*, 15 March 2013. Accessed 23 March 2016. http://justtv.wordpress.com/2013/03/15/veronica-mars-and-exchanges-of-value-revisited.

Molesworth, Richard. *Wiped! Doctor Who's Missing Episodes*. London: Telos Press, 2010.

Montola, Marcus. "Exploring the Edge of the Magic Circle: Defining Pervasive Games." Proceedings of DAC, Copenhagen, Denmark, 2005. Accessed 23 March 2016. http://www.iki.fi/montola/exploringtheedge.pdf.

Montola Marcus, Jaakko Stenros, and Annika Waern, eds. *Pervasive Games: Theory and Design*. Burlington, MA: Morgan Kaufmann, 2009.

Moores, Shaun. *Media/Theory*. London: Routledge, 2005.

Morgan, Thaïs. "Is There an Intertext in This Text?: Literary and Interdisciplinary Approaches to Intertextuality." *American Journal of Semiotics* 3, no. 1 (1985): 1–40.

Morrissey, Katherine E. "Fan/dom: People, Practices, and Networks." *Transformative Works and Cultures* 14 (2013). http://dx.doi.org/10.3983/twc.2013.0532.

Morson, Gary Saul, and Caryl Emerson. *Mikhail Bakhtin: Creation of a Prosaics*. Stanford, CA: Stanford University Press, 1990.

Mortensen, Torvill Elvira. "Me, the Other." In *Second Person: Role-Playing and Story in Games and Playable Media*, edited by Pat Harrigan and Noah Wardrip-Fruin, 297–306. Cambridge, MA: The MIT Press, 2007.

Murray, Janet. *Hamlet on the Holodeck: The Future of Narrative in Cyberspace*. Cambridge, MA: The MIT Press, 1997.

Newman, Michael Z. "Say 'Pulp Fiction' One More Goddamn Time: Quotation Culture and an Internet-Age Classic." *New Review of Film and Television Studies* 12, no. 2 (2014): 125–42.

Ng, Eve. "Reading the Romance of Fan Cultural Production: Music Videos of a Television Lesbian Couple." *Popular Communication* 6, no. 2 (2008): 103–21.

O'Neill, Patrick. *Fictions of Discourse: Reading Narrative Theory*. Toronto: University of Toronto Press, 1994.

O'Reilly, Tim. "What Is Web 2.0: Design Patterns and Business Models for the Next Generation of Software." *O'Reilly*, 30 September 2005. Accessed 23 March 2016. http://www.oreillynet.com/pub/a/oreilly/tim/news/2005/09/30/what-is-web-20.html.

obsession_inc. "Affirmational Fandom vs. Transformational Fandom," 01 June 2009. Accessed 23 March 2016. http://obsession-inc.dreamwidth.org/82589.html.

Oldnall, Andrew S. "Nursing as an Emerging Academic Discipline." *Journal of Advanced Nursing* 21 (1995): 605–12.

Ong, Walter. *Orality and Literacy*. London: Routledge, 2002.

Oren, Eyal. "SemperWiki: A Semantic Personal Wiki." Semantic Desktop Workshop 2005 @ ISWC. 2005. Accessed 23 March 2016. http://ceur-ws.org/Vol-175/27_oren_semperwiki_final.pdf.

Örnebring, Henrik. "Alternate Reality Gaming and Convergence Culture: The Case of *Alias*." *International Journal of Cultural Studies* 10, no. 4 (2007): 445–62.

Orr, Mary. *Intertextuality*. London: Polity, 2003.

Ostrom, Elinor. *Governing the Commons: The Evolution of Institutions for Collective Action*. Cambridge, UK: Cambridge University Press, 1990.

Ott, Brian L., and Cameron Walter. "Intertextuality: Interpretive Practice and Textual Strategy." *Critical Studies in Media Communication* 17, no. 4 (2000): 429–46.

Otto, Wayne, Sandra White, and Kay Camperell. "Text Comprehension Research to Classroom Application: Developing an Instructional Technique." *Reading Psychology* 1, no. 3 (1980): 184–91.

Pande, Rukmini. "Squee from the Margins: Racial/Cultural/Ethnic Identity in Global Media Fandom." In *Seeing Fans: Representations of Fandom in Media and Popular Culture*, edited by Lucy Bennett and Paul Booth, 209–20. New York: Bloomsbury, 2016.

Pariser, Eli. *The Filter Bubble: What the Internet Is Hiding from You*. London: Penguin UK, 2011.

Parkin, Lance. *AHistory: An Unauthorized History of the* Doctor Who *Universe*. Des Moines, IA: Mad Norwegian Press, 2006.

Pebler, Luke. "My Gigantic Issue with the *Veronica Mars* Kickstarter." *Revenge of the Fans* [Suzanne Scott], 13 March 2013. Accessed 23 March 2016. http://www.suzanne-scott.com/2013/03/15/guest-post-my-gigantic-issue-with-the-veronica-mars-kickstarter/.

Penley, Constance. *Nasa/Trek: Popular Science and Sex in America*. London: Verso, 1997.

Perez, Nistasha. "GIF Fics and the Rebloggable Canon of SuperWhoLock." In *Fan Phenomena: Doctor Who*, edited by Paul Booth, 148–57. Bristol, UK: Intellect, 2013.

Pérez-Reverte, Arturo. *The Club Dumas*, translated by Sonia Soto. New York: Vintage, 1997.

Perryman, Neil. "Doctor Who and the Convergence of Media: A Case Study in Transmedia Storytelling." *Convergence: The International Journal of Research into New Media Technologies* 14, no. 1 (2008): 21–39.

Pesce, Mark. "Piracy Is *Good*? New Models for the Distribution of Television Programming." Paper presented at the Australian Film Television and Radio School, Sydney, 6 May 2005. Accessed 23 March 2016. http://hyperreal.org/~mpesce/piracyisgood.pdf.

Petersen, Line Nybro. "*Sherlock* Fans Talk: Mediatized Talk on Tumblr." *Northern Lights* 12 (2014): 87–104.

Peterson, Karla. "A Shout-out to Blog Buzz: TV Shows Embrace the Passion (and Value) of Internet Fan Sites." *Union-Tribune*, 10 April 2006. Accessed 23 March 2016. http://legacy.sandiegouniontribune.com/news/features/20060410-9999-1c10karla.html.

Porter, Jennifer E. "Pilgrimage and the IDIC Ethic: Exploring *Star Trek* Convention Attendance as Pilgrimage." In *Intersecting Journeys: The Anthropology of Pilgrimage and Tourism*, edited by Ellen Badone and Sharon R. Roseman, 160–79. Champaign: University of Illinois Press, 2004.

Poster, Mark. *The Information Subject*. London: Taylor and Francis, 2001.

Postigo, Hector. "Video Game Appropriation through Modifications." *Convergence: The International Journal of Research into New Media Technologies* 14, no. 1 (2008): 59–74.

Potts, John. "Who's Afraid of Technological Determinism?: Another Look at Medium Theory." *Fibreculture* 12 (2008). Accessed 23 March 2016. http://twelve.fibreculturejournal.org/fcj-084-who%E2%80%99s-afraid-of-technological-determinism-another-look-at-medium-theory/.

Potts, Liza. "Amanda Palmer and the #LOFNOTC: How Online Fan Participation Is Rewriting Music Labels." *Participations* 9, no. 2 (2012). http://www.participations.org/Volume%209/Issue%202/20%20Potts.pdf.

Rafaeli, Sheizof. "Interactivity: From New Media to Communication." In *Sage Annual Review of Communication Research: Advancing Communication Science: Merging Mass and Interpersonal*

Processes, 16, edited by R. P. Hawkins, J. M. Wiemann, and S. Pingree, 110–34. Beverly Hills, CA: Sage, 1988.

Reid, Robin Anne. "Fan Studies." In *The Routledge Companion to Science Fiction*, edited by Mark Bould, Andrew Butler, Adam Roberts, and Sherryl Vint, 204–13. London: Routledge, 2009.

Rettberg, Jill Walker. *Blogging*. Cambridge, UK: Polity Press, 2008.

Rheingold, Howard. *The Virtual Community: Homesteading on the Electronic Frontier*. Reading, MA: Addison-Wesley, 1993.

Rosenbloom, A. "A Game Experience–in Every Application." *Communications of the ACH*, 46, no. 7 (2003): 28–31.

Rutherford, Jonathan. *Identity: Community, Culture, Difference*. London: Lawrence & Wishart, 1998.

Ryan, Marie-Laure. *Avatars of Story*. Minneapolis: University of Minnesota Press, 2006.

Ryan, Marie-Laure. "Beyond Myth and Metaphor: Narrative in Digital Media." *Poetics Today* 23, no. 4 (2002): 581–609.

Ryan, Marie-Laure. *Narrative as Virtual Reality*. Baltimore, MD: The Johns Hopkins University Press, 2001.

Sahlins, Marshall. "The Spirit of the Gift." In *The Logic of the Gift: Toward an Ethic of Generosity*, edited by Alan D. Schrift, 70–99. New York: Routledge, 1997.

Salas, Randy. "Enter the Dragon: Grab a Wand and Hunt All Manner of Fantasy Creatures at MagiQuest, a New Interactive Adventure at the Mall of America." *Star Tribune*, 13 October 2009. Accessed 23 March 2016. https://www.highbeam.com/doc/1G1-209623490.html.

Salen, Katie, and Eric Zimmerman. *Rules of Play: Game Design Fundamentals*. Cambridge, MA: The MIT Press, 2004.

Sandvoss, Cornel. *Fans: The Mirror of Consumption*. Malden, MA: Polity Press, 2005.

Sandvoss, Cornel, and Laura Kearns. "From Interpretive Communities to Interpretative Fairs: Ordinary Fandom, Textual Selection and Digital Media." In *Ashgate Research Companion to Fan Cultures*, edited by Linda Duits, Koos Zwaan, and Stijn Reijnders, 91–106. Farnham, UK: Ashgate, 2014.

Schneider, Jay, and Gerd Kortuem. "How to Host a Pervasive Game: Supporting Face-to-Face Interactions in Live-Action Roleplaying." UbiComp Workshop on Designing Ubiquitous Computing Games, 2000. Accessed 23 March 2016. http://citeseer.ist.psu.edu/viewdoc/summary?doi=10.1.1.29.3258.

Scholes, Robert, and Robert Kellogg. *The Nature of Narrative*. London: Oxford University Press, 1968.

Schoolfield, J. "Quest Masters." *Funworld* 22, no. 10 (2006): 42–51.

Schork, R. Joseph, "Acoustic Intratexts in *Aeneid* 7.122 and 4.408." *Classical Philology* 91, no. 1 (1996): 61–3.

Schrift, Alan D. "Introduction: Why Gift?" In *The Logic of the Gift: Toward an Ethic of Generosity*, edited by Alan D. Schrift, 1–22. New York: Routledge, 1997.

Schroeder, Randy. "Playspace Invaders: Huizinga, Baudrillard and Video Game Violence." *Journal of Popular Culture* 30, no. 3 (1996): 143–53.

Sconce, Jeffrey. "What If?: Charting Television's New Textual Boundaries." In *Television after TV*, edited by Lynn Spigel and Jan Olsson, 93–112. Durham, NC: Duke University Press, 2004.

Scott, A. O. "At Comic-Con, Bring out Your Fantasy and Fuel the Culture." *New York Times*, 15 July 2015. Accessed 23 March 2016. http://www.nytimes.com/2015/07/19/movies/at-comic-con-bring-out-your-fantasy-and-fuel-the-culture.html.

Scott, Suzanne. "The Moral Economy of Crowdfunding and the Transformative Capacity of Fan-ancing." *New Media and Society* 17, no. 2 (2015): 167–82.

Scott, Suzanne. "Repackaging Fan Culture: The Regifting Economy of Ancillary Content Models." *Transformative Works and Cultures* 3 (2014). http://doi:10.3983/twc.2009.0150.

Serfaty, Viviane. *The Mirror and the Veil: An Overview of American Online Diaries and Blogs*. Amsterdam: Rodopi, 2004.

Shave, Rachel. "Slash Fandom on the Internet, or Is the Carnival Over?" *Refractory* 6 (2004). Accessed 23 March 2016. http://blogs.arts.unimelb.edu.au/refractory/2004/06/17/slash-fandom-on-the-internet-or-is-the-carnival-over-rachel-shave/.

Shefrin, Elana. "*Lord of the Rings*, *Star Wars*, and Participatory Fandom: Mapping New Congruencies between the Internet and Media Entertainment Culture." *Critical Studies in Media Communication* 21, no. 3 (2004): 261–81.

Shen, Dan, and Dejin Xu. "Intratextuality, Extratextuality, Intertextuality: Unreliability in Autobiography Versus Fiction." *Poetics Today* 28, no. 1 (2007): 43–87.

Shepherd, David, ed., *Bakhtin: Carnival and Other Subjects: Selected Papers from the Fifth International Bakhtin Conference University of Manchester, July 1991*, Atlanta, GA: Rodopi, 1993.

Shiga, John. "Copy-and-Persist: The Logic of Mash-Up Culture." *Critical Studies in Media Communication* 24, no. 2 (2007): 93–114.

Smith, Craig. "By the Numbers: 17 MySpace Stats and Facts Then and Now," 14 August 2015. Accessed 23 March 2016. http://expandedramblings.com/index.php/myspace-stats-then-now/.

Smith, Marc A., and Peter Kollock, eds. *Communities in Cyberspace*. London: Routledge, 1999.

Soukup, Charles. "Hitching a Ride on a Star: Celebrity, Fandom, and Identification on the World Wide Web." *Southern Communication Journal* 71, no. 4 (2006): 319–37.

Sparks, Kimberly A. "Will Your Next Job Be in an Alternate Reality?" *Canadian Screenwriter* 11, no. 2 (2009).

Sperling, Nicole. "Melissa Joan Hart Says Kickstarter Failure 'Started out Wrong'." *LA Times*, 20 May 2013. Accessed 23 March 2016. http://articles.latimes.com/2013/may/20/entertainment/la-et-mn-melissa-joan-hart-kickstarter-failure-20130520.

Stallybrass, Peter, and Allon White. *The Politics and Poetics of Transgression*. Ithaca, NY: Cornell University Press, 1986.

Stanfill, Mel. "The Veronica Mars Kickstarter, Fan-ancing, and Austerity Logics." *Mel Stanfill*, 25 March 2013. Accessed 23 March 2016. http://www.melstanfill.com/the-veronica-mars-kickstarter-fan-ancing-and-austerity-logics/.

Stein, Louisa. *Millennial Fandom*. Iowa City: University of Iowa Press, 2015.

Stein, Louisa. "Sherlock, SuperWhoLock and Transfandom." Presentation at the Society for Cinema and Media Studies conference, April 2016, Atlanta, GA.

Stein, Louisa E. "'This Dratted Thing': Fannish Storytelling through New Media." In *Fan Fiction and Fan Communities in the Age of the Internet*, edited by Karen Hellekson and Kristina Busse, 245–60. Jefferson, NC: McFarland, 2006.

Stevens, Carolyn. "Buying Intimacy: Proximity and Exchange at a Japanese Rock Concert." In *Fanning the Flames: Fans and Consumer Culture in Japan*, edited by William W. Kelly, 59–77. New York: SUNY Press, 2004.

Stone, Alluquere Rosanne. *The War of Desire and Technology at the Close of the Mechanical Age*. Cambridge, MA: The MIT Press, 1996.

Sullivan, Andrew. "Comments." *The Daily Dish*, 03 March 2008. Accessed 23 March 2016. http://andrewsullivan.theatlantic.com/the_daily_dish/2008/03/comments.html.

Szadkowski, J. "Resorts' MagiQuest Actives Slick Fantasy." *Washington Times*, 25 August 2007. Accessed 23 March 2016.

Szulborski, Dave. *This Is Not a Game: A Guide to Alternate Reality Gaming*. New York: New Fiction Publishing, 2005.

Tapscott, Don, and Anthony Williams. *Wikinomics: How Mass Collaboration Changes Everything*. New York: Penguin, 2006.

Terranova, Tiziana. "Free Labor: Producing Culture for the Digital Economy." *Social Text* 63 18, no. 2 (2000): 33–58.

Théberge, Paul. "Everyday Fandom: Fan Clubs, Blogging, and the Quotidian Rhythms of the Internet." *Canadian Journal of Communication* 30, no. 4 (2005). Accessed 23 March 2016. http://cjc-online.ca/index.php/journal/article/viewArticle/1673/1810.

Thomas, Angela. "Fictional Blogs." In *Use of Blogs*, edited by Axel Bruns and Joanne Jacobs, 199–209. New York: Peter Lang, 2006.

Thompson, Kristin. *Storytelling in Film and Television*. Cambridge, MA: Harvard University Press, 2003.

Thorson, Kjerstin S., and Shelly Rodgers. "Relationships between Blogs and eWOM and Interactivity, Perceived Interactivity, and Parasocial Interaction." *Journal of Interactive Advertising* 6, no. 2 (2006): 39–50.

Toffler, Alvin. *The Third Wave*. New York: Bantam Books, 1980.

Tulloch, John, and Henry Jenkins. *Science Fiction Audiences: Watching 'Doctor Who' and 'Star Trek.'* London: Routledge, 1995.

Turk, Tisha, and Joshua Johnson. "Toward an Ecology of Vidding." In "Fan/Remix Video," edited by Francesca Coppa and Julie Levin Russo, special issue, *Transformative Works and Cultures* 9 (2012). doi:10.3983/twc.2012.0326.

Turkle, Sherry. *Life on the Screen: Identity in the Age of the Internet*. New York: Simon and Schuster, 1995.

Tushnet, Rebecca. "Copyright Law, Fan Practices, and the Rights of the Author." In *Fandom: Identities and Communities in a Mediated World*, edited by Jonathan Gray, Cornel Sandvoss, and C. Lee Harrington, 60–74. New York: New York University Press, 2007.

Tussey, Ethan. "Fixing Television by Funding a Movie: The Crowdfunding of *Veronica Mars*." In *Crowdfunding the Future: Media Industries, Ethics, and Digital Society*, edited by Lucy Bennett, Bertha Chin, and Bethan Jones, 157–73. New York: Peter Lang, 2015.

Uren, Tim. "Finding the Game in Improvised Theater." In *Second Person: Role-Playing and Story in Games and Playable Media*, edited by Pat Harrigan and Noah Wardrip-Fruin, 279–84. Cambridge, MA: The MIT Press, 2007.

van Kokswijk, Jacob. *Digital Ego: Social and Legal Aspects of Virtual Identity*, rev. ed. Delft, The Netherlands: Eburon Academic Publishers, 2008.

Veale, Kevin. "Capital, Dialogue, and Community Engagement—*My Little Pony: Friendship Is Magic* Understood as an Alternate Reality Game." *Transformative Works and Cultures* 14 (2013). http://journal.transformativeworks.org/index.php/twc/article/view/510/405.

Wagner, Mitch. "Lost Fans Find Internet Thrills Via Wikis, Games, Second Life." *InformationWeek*, 26 April 2008. Accessed 23 March 2016. http://www.informationweek.com/news/personal_tech/virtualworlds/showArticle.jhtml?articleID=207401542.

Walker, Jill. "Distributed Narratives: Telling Stories across Networks." Paper presented at AoIR 5.0, Brighton, 21 September 2004. Accessed 23 March 2016. http://jilltxt.net/txt/Walker-AoIR-3500words.pdf.

Walker, Rob. "Enterprising: Her *Star Trek* Conventions Harnessed the Power of Media Fandom Long before the Barons of Content Did." *New York Times Magazine*, 23 December 2008. Accessed 23 March 2016. http://www.nytimes.com/2008/12/28/magazine/28wwln-consumed-t.html.

Walsh, Richard. "Fabula and Fictionality in Narrative Theory." *Style* 35, no. 4 (2001): 592–606.

Wanzo, Rebecca. "African American Acafandom and Other Strangers: New Genealogies of Fan Studies." *Transformative Works and Cultures* 20 (2015). http://dx.doi.org/10.3983/twc.2015.0699.

Warnick, Barbara. *Rhetoric Online: Persuasion and Politics on the World Wide Web*. New York: Peter Lang, 2007.

Webb, Darren. "Bakhtin at the Seaside: Utopia, Modernity and the Carnivalesque." *Theory, Culture & Society* 22, no. 3 (2005): 121–38.

Wesch, Michael. "What Is Web 2.0? What Does It Mean for Anthropology?" *Anthropology News* (2007): 30 1.

"What Effect Have Fans Had?" *The Lurkers Guide to Babylon* 5, 19 March 2006. Accessed 21 March 2016. http://www.midwinter.com/lurk/resources/fans.html#effect.

WHCnelson. Review of "Battlestar Galactica 1980/1985 Finding Earth." *Fanfiction.net*, 20 November 2007. Accessed 23 March 2016. http://www.fanfiction.net/r/3777645/.

Williams, Raymond. *Culture and Materialism*. London: Verso, 1980.

Williams, Raymond. *Keywords: A Vocabulary of Culture and Society*. Oxford, UK: Oxford University Press, 1976.

Williams, Raymond. *Television: Technology and Cultural Form*. London: Routledge, 1990.

Williams, Rebecca. "'It's about Power': Spoilers and Fan Hierarchy." *Slayage: The International Journal of Buffy Studies* 3, no. 3–4 (2004). Accessed 23 March 2016. http://slayageonline.com/Numbers/slayage11_12.htm.

Williams, Robin, and David Edge. "The Social Shaping of Technology." Accessed 23 March 2016. https://www.researchgate.net/publication/222482133_The_Social_Shaping_of_Technology.

Wilson, Elizabeth. *Cultural Passions: Fans, Aesthetes and Tarot Readers*. London: IB Tauris, 2013.
Wood, Megan M., and Linda Baughman. "*Glee* Fandom and Twitter: Something New, or More of the Same Old Thing?" *Communication Studies* 63, no. 3 (2012): 328–44.
Žižek, Slavoj. "Cyberspace, or the Unbearable Closure of Being." In *The Plague of Fantasies*. London: Verso, 1997.
Žižek, Slavoj. *Looking Awry: An Introduction to Jacques Lacan through Popular Culture*. Cambridge, MA: The MIT Press, 1991.
Zubernis, Lynn, and Katherine Larsen. *Fandom at the Crossroads: Celebration, Shame, and Fan/Produer Relationships*. Newcastle, UK: Cambridge Scholars Press, 2012.
Zwaan, Koos, Linda Duits, and Stijn Reijnders. "Introduction to *The Ashgate Research Companion to Fan Cultures*," 1–9. Farnham, UK: Ashgate, 2014.

FAN WORK CONSULTED

Full Titles, authors, and commenters of Blog Fan Fiction:

- wyrdwritere, "Hubris, or the Downfall of Helena Cain, part 3 of 4." <http://community.livejournal.com/so_say_we_all/12310.html> (no comments)
- runawaynun, "Cylons, Shoes, and Squirrels." <http://community.livejournal.com/bsg2003fics/560052.html> (comments by freifraufischer, runawaynun (x2), projectjulie)
- lyssie, "Seelix/Cally, Lubing the Viper." <http://lyssie.livejournal.com/1176623.html> (comments by life_on_queen (x2), lyssie (x11), stargazercmc, timjr, runawaynun, falconsoko, redscribe (x2), en_hattifnatt, auroramama, sheepfairy, projectjulie)
- wyrdwritere, "Hubris, or the Downfall of Helena Cain, part 4 of 4." <http://community.livejournal.com/bsg2003fics/564526.html> (comments by millari, wyrdwritere (x4), elzed, pellamerethiel, artemis_90)
- helen_c, "Four Times Lee Adama Almost Died (4/4)." <http://helen-c.livejournal.com/108938.htm> (comments by siljamus, helen_c (x7), fredsmith518 (x2), suffolkgirl, daybreak777, marenfic, pellamerethiel, rose_griffes)
- rowanjade, "They Took My Towel! And Other Furry Tails." <http://community.livejournal.com/bsg2003fics/567327.html> (comments by life_on_queen (x2), rowanjade (x3), raincitygirl (x2), rebelliousrose)

- jeeshee, "Gaius Pulls It Off." <http://community.livejournal.com/bsg2003fics/567738.html> (no comments)
- jeeshee, "Reconnaissance." <http://community.livejournal.com/bsg2003fics/568890.html> (comments by lyssie, jeeshee)
- danniisupernova, "The Woods." <http://community.livejournal.com/sprnova_fixxx/17311.html> (comments by rap541 (x2), danniisupernova (x2))
- lyssie, "Kara/Anders, Uninterrupted Routine." <http://lyssie.livejournal.com/1187196.html> (comments by lyssie (x3), cynthia_arrow, prolix_allie, latteaddict)
- Chilipip & Asso, "The Missing Year ~Update!" <http://chilipip.wordpress.com/welcome-2/kin/> (no comments)

Downloaded Wikis:

Heroes:
- "Main Page" <http://Heroeswiki.com/Main_Page>
- "Spoiler: Powerless" <http://Hereoeswiki.com/Spoiler:Powerless>
- "Episode: Powerless" <http://Hereoeswiki.com/Episode:Powerless>
- "Index" <http://Hereoswiki.com/Index>
- "Create Your Own Hero" <http://heroeswiki.com/Create_Your_Hero>

Lost:
- "Main Page" <http://www.Lostpedia.com/wiki/Main_Page>
- "Season 4/Spoilers" <http://www.Lostpedia.com/wiki/Season_4/spoilers>
- "Jack Shephard" <http://www.Lostpedia.com/wiki/Jack >
- "John Locke" <http://www.Lostpedia.com/wiki/John_Locke>
- "James (Sawyer) Ford" <http://www.Lostpedia.com/wiki/Sawyer>
- "Lost: Mysterious Happenings" <http://www.Lostpedia.com/wiki/Portal:Mysterious>

MySpace Character Profiles (few of these exist anymore):
- captain_j_harkness: Myspace.com/captain_j_harkness. (2008). <http://profile.myspace.com/index.cfm?fuseaction=user.viewprofile&friendid=119494970>
- Closefriendofjamestkirk: MySpace.com/closefriendofjameskirk. (2008). <http://profile.myspace.com/index.cfm?fuseaction=user.viewprofile&friendID=310509123>
- coffeeluvr4life: Myspace.com/coffeeluvrlife. (2007). <http://profile.myspace.com/index.cfm?fuseaction=user.viewprofile&friendid=110833970>

- Doctor Who: Myspace.com/leerainford. (2009). <http://www.myspace.com/leerainford>
- gilmoregirls3: Myspace.com/gilmoregirls3. (2007). <http://profile.myspace.com/index.cfm?fuseaction=user.viewprofile&friendid=193528050>
- Gwen Cooper: Myspace.com/Gwen. (2009). <http://profile.myspace.com/index.cfm?fuseaction=user.viewProfile&friendID=122033336>
- Gwencooper: Myspace.com/gwencooper. (2008). <http://profile.myspace.com/index.cfm?fuseaction=user.viewprofile&friendid=90956481>
- Jack Harkness™: Myspace.com/122039243 (2009). <http://profile.myspace.com/index.cfm?fuseaction=user.viewProfile&friendID=122039243>
- Jim: Myspace/45053624 (2009). <http://profile.myspace.com/index.cfm?fuseaction=user.viewProfile&friendID=45053624>
- Veronica_Mars: Myspace.com/veronica_mars. (2006). <http://profile.myspace.com/index.cfm?fuseaction=user.viewprofile&friendid=14789920>
- Veronica Mars CW: Myspace.com/veronicamarscw. (2006). <http://profile.myspace.com/index.cfm?fuseaction=user.viewprofile&friendID=92251231>
- Veronica Marz: Myspace.com/mizzmarzx. (2006). <http://profile.myspace.com/index.cfm?fuseaction=user.viewprofile&friendid=130922322>
- Zachandstaceforever:Myspace.com/zachandstaceforever. (2007). <http://profile.myspace.com/index.cfm?fuseaction=user.viewprofile&friendid=72513851>

INDEX

A

B

C

D

E

F

K

L

M

N

O

P

U

V

W

General Editor: *Steve Jones*

Digital Formations is the best source for critical, well-written books about digital technologies and modern life. Books in the series break new ground by emphasizing multiple methodological and theoretical approaches to deeply probe the formation and reformation of lived experience as it is refracted through digital interaction. Each volume in **Digital Formations** pushes forward our understanding of the intersections, and corresponding implications, between digital technologies and everyday life. The series examines broad issues in realms such as digital culture, electronic commerce, law, politics and governance, gender, the Internet, race, art, health and medicine, and education. The series emphasizes critical studies in the context of emergent and existing digital technologies.

Other recent titles include:

Felicia Wu Song
Virtual Communities: Bowling Alone, Online Together

Edited by Sharon Kleinman
The Culture of Efficiency: Technology in Everyday Life

Edward Lee Lamoureux, Steven L. Baron, & Claire Stewart
Intellectual Property Law and Interactive Media: Free for a Fee

Edited by Adrienne Russell & Nabil Echchaibi
International Blogging: Identity, Politics and Networked Publics

Edited by Don Heider
Living Virtually: Researching New Worlds

Edited by Judith Burnett, Peter Senker & Kathy Walker
The Myths of Technology: Innovation and Inequality

Edited by Knut Lundby
Digital Storytelling, Mediatized Stories: Self-representations in New Media

Theresa M. Senft
Camgirls: Celebrity and Community in the Age of Social Networks

Edited by Chris Paterson & David Domingo
Making Online News: The Ethnography of New Media Production

To order other books in this series please contact our Customer Service Department:

(800) 770-LANG (within the US)
(212) 647-7706 (outside the US)
(212) 647-7707 FAX

To find out more about the series or browse a full list of titles, please visit our website:

WWW.PETERLANG.COM

www.ingramcontent.com/pod-product-compliance
Ingram Content Group UK Ltd.
Pitfield, Milton Keynes, MK11 3LW, UK
UKHW020201290926
15471UKWH00005B/996

9 781433 131509